To the new generation: Emma Elīza, Noah and Kai

B&SS – Bocconi & Springer Series

Series Editors:
Sandro Salsa (*Editor-in-Chief*) • Carlo A. Favero • Peter Müller • Lorenzo Peccati • Eckhard Platen
Wolfgang J. Runggaldier

Giovanni Peccati · Murad S. Taqqu

Wiener Chaos: Moments, Cumulants and Diagrams

A survey with computer implementation

BOCCONI UNIVERSITY PRESS

 Springer

Giovanni Peccati
Mathematics Research Unit
University of Luxembourg
Luxembourg
giovanni.peccati@gmail.com

Murad S. Taqqu
Department of Mathematics and Statistics
Boston University
Boston
murad@math.bu.edu

Additional material to this book can be downloaded from http://extras.springer.com
Password: 978-88-470-1678-1

B&SS – Bocconi & Springer Series

ISSN print edition: 2039-1471   ISSN electronic edition : 2039-148X

ISBN 978-88-470-1678-1   ISBN 978-88-470-1679-8 (eBook)

DOI 10.1007/978-88-470-1679-8

Library of Congress Control Number: 2010929482

Springer Milan Dordrecht Heidelberg London New York

© Springer-Verlag Italia 2011

9 8 7 6 5 4 3 2 1

Cover-Design: K Design, Heidelberg

Typesetting with LaTeX: PTP-Berlin, Protago TeX-Production GmbH, Germany (www.ptp-berlin.eu)

Springer-Verlag Italia srl – Via Decembrio 28 – 20137 Milano
Springer is a part of Springer Science+Business Media (www.springer.com)

Contents

Preface XI

1 Introduction 1
   1.1   Overview 1
   1.2   Some related topics 5

2 The lattice of partitions of a finite set 7
   2.1   Partitions of a positive integer 7
   2.2   Partitions of a set 9
   2.3   Partitions of a set and partitions of an integer 12
   2.4   Bell polynomials, Stirling numbers and Touchard polynomials 16
   2.5   Möbius functions and Möbius inversion on partitions ... 19
   2.6   Möbius functions on partially ordered sets 20
         2.6.1   Incidence algebras and general Möbius inversion 20
         2.6.2   Computing a Möbius function 24
         2.6.3   Further properties of Möbius functions and a proof of
                 (2.5.21) 25

3 Combinatorial expressions of cumulants and moments 31
   3.1   Cumulants 31
   3.2   Relations involving moments and cumulants 33
   3.3   Bell and Touchard polynomials and the Poisson distribution 40

4 Diagrams and multigraphs 45
   4.1   Diagrams 45
   4.2   Solving the equation $\sigma \wedge \pi = \hat{0}$ 52
   4.3   From Gaussian diagrams to multigraphs 53
   4.4   Summarizing diagrams and multigraphs 55

5   Wiener-Itô integrals and Wiener chaos 57
   5.1   Completely random measures 57
   5.2   Single Wiener-Itô integrals 64
   5.3   Infinite divisibility of single integrals 65
   5.4   Multiple stochastic integrals of elementary functions 76
   5.5   Wiener-Itô stochastic integrals 78
   5.6   Integral notation 84
   5.7   The role of diagonal measures 85
   5.8   Example: completely random measures on the real line 87
   5.9   Chaotic representation 91
   5.10   Computational rules 92
   5.11   Multiple Gaussian integrals of elementary functions 92
   5.12   Multiple Poisson integrals of elementary functions.............. 97
   5.13   A stochastic Fubini theorem................................ 106

6   Multiplication formulae 109
   6.1   The general case 109
   6.2   Contractions.. 115
   6.3   Symmetrization of contractions 117
   6.4   The product of two integrals in the Gaussian case 118
   6.5   The product of two integrals in the Poisson case 122

7   Diagram formulae 127
   7.1   Formulae for moments and cumulants 127
   7.2   MSets and MZeroSets...................................... 132
   7.3   The Gaussian case 133
   7.4   The Poisson case 140

8   From Gaussian measures to isonormal Gaussian processes.......... 145
   8.1   Multiple stochastic integrals as Hermite polynomials 145
   8.2   Chaotic decompositions 148
   8.3   Isonormal Gaussian processes 149
   8.4   Wiener chaos 152
   8.5   Contractions, products and some explicit formulae.............. 156

9   Hermitian random measures and spectral representations 159
   9.1   The isonormal approach 159
   9.2   The Gaussian measure approach 161
   9.3   Spectral representation 163
   9.4   Stochastic processes 164
   9.5   Hermite processes 165
   9.6   Further properties of Hermitian Gaussian measures 166
   9.7   Caveat about normalizations 167

10   Some facts about Charlier polynomials 171

11   Limit theorems on the Gaussian Wiener chaos 177
11.1   Some features of the laws of chaotic random variables
(Gaussian case) 177
11.2   A useful estimate 179
11.3   A general problem 180
11.4   One-dimensional CLTs in the Gaussian case 182
11.5   An application to Brownian quadratic functionals 187
11.6   Proof of the propositions 190
11.7   Further combinatorial implications 197
11.8   A multidimensional CLT 199

12   CLTs in the Poisson case: the case of double integrals 203

Appendix A. Practical implementation using Mathematica 207
A.1   General introduction 207
A.2   When Starting 209
A.3   Integer Partitions 210
      A.3.1   IntegerPartitions 210
      A.3.2   IntegerPartitionsExponentRepresentation 210
A.4   Basic Set Partitions 210
      A.4.1   SingletonPartition 211
      A.4.2   MaximalPartition 211
      A.4.3   SetPartitions 211
      A.4.4   PiStar 212
A.5   Operations with Set Partitions 212
      A.5.1   PartitionIntersection 212
      A.5.2   PartitionUnion 212
      A.5.3   MeetSolve 213
      A.5.4   JoinSolve 213
      A.5.5   JoinSolveGrid 214
      A.5.6   MeetAndJoinSolve 214
      A.5.7   CoarserSetPartitionQ 215
      A.5.8   CoarserThan 215
A.6   Partition Segments 215
      A.6.1   PartitionSegment 215
      A.6.2   ClassSegment 216
      A.6.3   ClassSegmentSolve 216
      A.6.4   SquareChoose 217
A.7   Bell and Touchard polynomials 217
      A.7.1   StirlingS2 217
      A.7.2   BellPolynomialPartial 218

A.7.3	BellPolynomialPartialAuto	....	219
A.7.4	BellPolynomialComplete	....	220
A.7.5	BellPolynomialCompleteAuto	....	220
A.7.6	BellB	....	221
A.7.7	TouchardPolynomial	....	222
A.7.8	TouchardPolynomialTable	....	223
A.7.9	TouchardPolynomialCentered	....	223
A.7.10	TouchardPolynomialCenteredTable	....	224

A.8   Mobius Formula 224
- A.8.1   MobiusFunction 224
- A.8.2   MobiusRecursionCheck 225
- A.8.3   MobiusInversionFormulaZero 225
- A.8.4   MobiusInversionFormulaOne 226

A.9   Moments and Cumulants 227
- A.9.1   MomentToCumulants 227
- A.9.2   MomentToCumulantsBell 228
- A.9.3   MomentToCumulantsBellTable 229
- A.9.4   CumulantToMoments 230
- A.9.5   CumulantToMomentsBell 231
- A.9.6   CumulantToMomentsBellTable 231
- A.9.7   MomentProductToCumulants 232
- A.9.8   CumulantVectorToMoments 233
- A.9.9   CumulantProductVectorToCumulantVectors............ 233
- A.9.10   GridCumulantProductVectorToCumulantVectors 233
- A.9.11   CumulantProductVectorToMoments 234

A.10   Gaussian Multiple Integrals 234
- A.10.1   GaussianIntegral 234

A.11   Poisson Multiple Integrals 236
- A.11.1   BOne 236
- A.11.2   BTwo 236
- A.11.3   PBTwo 237
- A.11.4   PoissonIntegral 237
- A.11.5   PoissonIntegralExceed 239

A.12   Contraction and Symmetrization 240
- A.12.1   ContractionWithSymmetrization 240
- A.12.2   ContractionIntegration 241
- A.12.3   ContractionIntegrationWithSymmetrization 241

A.13   Solving Partition Equations Involving π^* 241
- A.13.1   PiStarMeetSolve 242
- A.13.2   PiStarJoinSolve 242
- A.13.3   PiStarMeetAndJoinSolve 243

A.14 Product of Two Gaussian Multiple Integrals 243
    A.14.1 ProductTwoGaussianIntegrals 243
A.15 Product of Two Poisson Multiple Integrals 244
    A.15.1 ProductTwoPoissonIntegrals 244
A.16 MSets and MZeroSets....................................... 245
    A.16.1 MSets 245
    A.16.2 MSetsEqualTwo 245
    A.16.3 MSetsGreaterEqualTwo 245
    A.16.4 MZeroSets 245
    A.16.5 MZeroSetsEqualTwo 246
    A.16.6 MZeroSetsGreaterEqualTwo 246
A.17 Hermite Polynomials 247
    A.17.1 HermiteRho 247
    A.17.2 HermiteRhoGrid 247
    A.17.3 Hermite... 248
    A.17.4 HermiteGrid 248
    A.17.5 HermiteH 250
    A.17.6 HermiteHGrid 250
A.18 Poisson-Charlier Polynomials 251
    A.18.1 Charlier... 251
    A.18.2 CharlierGrid....................................... 251
    A.18.3 CharlierCentered 253
    A.18.4 CharlierCenteredGrid 254

Appendix B. Tables of moments and cumulants 255
Centered and scaled Cumulant to Moments.......................... 256
Centered Cumulant to Moments 257
Cumulant to Moments 258
Centered and scaled Moment to Cumulants.......................... 259
Centered Moment to Cumulants 260
Moment to Cumulants 261

References... 263

Index 271

Preface

The theoretical aspects of Time Series Analysis in the Gaussian context have been well understood since the end of the Second World War, more than 60 years ago. Linear transformations (or filters) preserve the Gaussian property and hence fit well in that framework. Norbert Wiener wanted to extend the theory to the non-linear world by considering non-linear transformations of Gaussian processes, while requesting that the output of the non-linear filters preserve the finite variance-covariance property, which is one the hallmarks of Gaussian processes. Kiyoshi Itô attempted to do the same thing in Japan. This extension is now known as the "Wiener chaos" and the corresponding stochastic integrals as "Wiener-Itô stochastic integrals". The versatility of the non-linear theory, however, turned out to be more limited than what was hoped for. It is not easy, for example, to write down the distributions of the random variables that live in this Wiener chaos. This mathematical challenge led several researchers to develop ad-hoc graphical devices, known as diagram formulae, allowing to derive moments and cumulants of chaotic random variables by means of combinatorial identities. Although these tools are not always easy to manipulate, they have been successfully used in order to develop not only new types of central limit theorems where the limit is Gaussian but also non-central limit theorems where the limit is non-Gaussian.

This is a book about combinatorial structures in the Wiener chaos. The combinatorial structures involved in our analysis are those of lattices of partitions of finite sets, over which we define incidence algebras, Möbius functions and associated inversion formulae. As discussed in the text, this combinatorial standpoint (which is originally due to Rota and Wallstrom [132]) provides an ideal framework in order to systematically deal with diagram formulae. Several applications are described, in particular recent limit theorems for chaotic random variables. An explicit computer implementation into the *Mathematica* language completes the text.

Chaotic random variables play now a crucial role in several areas of theoretical and applied probability. For instance, the fact that every functional of a Gaussian process or a Poisson measure can be written as a series of multiple integrals (the so-called "Wiener-Itô chaotic representation property") is one of the staples of Malliavin calculus and of its many applications, for example, to stochastic partial differential equations or stochastic calculus. In a recent series of papers (that are described and analyzed in the book), essentially written by Nourdin, Nualart, Peccati and Taqqu, it has been shown that the properties of chaotic random variables are the key elements for deriving very general (and strikingly simple) convergence results for non-linear functionals of Gaussian fields, Poisson measures or Lévy processes.

The goal of this monograph is to render this subject more accessible. We do this in a number of ways. We provide many examples to illustrate the theory and we also implement many of the formulas in *Mathematica*, so that the user can get a concrete feeling for the various topics. The theoretical exposition is rigorous. We have tried to fill in many of the steps in the proofs we provide, and when proofs are not given, we include detailed references. The bibliography, for example, is rather extensive, with more than 150 references. Our emphasis is on the combinatorial aspect of the subject because it is through combinatorics that the various objects are related. We start with describing partitions of a integer, of a set, the relations between them, we continue with moments and cumulants and cover a number of graphical descriptions of the various diagram formulae. When considering stochastic integrals, we do not always eliminate diagonals as is usually done, but we consider integrals where an arbitrary subset of diagonals has been eliminated, and we specify the explicit relations between them. The stochastic integrals include not only multiple integrals with respect to a Gaussian measure but also multiple integrals with respect to Poisson measures.

As anticipated, the subject is very much in flux with new limit theorems being developed and further applications, for example, to the Malliavin calculus. Although we do not cover these subjects, we provide an overview of various directions that are being pursued by researchers. This survey provides a basis for understanding the new developments.

The readership we have in mind includes researchers, but also graduate students who are either starting their research or are already working on a doctoral thesis. For a detailed description of the contents, please refer to the introduction and its subsection 1.1 "Overview". The Contents, which include the title of the sections and subsections, also provide a good overview of the covered topics.

Acknowledgements. Part of this book has been written while the authors were visiting the Department of Mathematics and Applied Statistics of Turin University, in June 2007. The authors heartily thank Pierpaolo de Blasi, Massimo Marinacci and Igor Prünster for their kind hospitality and support. Giovanni Peccati acknowledges support from ISI Foundation–Lagrange Project. Murad S. Taqqu acknowledges support by the National Science Foundation under grants DMS-0706786 and DMS-1007616 at Boston University. We thank Florent Benaych-Georges, Domenico Marinucci and Marc Yor for a careful reading of an earlier draft of this manuscript, as well as for valuable suggestions. Finally we want to thank David Jeffrey Hamrick and Mark Veillette who developed the *Mathematica* code, and without whom the computer applications would not have existed.

Luxembourg/Boston, August 2010   *Giovanni Peccati*
                                                            Murad S. Taqqu

1

Introduction

1.1 Overview

The aim of this work is to provide a unified treatment of moments and cumulants associated with non-linear functionals of completely random measures. A "completely random measure" (also called an "independently scattered random measure") is a measure φ with values in a space of random variables, such that $\varphi(A)$ and $\varphi(B)$ are independent random variables, whenever A and B are disjoint sets. Examples are Gaussian, Poisson or Gamma random measures. We will specifically focus on multiple stochastic integrals with respect to the random measure φ. These integrals are of the form

$$\int_{\sigma} f(z_1, ..., z_n)\varphi(dz_1) \cdots \varphi(dz_n) \tag{1.1.1}$$

and

$$\int_{\geq \sigma} f(z_1, ..., z_n)\varphi(dz_1) \cdots \varphi(dz_n), \tag{1.1.2}$$

where f is a symmetric function and φ is a completely random measure (for instance, Poisson or Gaussian) on the real line. The integration is not over all of $\mathbb{R}^n$, but over a "diagonal" subset of $\mathbb{R}^n$ defined by a partition σ of the integers $\{1, ..., n\}$ as illustrated below.

We shall mainly adopt a combinatorial point of view. Our main inspiration is a truly remarkable paper by Rota and Wallstrom [132], building (among many others) on earlier works by Itô [40], Meyer [76, 77] and, most importantly, Engel [26] (see also Bitcheler [9], Kussmaul [59], Linde [64], Masani [73], Neveu [79] and Tsilevich and Vershik [159] for related works). In particular, in [132] the authors point out a crucial connection between the machinery of multiple stochastic integration and the structure of the lattice of partitions of a finite set, with specific emphasis on the role played by the associated Möbius function (see e.g. [2], as well as Chapter 2 below). As we will see later on, the connection between multiple stochastic integration and partitions is given by the natural isomorphism between the partitions of the set $\{1, ..., n\}$ and the

G. Peccati, M.S. Taqqu: Wiener Chaos: Moments, Cumulants and Diagrams –
A survey with computer implementation.
© Springer-Verlag Italia 2011

diagonal sets associated with the Cartesian product of n measurable spaces (a diagonal set is just a subset of the Cartesian product consisting of points that have two or more coordinates equal).

For example, going back to (1.1.1), if $n = 3$ and

$$\sigma = \{\{1, 2\}, \{3\}\}$$

in (1.1.1), then one integrates over $z_1, z_2, z_3 \in \mathbb{R}^3$ with

$$z_1 = z_2, \quad z_2 \neq z_3.$$

If

$$\sigma = \{\{1\}, \{2\}, \{3\}\},$$

then the integration is over

$$z_1 \neq z_2, \quad z_1 \neq z_3, \quad z_2 \neq z_3$$

(commonly denoted $z_1 \neq z_2 \neq z_3$). In (1.1.2), one integrates over "$\geq \sigma$", that is, over all partitions that are coarser than σ. For example, if $n = 3$ and $\sigma = \{\{1, 2\}, \{3\}\}$, then "$\geq \sigma$" indicates that one has not only to integrate over $z_1 = z_2, z_2 \neq z_3$, but also on the hyperdiagonal $z_1 = z_2 = z_3$ (corresponding to the one-block partition $\{\{1, 2, 3\}\}$). If $n \geq 2$, the integrals (1.1.1) and (1.1.2) are non-linear functionals of φ, and thus, even if φ is Gaussian, these integrals are non-Gaussian random variables.

As we will see in Chapter 5, a crucial element of our analysis is given by random variables of the type (1.1.1), where the integration is performed over a set *without diagonals*, that is, where all the coordinates z_i are different. Random variables of this type belong to the so-called **Wiener chaos** associated with the random measure φ. Chaotic random variables play now a crucial role in several areas of theoretical and applied probability. For instance, we will see that they enjoy several remarkable connections with orthogonal polynomials, such as Hermite and Charlier polynomials; also, we will show that every square-integrable functional of a Gaussian process or of a Poisson measure can be written as a series of multiple integrals over non-diagonal sets (the so-called "Wiener-Itô chaotic representation property"). This last fact is one of the staples of Malliavin calculus (see [93, 94, 21]) and of its many applications e.g. to stochastic partial differential equations or stochastic calculus. In a recent series of papers (that are described and analyzed in the last chapters of this book), it has been shown that the properties of chaotic random variables are fundamental in deriving convergence results for non-linear functionals of Gaussian fields, Poisson measures or Lévy processes.

The best description of the approach to stochastic integration followed in the present work is still given by the following sentences, taken from [132]:

> The basic difficulty of stochastic integration is the following. We are given a measure φ on a set S, and we wish to extend such a measure to the product set S^n. There is a well-known and established way of carrying out such

an extension, namely, taking the product measure. While the product measure is adequate in most instances dealing with a scalar valued measure, it turns out to be woefully inadequate when the measure is vector-valued, or, in the case dealt with presently, random-valued. The product measure of a nonatomic scalar measure will vanish on sets supported by lower-dimensional linear subspaces of S^n. This is not the case, however, for random measures. The problem therefore arises of modifying the definition of product measure of a random measure in such a way that the resulting measure will vanish on lower-dimensional subsets of S^n, or diagonal sets, as we call them.

As pointed out in [132], as well as in Chapter 5 below, the combinatorics of partition lattices provide the correct framework in order to define a satisfactory stochastic product measure.

As discussed in detail in Chapter 8, part of the results presented in this work extend to the case of *isonormal Gaussian processes*, that is, centered Gaussian families whose covariance structure is isomorphic to some (real or complex) Hilbert space. Note that isonormal Gaussian processes have gained enormous importance in recent years, for instance in connection with fractional processes (see e.g. the second edition of Nualart's book [94]), or as limit objects (known as *Gaussian Free Fields*) appearing e.g. in the theory of random matrices and random surfaces (see [129] and [139] for some general discussions in this direction).

To make the notions we introduce concrete, we have included an Appendix to this survey. It is devoted to a *Mathematica*[1] implementation of various formulae. We stress, however, that no knowledge of the *Mathematica* language is required. This is because we provide in the Appendix detailed instructions and examples. This unusual addendum may be welcome, in particular, by graduate students and new researchers in this area.

As apparent from the title, in the subsequent chapters a prominent role will be played by moments and cumulants. In particular, the principal aims of our work are the following:

– **Put diagram formulae in a proper algebraic setting.** Diagram formulae are mnemonic devices, allowing to compute moments and cumulants associated with one or more random variables. These tools have been developed and applied in a variety of frameworks: see e.g. [141, 151] for diagram formulae associated with general random variables; see [10, 12, 34, 68] for non-linear functionals of Gaussian fields; see [150] for non-linear functionals of Poisson measures. They can be quite useful in the obtention of Central Limit Theorems (CLTs) by means of the so-called *method of moments and cumulants* (see e.g. [66]). Inspired by the works by McCullagh [74], Rota and Shen [131] and Speed [145], we shall show that all diagram formulae quoted above can be put in a unified framework, based on the

---

[1] *Mathematica* is a computational software program developed by Wolfram Research, which is used widely in scientific and mathematical fields.

use of partitions of finite sets. Although somewhat implicit in the previously quoted references, this clear algebraic interpretation of diagrams is new. In particular, in Chapter 4 we will show that all diagrams encountered in the probabilistic literature (such as Gaussian, non-flat and connected diagrams) admit a neat translation in the combinatorial language of partition lattices.

- **Illustrate the Engel-Rota-Wallstrom theory.** We shall show that the theory developed in [26] and [132] allows to recover several crucial results of stochastic analysis, such as multiplication formulae for multiple Gaussian and Poisson integrals see [49, 94, 150]. This extends the content of [132], which basically dealt with product measures. See also [28] for other results in this direction.

- **Shed light the combinatorial implications of new CLTs.** In a recent series of papers (see [69, 82, 86, 87, 90, 95, 98, 103, 106, 109, 110, 111]), a new set of tools has been developed, allowing to deduce simple CLTs involving random variables having the form of multiple stochastic integrals. All these results can be seen as simplifications of the method of moments and cumulants. In Chapter 11, we will illustrate these results from a combinatorial standpoint, by providing some neat interpretations in terms of diagrams and graphs. In particular, we will prove that in these limit theorems a fundamental role is played by the so-called *circular diagrams*, that is, connected Gaussian diagrams whose edges only connect subsequent rows.

We will develop the necessary combinatorial tools related to partitions, diagram and graphs from first principles in Chapter 2 and Chapter 4. Chapter 3 provides a self-contained treament of moments and cumulants from a combinatorial point of view. Stochastic integration is introduced in Chapter 5. Chapter 6 and Chapter 7 deal, respectively, with product formulae and diagram formulae. Chapter 8 deals with Gaussian random measures, isonormal Gaussian processes and the relationship between corresponding multiple integrals and Hermite polynomials. In Chapter 9 we describe Hermitian random measures and spectral representations and define the Hermite processes. In Chapter 10 we introduce Charlier polynomials and relate them to multiple integrals with respect to a Poisson random measure. Chapter 11 and Chapter 12 deal with CLTs on Wiener and Poisson chaos respectively. There are two appendices[2]. Appendix A describes the *Mathematica* commands. These are also listed in the Contents. Finally, Appendix B contains tables of moments and cumulants.

---

[2] The index does not include the appendices.

1.2 Some related topics

In this survey, we choose to follow a very precise path, namely starting with the basic properties of partition lattices and diagrams, and develop from there as many as possible of the formulae associated with products, moments and cumulants in the theory of stochastic integration with respect to completely random measures. In order to keep the length of the present work within bounds, several crucial topics are not included (or are just mentioned) in the discussion to follow. One remarkable omission is of course a complete discussion of the connections between multiple stochastic integrals and orthogonal polynomials. This topic is partially treated in Chapters 8 and 10 below, in the particular case of Gaussian and Poisson fields. For recent references on more general stochastic processes (such as Lévy processes), see e.g. the monograph by Schoutens [137] and the two papers by Solé and Utzet [143, 144]. Other related (and missing) topics are detailed in the next list, whose entries are followed by a brief discussion.

- *Wick products.* Wick products are intimately related to chaotic expansions. A complete treatment of this topic can be found e.g. in Janson's book [46].
- *Malliavin calculus.* See the two monographs by Nualart [93, 94] for Malliavin calculus in a Gaussian setting. The monograph by Di Nunno, Øksendal and Proske [21] provides an introduction to Malliavin calculus with applications to finance. A good introduction to Malliavin calculus for Poisson measures is contained in the classic papers by Nualart and Vives [100], Privault [120] and Privault and Wu [125], as well as in the recent monograph by Privault [122]. A fundamental connection between Malliavin operators and limit theorems has been first pointed out in [96]. See [82, 86, 106] for further developments.
- *Hu-Meyer formulae.* Hu-Meyer formulae connect Stratonovich multiple integrals and multiple Wiener-Itô integrals. See [94] for a standard discussion of this topic in a Gaussian setting. Hu-Meyer formulae for general Lévy processes can be naturally obtained by means of the theory described in the present book: see the excellent paper by Farré *et al.* [28] for a complete treatment of this point.
- *Stein's method.* Stein's method for normal and non-normal approximation can be a very powerful tool in order to obtain central and non-central limit theorems for non-linear functionals of random fields. In Chapter 3, we will only scratch the surface of this topic, by proving two basic results related to Stein's method, namely the Stein's Lemma for the normal distribution, and the Chen-Stein Lemma for the Poisson distribution. Some further discussion is contained in Chapter 11. See [147] for a classic reference on the subject and [15] for an exhaustive recent monograph. See [86, 87, 90] for several limit theorems involving functionals of Gaussian fields, obtained by means of Stein's method and Malliavin calculus. See [106, 114] for applications of Stein's method to functionals of Poisson measures.
- *Free probability.* The properties of the lattice of (non-crossing) partitions and the corresponding Möbius function are crucial in free probability. See the monograph by Nica and Speicher [80] for a valuable introduction to the combinatorial aspects

of free probability. See Anshelevich [3, 4] for some instances of a "free" theory of multiple stochastic integration. The paper [53], by Kemp *et al.*, establishes some explicit connections between limit theorems in free probability and the topics discussed in Chapter 11 below.

2

The lattice of partitions of a finite set

In this chapter we recall some combinatorial results concerning the lattice of partitions of a finite set. These objects play an important role in the obtention of the *diagram formulae* presented in Chapter 5. The reader is referred to Stanley [146, Ch. 3] and Aigner [2] for a detailed presentation of (finite) partially ordered sets and Möbius inversion formulae.

2.1 Partitions of a positive integer

Given an integer $n \geq 1$, we define the set $\Lambda(n)$ of *partitions* of n as the collection of all vectors of the type $\lambda = (\lambda_1, ..., \lambda_k)$ $(k \geq 1)$, where:

> (i) λ_j is an integer for every $j = 1, ..., k$;
> (ii) $\lambda_1 \geq \lambda_2 \geq \cdots \geq \lambda_k \geq 1$; $\hspace{2cm}$ (2.1.1)
> (iii) $\lambda_1 + \cdots + \lambda_k = n$.

We call k the *length* of λ. It is sometimes convenient to write a partition $\lambda = (\lambda_1, ..., \lambda_k) \in \Lambda(n)$ in the form

$$\lambda = (1^{r_1} 2^{r_2} \cdots n^{r_n}).$$

This representation (which encodes all information about λ) simply indicates that, for every $i = 1, ..., n$, the vector λ contains exactly r_i (≥ 0) components equal to i. Clearly, if $\lambda = (\lambda_1, ..., \lambda_k) = (1^{r_1} 2^{r_2} \cdots n^{r_n}) \in \Lambda(n)$, then

$$1 r_1 + \cdots + n r_n = n \hspace{2cm} (2.1.2)$$

and $r_1 + \cdots + r_n = k$. We will sometimes use the (more conventional) notation

$$\lambda \vdash n \quad \text{instead of} \quad \lambda \in \Lambda(n).$$

G. Peccati, M.S. Taqqu: Wiener Chaos: Moments, Cumulants and Diagrams –
A survey with computer implementation.
© Springer-Verlag Italia 2011

Example 2.1.1   (i) If $n = 5$, one can, for example, have $5 = 4 + 1$ or $5 = 1 + 1 + 1 + 1 + 1$. In the first case the length is $k = 2$, with $\lambda_1 = 4$ and $\lambda_2 = 1$, and the partition is $\lambda = (1^1 2^0 3^0 4^1 5^0)$. In the second case, the length is $k = 5$ with $\lambda_1 = \ldots = \lambda_5 = 1$, and the partition is $\lambda = (1^5 2^0 3^0 4^0 5^0)$.

  (ii) One can go easily from one representation to the other. Thus $\lambda = (1^2 2^3 3^0 4^2)$ corresponds to

$$n = (1 \times 2) + (2 \times 3) + (3 \times 0) + (4 \times 2) = 16,$$

that is, to the decomposition $16 = 4 + 4 + 2 + 2 + 2 + 1 + 1$, and thus to

$$\lambda = (\lambda_1, \lambda_2, \lambda_3, \lambda_4, \lambda_5, \lambda_6, \lambda_7) = (4, 4, 2, 2, 2, 1, 1).$$

Remark. Fix $n \geq 2$ and $k \in \{1, \ldots, n\}$. Consider a partition of n of length k, say $\lambda = (\lambda_1, \ldots, \lambda_k)$, and write it in the form $\lambda = (1^{r_1} 2^{r_2} \cdots n^{r_n})$. Then, one has necessarily that $r_j = 0$ for every $j > n - k + 1$ (or, equivalently, $n \leq j + k - 2$). Indeed, if $r_j > 0$ for such a j, then there would exist $\lambda_{a^*} \in \lambda$ such that $n \leq \lambda_{a^*} + k - 2$. This contradicts the inequality $n = \lambda_1 + \cdots \lambda_k \geq \lambda_{a^*} + k - 1$, which results from the fact that λ is composed of k strictly positive integers whose sum equals n. This fact implies that, for fixed $n \geq 2$ and $k \in \{1, \ldots, n\}$, every partition $\lambda \in \Lambda(n)$ of length k has the form

$$\lambda = \left(1^{r_1} 2^{r_2} \cdots (n - k + 1)^{r_{n-k+1}} (n - k + 2)^0 \cdots n^0\right), \qquad (2.1.3)$$

thus yielding immediately the following statement.

Proposition 2.1.2 *There exists a bijection, say β, between the subset of $\Lambda(n)$ composed of partitions with length k and the collection of all vectors $(r_1, \ldots, r_{n-k+1})$ of nonnegative integers such that*

$$r_1 + \cdots + r_{n-k+1} = k \quad and \quad 1 r_1 + 2 r_2 + \cdots + (n - k + 1) r_{n-k+1} = n. \quad (2.1.4)$$

The bijection is obtained as follows: if $(r_1, \ldots, r_{n-k+1})$ verifies (2.1.4), then β^{-1} $(r_1, \ldots, r_{n-k+1})$ is the element of $\Lambda(n)$ of length k given by (2.1.3).

A vector of nonnegative integers $(r_1, \ldots, r_m)$ verifying $r_1 + \cdots + r_m = k$ is customarily called a *weak m-composition* of k (see for example Stanley [146, p. 15]).

Example 2.1.3 Consider the case $n = 4$ and $k = 2$. Then, there exist only two vectors (r_1, r_2, r_3) of nonnegative integers satisfying (2.1.4) (that is, such that $r_1 + r_2 + r_3 = 2$ and $r_1 + 2 r_2 + 3 r_3 = 4$), namely $(1, 0, 1)$ and $(0, 2, 0)$. These vectors correspond respectively to the partitions $\lambda_1 = (3, 1) = (1^1 2^0 3^1 4^0)$ and $\lambda_2 = (2, 2) = (1^0 2^2 3^0 4^0)$. Proposition 2.1.2 implies that λ_1 and λ_2 are the only elements of $\Lambda(4)$ having length 2.

2.2 Partitions of a set

Let b denote a finite nonempty set and let

$$\mathcal{P}(b) \text{ be the set of } \textit{partitions} \text{ of } b.$$

By definition, an element π of $\mathcal{P}(b)$ is a collection of nonempty and disjoint subsets of b (called *blocks*), such that their union equals b. The symbol $|\pi|$ indicates the number of blocks (or the *size*) of the partition π.

Notation. For each pair $i, j \in b$ and for each $\pi \in \mathcal{P}(b)$, we write

$$i \sim_\pi j$$

whenever i and j belong to the same block of π.

We now define a partial ordering on $\mathcal{P}(b)$. For every $\sigma, \pi \in \mathcal{P}(b)$, we write

$$\sigma \leq \pi$$

if and only if

$$\text{each block of } \sigma \text{ is contained in a block of } \pi.$$

Thus,

$$\text{If } \quad \sigma \leq \pi, \quad \text{ then } \quad |\sigma| \geq |\pi|.$$

Borrowing from the terminology used in topology one also says that π is *coarser* than σ. It is clear that $\leq$ is a *partial ordering relation*, that is, $\leq$ is a *binary* relation on $\mathcal{P}(b)$, which is also *reflexive*, *transitive* and *antisymmetric*, that is:

(i) $\sigma \leq \sigma$, for every $\sigma \in \mathcal{P}(b)$ (reflexivity);
(ii) if $\sigma \leq \pi$ and $\pi \leq \rho$, then $\sigma \leq \rho$ (transitivity);
(iii) if $\sigma \leq \pi$ and $\pi \leq \sigma$, then $\sigma = \pi$ (antisymmetry);

(see also Stanley [146, pp. 97-98]).

Example 2.2.1   (i) If $b = \{1, 2, 3, 4, 5\}$, $\pi = \{\{1, 2, 3\}, \{4, 5\}\}$ and $\sigma = \{\{1, 2\}, \{3\}, \{4, 5\}\}$, then, $\sigma \leq \pi$ because each block of σ is contained in a block of π. We have $3 = |\sigma| > |\pi| = 2$.
(ii) If $\pi = \{\{1, 2, 3\}, \{4, 5\}\}$ and $\sigma = \{\{1, 2\}, \{3, 4, 5\}\}$, then π and σ are not ordered.

Moreover, the relation $\leq$ induces on $\mathcal{P}(b)$ a *lattice* structure. Recall that a lattice is a partially ordered set such that each pair of elements has a least upper bound and a greatest lower bound (see the forthcoming remark for a definition of these two notions, as well as [146, p. 102]). In particular, the partition

$$\sigma \wedge \pi, \quad \sigma, \pi \in \mathcal{P}(b),$$

called *meet* of σ and π, is the partition of b such that each block of $\sigma \wedge \pi$ is a nonempty intersection between one block of σ and one block of π. On the other hand, the partition

$$\sigma \vee \pi, \quad \sigma, \pi \in \mathcal{P}(b),$$

called *join* of σ and π, is the element of $\mathcal{P}(b)$ whose blocks are constructed by taking the non-disjoint unions of the blocks of σ and π, that is, by taking the union of those blocks that have at least one element in common.

Remarks. (a) Whenever $\pi_1 \leq \pi_2$, one has $|\pi_1| \geq |\pi_2|$. In particular, $|\sigma \wedge \pi| \geq |\sigma \vee \pi|$.

(b) The partition $\sigma \wedge \pi$ is the greatest lower bound associated with the pair (σ, π). As such, $\sigma \wedge \pi$ is completely characterized by the property of being the unique element of $\mathcal{P}(b)$ such that: (i) $\sigma \wedge \pi \leq \sigma$, (ii) $\sigma \wedge \pi \leq \pi$, and (iii) $\rho \leq \sigma \wedge \pi$ for every $\rho \in \mathcal{P}(b)$ such that $\rho \leq \sigma, \pi$.

(c) Analogously, the partition $\sigma \vee \pi$ is the least upper bound associated with the pair (σ, π). It follows that $\sigma \vee \pi$ is completely characterized by the property of being the unique element of $\mathcal{P}(b)$ such that: (i) $\sigma \leq \sigma \vee \pi$, (ii) $\pi \leq \sigma \vee \pi$, and (iii) $\sigma \vee \pi \leq \rho$ for every $\rho \in \mathcal{P}(b)$ such that $\sigma, \pi \leq \rho$.

Example 2.2.2   (i) Take $b = \{1, 2, 3, 4, 5\}$. If $\pi = \{\{1, 2, 3\}, \{4, 5\}\}$ and $\sigma = \{\{1, 2\}, \{3\}, \{4, 5\}\}$, then, as noted above, $\sigma \leq \pi$ and we have

$$\sigma \wedge \pi = \sigma \quad \text{and} \quad \sigma \vee \pi = \pi.$$

A graphical representation of π, σ, $\sigma \wedge \pi$ and $\sigma \vee \pi$ is:

$\pi =$	1 2 3	4 5

$\sigma =$	1 2	3	4 5

$\sigma \wedge \pi =$	1 2	3	4 5

$\sigma \vee \pi =$	1 2 3	4 5

(ii) If $\pi = \{\{1, 2, 3\}, \{4, 5\}\}$ and $\sigma = \{\{1, 2\}, \{3, 4, 5\}\}$, then π and σ are not ordered and

$$\sigma \wedge \pi = \{\{1, 2\}, \{3\}, \{4, 5\}\} \quad \text{and} \quad \sigma \vee \pi = \{b\} = \{\{1, 2, 3, 4, 5\}\}.$$

A graphical representation of π, σ, $\sigma \wedge \pi$ and $\sigma \vee \pi$ is:

$$\pi = \boxed{\begin{array}{c|c} 1\ 2\ 3 & 4\ 5 \end{array}}$$

$$\sigma = \boxed{\begin{array}{c|c} 1\ 2 & 3\ 4\ 5 \end{array}}$$

$$\sigma \wedge \pi = \boxed{\begin{array}{c|c|c} 1\ 2 & 3 & 4\ 5 \end{array}}$$

$$\sigma \vee \pi = \boxed{\begin{array}{c} 1\ 2\ 3\ 4\ 5 \end{array}}$$

(iii) A convenient way to build $\sigma \vee \pi$ is to do it in successive steps. Take the union of two blocks with a common element and look at it as a new block of π. See if it shares an element with another block of σ. If yes, repeat. For instance, suppose that $\pi = \{\{1,2\}, \{3\}, \{4\}\}$ and $\sigma = \{\{1,3\}, \{2,4\}\}$. Then, π and σ are not ordered and

$$\sigma \wedge \pi = \{\{1\}, \{2\}, \{3\}, \{4\}\} \quad \text{and} \quad \sigma \vee \pi = \{\{1,2,3,4\}\}.$$

One now obtains $\sigma \vee \pi$ by noting that the element 2 is common to $\{1,2\} \in \pi$ and $\{2,4\} \in \sigma$, and the "merged" block $\{1,2,4\}$ shares the element 1 with the block $\{1,3\} \in \sigma$, thus implying the conclusion. A graphical representation of π, σ, $\sigma \wedge \pi$ and $\sigma \vee \pi$ is:

$$\pi = \boxed{\begin{array}{c|c|c} 1\ 2 & 3 & 4 \end{array}}$$

$$\sigma = \boxed{\begin{array}{c|c} 1\ 3 & 2\ 4 \end{array}}$$

$$\sigma \wedge \pi = \boxed{\begin{array}{c|c|c|c} 1 & 2 & 3 & 4 \end{array}}$$

$$\sigma \vee \pi = \boxed{\begin{array}{c} 1\ 2\ 3\ 4 \end{array}}$$

Notation. When displaying a partition π of $\{1, ..., n\}$ $(n \geq 1)$, the blocks $b_1, ..., b_k \in \pi$ will always be listed in the following way: b_1 will always contain the element 1, and

$$\min\{i : i \in b_j\} < \min\{i : i \in b_{j+1}\}, \ j = 1, ..., k-1.$$

Also, the elements within each block will be always listed in increasing order. For instance, if $n = 6$ and the partition π involves the blocks $\{2\}, \{4\}, \{1,6\}$ and $\{3,5\}$, we will write $\pi = \{\{1,6\}, \{2\}, \{3,5\}, \{4\}\}$.

Definition 2.2.3 *The **maximal element** of $\mathcal{P}(b)$ is the trivial partition*

$$\boxed{\hat{1} = \{b\}.}$$

*The **minimal element** of $\mathcal{P}(b)$ is*

> *the partition $\hat{0}$, such that each block of $\hat{0}$ contains exactly one element of b.*

Observe that $\left|\hat{1}\right| = 1$ and $\left|\hat{0}\right| = |b|$, and also $\hat{0} \leq \hat{1}$. Thus if $b = \{1, 2, 3\}$, then

$$\hat{0} = \{\{1\}, \{2\}, \{3\}\} \quad \text{and} \quad \hat{1} = \{1, 2, 3\}.$$

Definition 2.2.4 *The **partition segment** (or **interval**) $[\sigma, \pi]$ in $\mathcal{P}(b)$, with $\sigma \leq \pi$, is the following subset of partitions of b:*

> $$[\sigma, \pi] = \{\rho \in \mathcal{P}(b) : \sigma \leq \rho \leq \pi\}.$$

Plainly,

$$\mathcal{P}(b) = \left[\hat{0}, \hat{1}\right].$$

2.3 Partitions of a set and partitions of an integer

We now focus on the notion of *class*, which associates to a segment of partitions a partition of an integer.

Definition 2.3.1 *The **class** of a segment $[\sigma, \pi]$ ($\sigma \leq \pi$), denoted $\lambda(\sigma, \pi)$, is defined as the partition of the integer $|\sigma|$ given by*

$$\lambda(\sigma, \pi) = (1^{r_1} 2^{r_2} \cdots |\sigma|^{r_{|\sigma|}}), \tag{2.3.5}$$

where r_i, $i = 1, ..., |\sigma|$, indicates the number of blocks of π that contain exactly i blocks of σ. We stress that necessarily $|\sigma| \geq |\pi|$, and also

$$|\sigma| = 1r_1 + 2r_2 + \cdots + |\sigma|r_{|\sigma|} \quad \text{and} \quad |\pi| = r_1 + \cdots + r_{|\sigma|}.$$

The lenght of $\lambda(\sigma, \pi)$ equals $|\pi|$.

Example 2.3.2 (i) If $\pi = \{\{1, 2, 3\}, \{4, 5\}\}$ and $\sigma = \{\{1, 2\}, \{3\}, \{4, 5\}\}$, then since $\{1, 2\}$ and $\{3\}$ are contained in $\{1, 2, 3\}$ and $\{4, 5\}$ in $\{4, 5\}$, we have 1 block of π (namely $\{4, 5\}$) containing 1 block of σ and 1 block of π (namely $\{1, 2, 3\}$) containing 2 blocks of σ. Thus $r_1 = 1$, $r_2 = 1$, $r_3 = 0$, that is, $\lambda(\sigma, \pi) = (1^1 2^1 3^0)$, corresponding to the partition of the integer $3 = 2 + 1$.
(ii) In view of (2.1.1), one may suppress the terms $r_i = 0$ in (2.3.5), and write for instance $\lambda(\sigma, \pi) = (1^1 2^0 3^2) = (1^1 3^2)$ for the class of the segment $[\sigma, \pi]$, associated with the two partitions $\sigma = \{\{1\}, \{2\}, \{3\}, \{4\}, \{5\}, \{6\}, \{7\}\}$ and $\pi = \{\{1\}, \{2, 3, 4\}, \{5, 6, 7\}\}$.

From now on, we let

$$[n] = \{1, \cdots, n\}, \quad n \geq 1. \tag{2.3.6}$$

With this notation, the maximal and minimal element of the set $\mathcal{P}([n])$ are given, respectively, by

$$\hat{1} = \{[n]\} = \{\{1, ..., n\}\} \quad \text{and} \quad \hat{0} = \{\{1\}, ..., \{n\}\}. \tag{2.3.7}$$

Now fix a set b, say $b = [n] = \{1, \cdots, n\}$ with $n \geq 1$ and consider the partition $\hat{0} = \{\{1\}, ..., \{n\}\}$. Then, for a fixed $\lambda = (1^{r_1} 2^{r_2} \cdots n^{r_n}) \vdash n$, the number of partitions $\pi \in \mathcal{P}(b)$ such that $\lambda\left(\hat{0}, \pi\right) = \lambda$ is given by

$$\begin{bmatrix} n \\ \lambda \end{bmatrix} \triangleq \begin{bmatrix} n \\ r_1, ..., r_n \end{bmatrix} = \frac{n!}{(1!)^{r_1} r_1! (2!)^{r_2} r_2! \cdots (n!)^{r_n} r_n!}. \tag{2.3.8}$$

This is the number of partitions π containing exactly r_1 blocks of size 1, r_2 blocks of size 2, ..., r_n blocks of size n. Equation (2.3.8) follows from the following fact. Fix $i = 1, ..., n$. If π contains r_i blocks of size i, then the elements in each block can be permuted within the block, yielding $(i!)^{r_i}$ possibilities and, in addition, the posiiton of the r_i blocks can be permuted as well, yielding $r_i!$ possiblities (see, for example, [146] for more details). The requirement that $\lambda\left(\hat{0}, \pi\right) = \lambda = (1^{r_1} 2^{r_2} \cdots n^{r_n})$ simply means that, for each $i = 1, ..., n$, the partition π must have exactly r_i blocks containing i elements of b. Recall that the integers $r_1, ..., r_n$ must satisfy (2.1.2), namely $1r_1 + \cdots + nr_n = n$.

Example 2.3.3   (i) For any finite set b, one has always that

$$\lambda\left(\hat{0}, \hat{1}\right) = \left(1^0 2^0 \cdots |b|^1\right),$$

because $\hat{1}$ has only one block, namely b, and that block contains $|b|$ blocks of $\hat{0}$.
(ii) Fix $k \geq 1$ and let b be such that $|b| = n \geq k+1$. Consider $\lambda = (1^{r_1} 2^{r_2} \cdots n^{r_n}) \vdash n$ be such that $r_k = r_{n-k} = 1$ and $r_j = 0$ for every $j \neq k, n - k$. For instance, if $n = 5$ and $k = 2$, then $\lambda = \left(1^0 2^1 3^1 4^0 5^0\right)$. Then, each partition $\pi \in \mathcal{P}(b)$ such that $\lambda\left(\hat{0}, \pi\right) = \lambda$ has only one block of k elements and one block of $n - k$ elements. To construct such a partition, it is sufficient to specify the block of k elements. This implies that there exists a bijection between the set of partitions $\pi \in \mathcal{P}(b)$ such that $\lambda\left(\hat{0}, \pi\right) = \lambda$ and the collection of the subsets of b having exactly k elements. In particular, (2.3.8) gives

$$\begin{bmatrix} n \\ \lambda \end{bmatrix} = \binom{n}{k} = \frac{n!}{k! (n - k)!}.$$

(iii) Let $b = [7] = \{1, ..., 7\}$ and $\lambda = (1^1 2^3 3^0 4^0 5^0 6^0 7^0)$. Then, (2.3.8) implies that there are exactly $\frac{7!}{3!(2!)^3} = 105$ partitions $\pi \in \mathcal{P}(b)$, such that $\lambda(\hat{0}, \pi) = \lambda$. One of these partitions is $\{\{1\}, \{2, 3\}, \{4, 5\}, \{6, 7\}\}$. Another is $\{\{1, 7\}, \{2\}, \{3, 4\}, \{5, 6\}\}$.

(iv) Let $b = [5] = \{1, ..., 5\}$, $\sigma = \{\{1\}, \{2\}, \{3\}, \{4, 5\}\}$ and $\pi = \{\{1, 2, 3\}, \{4, 5\}\}$. Then, $\sigma \leq \pi$ and the set of partitions defined by the interval $[\sigma, \pi]$ is $\{\sigma, \pi, \rho_1, \rho_2, \rho_3\}$, where

$$\rho_1 = \{\{1, 2\}, \{3\}, \{4, 5\}\}$$
$$\rho_2 = \{\{1, 3\}, \{2\}, \{4, 5\}\}$$
$$\rho_3 = \{\{1\}, \{2, 3\}, \{4, 5\}\}.$$

The partitions ρ_1, ρ_2 and ρ_3 are not ordered (i.e., for every $1 \leq i \neq j \leq 3$, one cannot write $\rho_i \leq \rho_j$), and are built by taking unions of blocks of σ in such a way that they are contained in blocks of π. Moreover, $\lambda(\sigma, \pi) = (1^1 2^0 3^1 4^0 5^0)$, since there is exactly one block of π containing one block of σ, and one block of π containing three blocks of σ.

(v) This example is related to the techniques developed in Chapter 6. Fix $n \geq 2$, as well as a partition $\gamma = (\gamma_1, ..., \gamma_k) \in \Lambda(n)$ such that $\gamma_k \geq 1$. Recall that, by definition, one has that $\gamma_1 \geq \gamma_2 \geq \cdots \geq \gamma_k$ and $\gamma_1 + \cdots + \gamma_k = n$. Now consider the segment $[\hat{0}, \pi]$, where

$$\hat{0} = \{\{1\}, \{2\}, ..., \{n\}\}, \text{ and}$$
$$\pi = \{\{1, ..., \gamma_1\}, \{\gamma_1 + 1, ..., \gamma_1 + \gamma_2\}, ..., \{\gamma_1 + \cdots + \gamma_{k-1} + 1, ..., n\}\}.$$

Then, the jth block of π contains exactly γ_j blocks of $\hat{0}$, for every $j = 1, ..., k$, implying that the class of the segment $[\hat{0}, \pi]$ coincides with $(\gamma_1, ..., \gamma_k)$. For instance, when $\gamma_1 > \gamma_2 > ... > \gamma_k$ (that is, all the γ_i's are different), one has that the class $\lambda(\hat{0}, \pi)$ is such that $\lambda(\hat{0}, \pi) = (\gamma_k^1 \gamma_{k-1}^1 \cdots \gamma_1^1) = \gamma$, after suppressing the indicators of the type r^0.

The following statement is a consequence of (2.3.8) and of Proposition 2.1.2.

Proposition 2.3.4 *Let $n \geq 1$ and $k \in \{1, ..., n\}$. Then, the number of partitions of $[n]$ having exactly k blocks is given by*

$$S(n, k) \triangleq \sum_{r_1, ..., r_{n-k+1}} \frac{n!}{(1!)^{r_1} r_1! (2!)^{r_2} r_2! \cdots (n - k + 1)!^{r_{n-k+1}} r_{n-k+1}!}, \quad (2.3.9)$$

where the sum runs over all vectors on nonnegative integers $(r_1, ..., r_{n-k+1})$ satisfying $r_1 + \cdots + r_{n-k+1} = k$ and $1r_1 + 2r_2 \cdots + (n - k + 1) r_{n-k+1} = n$.

Proof. By Definition 2.3.1, a partition $\pi \in \mathcal{P}([n])$ has k blocks if and only if $\lambda\left(\hat{0}, \pi\right)$ has length k. Since $\lambda\left(\hat{0}, \pi\right)$ is a partition of the integer n, then using Proposition 2.1.2 one deduces that $\pi \in \mathcal{P}([n])$ has k blocks if and only if $\lambda\left(\hat{0}, \pi\right)$ has the form of the right-hand side of (2.1.3), for some vector on nonnegative integers $(r_1, ..., r_{n-k+1})$ satisfying $r_1 + \cdots + r_{n-k+1} = k$ and $1r_1 + 2r_2 \cdots + (n-k+1) r_{n-k+1} = n$. The proof is concluded by using (2.3.8). ■

Remark. One defines customarily $S(0,0) = 1$ and, for $n \geq 1$, $S(n,0) = S(n,k) = 0$, for every $k > n$. The integers

$$S(n,k), \quad n, k \geq 0, \tag{2.3.10}$$

defined by these conventions and by (2.3.9), are called the *Stirling numbers of the second kind*. See, for example, [158, Ch. 8] and [146, p. 33] for some exhaustive presentations of the properties of Stirling numbers. See Section 2.4 for a connection with Bell and Touchard polynomials.

Example 2.3.5 (i) For every $n \geq 1$, one has that $S(n, 1) = 1$, that is, there exists only one partition of $[n]$ containing exactly one block (i.e., the trivial partition $\{[n]\}$). To see that this is consistent with (2.3.9) in the case $k = 1$, observe that the only integer solution to the system

$$r_1 + \cdots + r_n = 1, \quad \text{and} \quad 1r_1 + 2r_2 \cdots + n r_n = n,$$

is given by $r_1 = \cdots = r_{n-1} = 0$ and $r_n = 1$, and in this case

$$\frac{n!}{(1!)^{r_1} r_1! (2!)^{r_2} r_2! \cdots (n!)^{r_n} r_n!} = 1.$$

By a similar route, one also checks that $S(n, n) = 1$.

(ii) Fix $n \geq 3$. We want to compute $S(n, 2)$, that is, the number of partitions of $[n]$ containing exactly two blocks. This case corresponds to $k = 2$, and one has therefore to consider the system

$$r_1 + \cdots + r_{n-1} = 2, \quad \text{and} \quad 1r_1 + 2r_2 \cdots + (n-1) r_{n-1} = n.$$

When n is even, this system has exactly $n/2$ solutions, obtained by choosing either $r_{n/2} = 2$ and $r_l = 0$ elsewhere, or $r_j = r_{n-j} = 1$ and $r_l = 0$ elsewhere, for some $j = 1, ..., n/2 - 1$. On the other hand, when n is odd the system has exactly $(n-1)/2$ solutions, obtained by choosing $r_j = r_{n-j} = 1$ and $r_l = 0$ elsewhere, for some $j = 1, ..., (n-1)/2$. Using (2.3.9), we therefore deduce that

$$S(n, 2) = \sum_{j=1}^{(n/2)-1} \binom{n}{j} + \frac{1}{2} \binom{n}{n/2}, \quad \text{if } n \text{ is even,}$$

$$S(n, 2) = \sum_{j=1}^{(n-1)/2} \binom{n}{j}, \quad \text{if } n \text{ is odd.}$$

For instance $S(3,2) = \binom{3}{1} = 3$, $S(4,2) = \binom{4}{1} + \frac{1}{2}\binom{4}{2} = 4 + 3 = 7$, and

$$S(5,2) = \binom{5}{1} + \binom{5}{2} = 5 + 10 = 15.$$

The following statement contains a useful identity.

Proposition 2.3.6 *Fix $n \geq 1$. Let f be a function on $\mathcal{P}([n])$ such that there exists a function h on $\Lambda(n)$ (the set of the partitions of n) verifying $f(\pi) = h(\lambda(\hat{0}, \pi))$ (that is, f only depends on the class $\lambda(\hat{0}, \pi)$), one has*

$$\sum_{\pi = \{b_1, \ldots, b_k\} \in \mathcal{P}([n])} f(\pi) = \sum_{\lambda = (1^{r_1} 2^{r_2} \cdots n^{r_n}) \vdash n} \begin{bmatrix} n \\ \lambda \end{bmatrix} h(\lambda).$$

The proof of Proposition 2.3.6 is elementary and left to the reader.

2.4 Bell polynomials, Stirling numbers and Touchard polynomials

We will now connect some of the objects presented in the previous sections (in particular, the Stirling numbers of the second kind introduced in (2.3.9)) to the remarkable classes of Bell and Touchard polynomials. These polynomials will appear later in the book, as they often provide a neat way to express combinatorial relations between moment and cumulants of random variables. See, for example, [13, Ch. 11] for an exhaustive discussion of these objects, as well as [119, Ch. 1] and [146, p. 33] and the references therein.

Definition 2.4.1 *Fix $n \geq 1$. For every $k \in \{1, \ldots, n\}$, the* **partial Bell polynomial** *of index (n, k) is the polynomial in $n - k + 1$ variables given by*

$$B_{n,k}(x_1, \ldots, x_{n-k+1}) = \sum_{r_1, \ldots, r_{n-k+1}} \frac{n!}{r_1! r_2! \cdots r_{n-k+1}!} \left(\frac{x_1}{1!}\right)^{r_1} \cdots$$

$$\left(\frac{x_{n-k+1}}{(n-k+1)!}\right)^{r_{n-k+1}} \tag{2.4.11}$$

$$= \sum_{\substack{\lambda = (1^{r_1} 2^{r_2} \cdots n^{r_n}) \vdash n \\ \lambda \text{ has length } k}} \begin{bmatrix} n \\ \lambda \end{bmatrix} x_1^{r_1} \times \cdots \times x_{n-k+1}^{r_{n-k+1}}, \tag{2.4.12}$$

where the first sum runs over all vectors on nonnegative integers $(r_1, \ldots, r_{n-k+1})$ satisfying $r_1 + \cdots + r_{n-k+1} = k$ and $1r_1 + 2r_2 \cdots + (n - k + 1) r_{n-k+1} = n$, and the symbol $\begin{bmatrix} n \\ \lambda \end{bmatrix}$ is defined in (2.3.8). The nth **complete Bell polynomial** *is the polynomial* *in n variables given by*

$$B_n(x_1, ..., x_n) = \sum_{k=1}^{n} B_{n,k}(x_1, ..., x_{n-k+1}) \tag{2.4.13}$$

$$= \sum_{\lambda=(1^{r_1}2^{r_2}\cdots n^{r_n})\vdash n} \begin{bmatrix} n \\ \lambda \end{bmatrix} x_1^{r_1} \times \cdots \times x_n^{r_n}. \tag{2.4.14}$$

Finally, the collection $\{T_n : n \geq 0\}$ *of the* **Touchard** *(or* **exponential***) polynomials (in one variable) is defined as:* $T_0 = 1$ *and*

$$T_n(x) = \sum_{k=1}^{n} S(n, k)x^k, \quad n \geq 1, \tag{2.4.15}$$

where the Stirling number of the second kind $S(n, k)$ *is defined in* (2.3.9).

Remarks. (a) In view of (2.3.8), the polynomial $B_{n,k}$ in (2.4.11) is obtained by multiplying each monomial $x_1^{r_1} \times \cdots \times x_{n-k+1}^{r_{n-k+1}}$ by a coefficient equal to the number of partitions $\pi \in \mathcal{P}([n])$ such that $\lambda(\hat{0}, \pi) = \left(1^{r_1}2^{r_2}\cdots(n-k+1)^{r_{n-k+1}}(n-k+2)^0 \cdots n^0\right)$. Due to this construction, equalities (2.4.12) and (2.4.14) are direct consequences of Proposition 2.1.2.

(b) A more general class of Touchard polynomials (in several variables) is discussed, for example, in [13, Sect. 11.7].

Example 2.4.2 (i) By reasoning as in Example 2.3.5-(i), one deduces that, for every $n \geq 1$, $B_{n,1}(x_1, ..., x_n) = x_n$ and $B_{n,n}(x_1, ..., x_n) = x_1^n$.

(ii) By reasoning as in Example 2.3.5-(ii), one sees that, for every $n \geq 3$,

$$B_{n,2}(x_1, ..., x_n) = \sum_{j=1}^{(n/2)-1} \binom{n}{j} x_j x_{n-j} + \frac{1}{2}\binom{n}{n/2} x_{n/2}^2,$$

if n is even, and

$$B_{n,2}(x_1, ..., x_n) = \sum_{j=1}^{(n-1)/2} \binom{n}{j} x_j x_{n-j},$$

if n is odd.

(iii) Putting Points (i) and (ii) together, we can compute the third complete Bell polynomial $B_3(x_1, x_2, x_3)$, namely

$$B_3(x_1, x_2, x_3) = B_{3,1}(x_1, x_2, x_3) + B_{3,2}(x_1, x_2, x_3) + B_{3,3}(x_1, x_2, x_3)$$
$$= x_3 + 3x_1 x_2 + x_1^3.$$

The following statement establishes some relations between Bell and Touchard polynomials, as well as between these polynomials and Bell numbers. The proof follows almost immediately from Definition 2.4.1 and relation (2.3.9), and is left to the reader as an easy exercise.

Proposition 2.4.3 *For every $n \geq 1$, one has that, for every $k \in \{1, ..., n\}$,*

$$B_{n,k}(x, ..., x) = S(n, k)x^k. \tag{2.4.16}$$

Moreover,

$$T_n(x) = B_n(x, ..., x), \tag{2.4.17}$$

and

$$T_n(1) = B_n, \tag{2.4.18}$$

*where $B_n \triangleq \sum_{k=1}^{n} S(n, k) = |\mathcal{P}([n])|$ is the so-called nth **Bell number**, counting the number of partitions of $[n]$.*

To conclude, we present two remarkable relations involving Touchard polynomials. The proof is deferred to Section 3.3, where they will be deduced (quite elegantly) from moments and cumulants computations (see Proposition 3.3.2).

Proposition 2.4.4 *The class of Touchard polynomials enjoys the following relations: for every $n \geq 0$,*

$$T_{n+1}(x) = x \sum_{k=0}^{n} \binom{n}{k} T_k(x) = x \sum_{k=0}^{n} \binom{n}{k} T_{n-k}(x), \tag{2.4.19}$$

and

$$T_n(x + y) = \sum_{k=0}^{n} \binom{n}{k} T_k(x) T_{n-k}(y). \tag{2.4.20}$$

The second inequality in (2.4.19) follows from the identity $\binom{n}{k} = \binom{n}{n-k}$. Observe that relation (2.4.20) is equivalent to saying that the sequence of polynomials $\{T_n : n \geq 0\}$ is *of the binomial type*. See, for example, [56, Sect. 4.3], and the references therein, for an introduction to polynomial sequences of the binomial type.

Example 2.4.5 (i) We use relation (2.4.19) in order to compute the first few Touchard polynomials. One has: $T_0(x) = 1$, and

$$
\begin{aligned}
T_1(x) &= xT_0(x) = x \\
T_2(x) &= x(T_0(x) + T_1(x)) = x + x^2 \\
T_3(x) &= x(T_0(x) + 2T_1(x) + T_2(x)) = x + 3x^2 + x^3.
\end{aligned}
$$

In view of (2.4.18), this also provides the explicit value of the first Bell numbers, namely $B_1 = 1$, $B_2 = 2$ and $B_3 = 5$. This means that there is exactly one partition of the set [1], and that there are exactly 2 and 5 partitions, respectively, of [2] and [3]. In particular, the partitions of [2] are $\{\{1\}, \{2\}\}$ and $\{[2]\}$, whereas the partitions of [3] are

$$\{\{1\}, \{2\}, \{3\}\}, \quad \{\{1\}, \{2,3\}\}, \quad \{\{1,2\}, \{3\}\}, \quad \{\{1,3\}, \{2\}\} \text{ and } \{[3]\}.$$

(ii) Thanks to (2.4.18), equation (2.4.19) (in the case $x = 1$) yields a well-known recursion formula for Bell numbers, i.e.

$$B_{n+1} = \sum_{k=0}^{n} \binom{n}{k} B_k, \quad n \geq 0,$$

where one sets by convention $B_0 = 1$.

2.5 Möbius functions and Möbius inversion on partitions

We now state several properties of the *Möbius function* associated with the lattice of partitions of a finite set. These properties are used throughout the book. The next Section 2.6 contains some general facts on Möbius functions associated with arbitrary (finite) partially ordered sets.

For $\sigma, \pi \in \mathcal{P}(b)$, we denote by $\mu(\sigma, \pi)$ the *Möbius function* associated with the lattice $\mathcal{P}(b)$. It is defined as follows. If $\sigma \not\leq \pi$ (that is, if the relation $\sigma \leq \pi$ does not hold), then $\mu(\sigma, \pi) = 0$. If $\sigma \leq \pi$, then the quantity $\mu(\sigma, \pi)$ depends only on the class $\lambda(\sigma, \pi)$ of the segment $[\sigma, \pi]$, and is given by

$$\mu(\sigma, \pi) = (-1)^{n-r} (2!)^{r_3} (3!)^{r_4} \cdots ((n-1)!)^{r_n} \qquad (2.5.21)$$

$$= (-1)^{n-r} \prod_{j=0}^{n-1} (j!)^{r_{j+1}}, \qquad (2.5.22)$$

where $n = |\sigma|$, $r = |\pi|$, and $\lambda(\sigma, \pi) = (1^{r_1} 2^{r_2} \cdots n^{r_n})$ (that is, there are exactly r_i blocks of π containing exactly i blocks of σ). Since $0! = 1! = 1$, expressions (2.5.21) and (2.5.22) do not depend on the specific values of r_1 (the number of blocks of π containing exactly 1 block of σ) and r_2 (the number of blocks of π containing exactly two blocks of σ).

Example 2.5.1   (i) If $|b| = n \geq 1$ and $\sigma \in \mathcal{P}(b)$ is such that $|\sigma| = k \, (\leq n)$, then

$$\mu\left(\sigma, \hat{1}\right) = (-1)^{k-1} (k-1)!. \qquad (2.5.23)$$

Indeed, in (2.5.21) $r_k = 1$, since $\hat{1}$ has a single block which contains the k blocks of σ. In particular, $\mu\left(\hat{0}, \{b\}\right) = \mu\left(\hat{0}, \hat{1}\right) = (-1)^{n-1} (n-1)!$.

(ii) For every $\pi \in \mathcal{P}(b)$, one has $\mu(\pi, \pi) = 1$. Indeed, since (trivially) each element of π contains exactly one element of π, one has $\lambda(\pi, \pi) = \left(1^{|\pi|}2^0 3^0 \cdots n^0\right)$.

(iii) Fix $n \geq 2$ and let $b \subset [n]$ be such that $0 < |b| < n$. Consider a partition $\pi \in \mathcal{P}([n])$ such that b is a block of π, and write π' in order to indicate the partition of $[n] \backslash b$ (i.e. of the complement of b) composed of the blocks of π that are different from b. Then, denoting by μ and μ', respectively, the Möbius functions associated with $\mathcal{P}([n])$ and $\mathcal{P}([n] \backslash \pi)$, one has that

$$\mu(\hat{0}, \pi) = (-1)^{|b|-1}(|b| - 1)! \times \mu'(\hat{0}, \pi'). \qquad (2.5.24)$$

Equality (2.5.24) can be deduced by a direct inspection of formula (2.5.22) or, alternatively, as an application of the forthcoming Proposition 2.6.6.

The next result is crucial for the obtention of the combinatorial formulae found in Chapter 3 and Chapter 5 below. For every pair of functions $G, F : \mathcal{P}(b) \to \mathbb{C}$ one has that, $\forall \sigma \in \mathcal{P}(b)$,

$$\left| \; G(\sigma) = \sum_{\hat{0} \leq \pi \leq \sigma} F(\pi) \quad (\text{resp. } G(\sigma) = \sum_{\sigma \leq \pi \leq \hat{1}} F(\pi)) \; \right| \qquad (2.5.25)$$

if and only if, $\forall \pi \in \mathcal{P}(b)$,

$$\left| \; F(\pi) = \sum_{\hat{0} \leq \sigma \leq \pi} \mu(\sigma, \pi) G(\sigma) \quad (\text{resp. } F(\pi) = \sum_{\pi \leq \sigma \leq \hat{1}} \mu(\pi, \sigma) G(\sigma)), \; \right|$$

$$(2.5.26)$$

where $\mu(\cdot, \cdot)$ is the Möbius function given in (2.5.21). Relation (2.5.26) is known as the *Möbius inversion formula* associated with $\mathcal{P}(b)$. For a proof of the equivalence between (2.5.25) and (2.5.26), see the next section (general references are, for example, [146, Sect. 3.7] and [2]). To understand (2.5.26) as inversion formulae, one can interpret the sum $\sum_{\hat{0} \leq \pi \leq \sigma} F(\pi)$ as an integral of the type $\int_{\hat{0}}^{\sigma} F(\pi) \, d\pi$ (and analogously for the other sums appearing in (2.5.25) and (2.5.26)).

2.6 Möbius functions on partially ordered sets

We now present a short discussion on Möbius functions associated with general partially ordered finite sets. As anticipated, this discussion implicitly provides a proof of the inversion formula (2.5.26), as well as some useful additional identities. The reader is referred to the already quoted references [2] and [146] for further details.

2.6.1 Incidence algebras and general Möbius inversion

Let P be a finite partially ordered set with partial order $\preceq$. We write $x \prec y$ to indicate that $x \preceq y$ and $x \neq y$. For instance, P could be a set of subsets, with $\preceq$ equal to the inclusion relation $\subseteq$. In our context, $P = \mathcal{P}(b)$, the set of partitions of b, and $\preceq$ is the partial order $\leq$ considered above. The *incidence algebra* associated with P, denoted by $\mathcal{I}(P)$ is the collection of all those functions $f : P \times P \rightarrow \mathbb{C}$ such that $f(x, y) = 0$ whenever $x \not\preceq y$, endowed with the convolution operation

$$
(f * g)(x, y) = \sum_{z \,:\, x \preceq z \preceq y} f(x, z) g(z, y). \tag{2.6.27}
$$

Two elements of $\mathcal{I}(P)$, denoted by δ and ζ, play special roles in our discussion. The *delta function* δ is defined by the relation: $\delta(x, y) = 1$ if and only if $x = y$. Note that δ is the (left and right) *neutral element* with respect to the convolution $*$, meaning that $f * \delta = \delta * f = f$ for every $f \in \mathcal{I}(P)$. The *zeta function* ζ is defined as: $\zeta(x, y) = 1$ if $x \preceq y$, and $= 0$ otherwise.

In Section 2.5 the Möbius function μ was defined through relations (2.5.21)–(2.5.22). We define here through Proposition 2.6.1, and we will show that (2.5.21)–(2.5.22) hold.

Proposition 2.6.1 *The function ζ admits a (right and left) convolutional inverse, that is, there exists a unique function $\mu \in \mathcal{I}(P)$ such that*

$$
\zeta * \mu = \mu * \zeta = \delta. \tag{2.6.28}
$$

Moreover, the function μ is completely determined by the following recursive relations:

$$
\begin{aligned}
&\mu(x, x) = 1 && \forall x \in P, \\
&\mu(x, y) = -\sum_{x \preceq z \prec y} \mu(x, z), && \forall \, x, y \in P : x \prec y, \\
&\mu(x, y) = -\sum_{x \prec z \preceq y} \mu(z, y), && \forall \, x, y \in P : x \prec y, \\
&\mu(x, y) = 0 && \forall \, x, y \in P : x \not\preceq y.
\end{aligned} \tag{2.6.29}
$$

Proof. It is clear that (2.6.29) univocally defines an element of $\mathcal{I}(P)$ (see Section 2.6.2 below for more details on this point). It is therefore sufficient to prove that a function $\mu \in \mathcal{I}(P)$ verifies relation (2.6.29) if and only if it satisfies (2.6.28). In order to prove this double implication, the crucial observation is that, by virtue of (2.6.27), for every $f \in \mathcal{I}(P)$ and for every $x \preceq y$,

$$
\sum_{x \preceq z \preceq y} f(x, z) = (f * \zeta)(x, y)
$$

and

$$
\sum_{x \preceq z \preceq y} f(z, y) = (\zeta * f)(x, y),
$$

and, in particular, $(f * \zeta)(x, x) = (\zeta * f)(x, x) = f(x, x)$. It follows from these relations that a function $\mu \in \mathcal{I}(P)$ verifies (2.6.29) if and only if both $(\mu * \zeta)(x, y)$ and $(\zeta * \mu)(x, y)$ equal 1 or 0, according as $x = y$ or $x \neq y$. This is equivalent to saying that μ verifies (2.6.29) if and only if $\zeta * \mu = \mu * \zeta = \delta$, thus yielding the desired conclusion.   ∎

Remark. It is not difficult to prove (see, for example, [146, Proposition 3.6.2, p. 114]) that for a function $f \in \mathcal{I}(P)$ the following four properties are equivalent:

  (i) $f(x, x) \neq 0$ for every $x \in P$;
 (ii) f has a right inverse in $\mathcal{I}(P)$;
(iii) f has a left inverse in $\mathcal{I}(P)$, and
 (iv) f has a two-sided inverse in $\mathcal{I}(P)$.

Property (iv) means that there exists a function $g \in \mathcal{I}(P)$ such that g is both the left and right inverse of f. It is easily seen that such a function g is unique.

The convolutional inverse μ appearing in formulae (2.6.28)–(2.6.29) is called the *Möbius function* associated with P. In particular, when specializing (2.6.29) to the case where P is equal to $\mathcal{P}(b)$ (that is, to the collection of all partitions of the finite set b, endowed with the partial order $\leq$), one obtains that the Möbius function μ associated with $\mathcal{P}(b)$ is completely characterized by the relations

$$
\begin{aligned}
\mu(\sigma, \sigma) &= 1 & \forall \sigma \in \mathcal{P}(b), \\
\mu(\sigma, \pi) &= -\sum_{\sigma \leq \rho < \pi} \mu(\sigma, \rho), & \forall \sigma, \pi \in \mathcal{P}(b) : \sigma < \pi, \\
\mu(\sigma, \pi) &= -\sum_{\sigma < \rho \leq \pi} \mu(\rho, \pi), & \forall \sigma, \pi \in \mathcal{P}(b) : \sigma < \pi, \\
\mu(\sigma, \pi) &= 0 & \forall \sigma, \pi \in \mathcal{P}(b) : \sigma \nleq \pi,
\end{aligned}
\tag{2.6.30}
$$

where we write $\sigma < \pi$ to indicate that $\sigma \leq \pi$ and $\sigma \neq \pi$ (and similarly for $\rho < \pi$). It is now a routine combinatorial computation to check that the function $\mu : \mathcal{P}(b) \times \mathcal{P}(b) \to \mathbb{C}$, defined in (2.5.21)–(2.5.22), verifies (2.6.30), and therefore coincides with the convolutional inverse of zeta function associated with $\mathcal{P}(b)$. See Section 2.6.3 below for a direct proof of the fact that (2.5.21) the Möbius function of $\mathcal{P}(b)$.

Example 2.6.2 Consider the set $b = \{1, 2, 3\}$. In this case, one has that $\mathcal{P}(b)$ contains exactly five elements, namely: $\hat{0} = \{\{1\}, \{2\}, \{3\}\}$, $\hat{1} = \{b\}$, $\pi_1 = \{\{1, 2\}, \{3\}\}$, $\pi_2 = \{\{1\}, \{2, 3\}\}$ and $\pi_3 = \{\{1, 3\}, \{2\}\}$. Note that the three non-trivial partitions π_1, π_2, π_3 are not ordered. According to the definition of the Möbius function given in Section 2.5, one has that $\mu(\hat{0}, \hat{1}) = (-1)^2 2! = 2$ and $\mu(\pi_i, \hat{1}) = \mu(\hat{0}, \pi_i) = -1$ $(i = 1, 2, 3)$, yielding the relations

$$
\begin{aligned}
\mu(\hat{0}, \hat{1}) &= -[\mu(\pi_1, \hat{1}) + \mu(\pi_2, \hat{1}) + \mu(\pi_3, \hat{1}) + \mu(\hat{1}, \hat{1})] \\
&= -[\mu(\hat{0}, \pi_1) + \mu(\hat{0}, \pi_2) + \mu(\hat{0}, \pi_3) + \mu(\hat{0}, \hat{0})] \\
\mu(\hat{0}, \pi_i) &= \mu(\pi_i, \hat{1}) = -\mu(\pi_i, \pi_i) \quad (i = 1, 2, 3),
\end{aligned}
$$

that are consistent with (2.6.29).

Relation (2.6.28) yields a general version of the inversion formulae stated in (2.5.26).

Proposition 2.6.3 (General Möbius inversion) *Let P be a partially ordered set. For every pair of functions* $G, F : P \to \mathbb{C}$ *one has that,* $\forall y \in P$,

$$G(y) = \sum_{x \preceq y} F(x) \quad (\text{resp. } G(y) = \sum_{y \preceq x} F(x)) \tag{2.6.31}$$

if and only if, $\forall x \in P$,

$$F(x) = \sum_{y \preceq x} \mu(y, x) G(y) \quad (\text{resp. } F(x) = \sum_{x \preceq y} \mu(x, y) G(y)), \tag{2.6.32}$$

where μ *is the Möbius function associated with* P.

Proof. The fact that the left hand side of (2.6.31) is equivalent to the left hand side of (2.6.32) is a consequence of the relations

$$\sum_{y \preceq x} \mu(y, x) \sum_{u \preceq y} F(u) = \sum_{u \preceq x} F(u) \sum_{u \preceq y \preceq x} \mu(y, x) = \sum_{u \preceq x} F(u)(\zeta * \mu)(u, x) = F(x),$$

and

$$\sum_{x \preceq y} \sum_{u \preceq x} \mu(u, x) G(u) = \sum_{u \preceq y} G(u) \sum_{u \preceq x \preceq y} \mu(u, x) = \sum_{u \preceq y} G(u)(\mu * \zeta)(u, y) = G(y),$$

where we used the fact that $\mu * \zeta = \zeta * \mu = \delta$. Analogously, one has that

$$\sum_{x \preceq y} \mu(x, y) \sum_{y \preceq u} F(u) = \sum_{x \preceq u} F(u) \sum_{x \preceq y \preceq u} \mu(x, y) = \sum_{x \preceq u} F(u)(\mu * \zeta)(x, u) = F(x),$$

and

$$\sum_{y \preceq x} \sum_{x \preceq u} \mu(x, u) G(u) = \sum_{y \preceq u} G(u) \sum_{y \preceq x \preceq u} \mu(x, u) = \sum_{y \preceq u} G(u)(\zeta * \mu)(y, u) = G(y),$$

thus proving the remaining parts of the statement. ∎

Observe that relation (2.6.29) can be reformulated as follows: for each $x \preceq y$,

$$\sum_{x \preceq z \preceq y} \mu(z, y) = \sum_{x \preceq z \preceq y} \mu(x, z) = \begin{cases} 0 & \text{if } x \neq y, \\ \mu(x, x)\,(= 1) & \text{if } x = y, \end{cases} \tag{2.6.33}$$

which will be used in the sequel. In the next section, we shall demonstrate that either one of the two equalities appearing in (2.6.33) completely determines (by recursion) the function μ.

2.6.2 Computing a Möbius function

Let P be a finite partially ordered set, with partial order $\preceq$ and Möbius function μ. In this section, we shall describe a recursive procedure which allows to deduce the values of μ from either one of the two systems

$$\sum_{x \preceq z \preceq y} \mu\,(z, y) = \begin{cases} 0 \text{ if } x \neq y, \\ 1 \text{ if } x = y, \end{cases} \tag{2.6.34}$$

and

$$\sum_{x \preceq z \preceq y} \mu\,(x, z) = \begin{cases} 0 \text{ if } x \neq y, \\ 1 \text{ if } x = y. \end{cases} \tag{2.6.35}$$

As already done in Definition 2.2.4 (in the special case $P = \mathcal{P}(b)$), given $x \preceq y$ we define the *segment* (or *interval*) $[x, y]$ as

$$[x, y] \triangleq \{z \in P : x \preceq z \preceq y\}.$$

We also define the *length* of $[x, y]$ as the largest integer l such that there exist $\rho_0, \rho_1, ..., \rho_l \in P$ verifying

$$x = \rho_0 \prec \rho_1 \prec \rho_2 \prec \cdots \prec \rho_l = y.$$

If $x = y$, then the length of $[x, y] = \{x\}$ is defined to be zero by convention. Note that the length of a segment is always well-defined, since the partially ordered set P is finite; moreover the length of a segment is always $\leq |P| - 1$. If the length of $[x, y]$ is 1, then one says that y *covers* x (that is, y covers x if and only if $x \prec y$ and there is no $z \in P$ such that $x \prec z \prec y$).

Example 2.6.4 Let $P = \mathcal{P}([4])$ and $\pi = \{\{1\}, \{2\}, \{3, 4\}\} \in \mathcal{P}([4])$. Then, the interval $[\pi, \hat{1}]$ (recall that $\hat{1} = \{[4]\}$ is the maximal partition) contains $\pi, \hat{1}$, as well as the three partitions

$$\pi_1 = \{\{1, 2\}, \{3, 4\}\}, \quad \pi_2 = \{\{1\}, \{2, 3, 4\}\}, \quad \pi_3 = \{\{1, 3, 4\}, \{2\}\}.$$

Since π_1, π_2 and π_3 are not ordered, one deduces that the length of $[\pi, \hat{1}]$ is 2. Observe that $\hat{1}$ covers π_i, for $i = 1, 2, 3$.

A recursive procedure for computing μ. We focus on (2.6.34). By construction, one has that $\mu(x, x) = 1$ for every $x \in P$. Moreover, if y covers x, then (2.6.34) becomes $\mu(x, y) + \mu(y, y) = 0$, thus yielding $\mu(x, y) = -1$. Now fix an integer $l \geq 1$, assume that the values of $\mu(x, y)$ are known for every $x \preceq y$ such that the length of $[x, y]$ is $\leq l$, and consider a pair $u \preceq v$ such that $[u, v]$ has length $l + 1$. Then, (2.6.34) implies that

$$\mu(u, v) = - \sum_{u \prec z \preceq v} \mu(z, v).$$

Since the right hand side of the preceding equation only involves pairs $z \preceq v$ such that $[z, v]$ has length at most l, this procedure completely determines the value of $\mu(u, v)$. By recursion on l, it also determines the whole Möbius function μ. Note that one could equivalently use relation (2.6.35).

2.6.3 Further properties of Möbius functions and a proof of (2.5.21)

Consider two finite partially ordered sets P, Q, whose order relations are denoted, respectively, by $\preceq_P$ and $\preceq_Q$. The *product* of P and Q is defined as the Cartesian product $P \times Q$, endowed with the following partial order relation: $(x, y) \preceq_{P \times Q} (x', y')$ if, and only if, $x \preceq_P x'$ and $y \preceq_Q y'$. Products of more than two partially ordered sets are defined analogously. Note that product of lattices are again lattices. We say (see, for example, [146, p. 98]) that P and Q are *isomorphic* if there exists a bijection $\psi : P \to Q$ which is order-preserving and such that the inverse of ψ is also order-preserving; this requirement on the bijection ψ is equivalent to saying that, for every $x, x' \in P$,

$$x \preceq_P x' \text{ if and only if } \psi(x) \preceq_Q \psi(x'). \tag{2.6.36}$$

Of course, two isomorphic partially ordered sets have the same cardinality. The following result is quite useful for explicitly computing Möbius functions. It states that the Möbius function is invariant under isomorphisms, and that the Möbius function of a lattice product is the product of the associated Möbius functions. See, for example, [146, Sect. 3.8] for more applications of these results.

Proposition 2.6.5 *Let P, Q be two partially ordered sets, and let μ_P and μ_Q denote their Möbius functions. Then,*

1 If P and Q are isomorphic, then

$$\mu_P(x, y) = \mu_Q(\psi(x), \psi(y))$$

for every $x, y \in P$, where ψ is the bijection appearing in (2.6.36).
2 The Möbius function associated with the partially ordered set $P \times Q$ is given by:

$$\mu_{P \times Q}[(x, y), (x', y')] = \mu_P(x, x') \times \mu_Q(y, y').$$

Proof. Point 1 is an immediate consequence of (2.6.29). Point 2 follows from the fact that, for every pair of partially ordered sets P and Q with zeta functions ζ_P and ζ_Q, one has that (by definition)

$$\zeta_{P \times Q}[(x,y),(x',y')] = \zeta_P(x,x')\zeta_Q(y,y'),$$

whenever $(x,y) \preceq_{P \times Q} (x',y')$.   ∎

We will implicitly apply Proposition 2.6.5 to deduce formula (2.5.21), as well as in the proof of Theorem 6.1.1. In both instances, it will be used in combination with the next statement. Its proof will be illuminated by an example.

Proposition 2.6.6 *Let b be a finite set, and let $\pi, \sigma \in \mathcal{P}(b)$ be such that:*

  (i) *$\sigma \leq \pi$;*
  (ii) *the segment $[\sigma, \pi]$ has class $(\lambda_1, ..., \lambda_k) \vdash |\sigma|$.*

Then, $[\sigma, \pi]$ is a partially ordered set isomorphic to the lattice product of the k sets $\mathcal{P}([\lambda_i])$, $i = 1, ..., k$.

Proof. To prove the statement, we shall use the fact that each partition in $[\sigma, \pi]$ is obtained by taking unions of the blocks of σ that are contained in the same block of π. Focus on a partition $\rho \in [\sigma, \pi]$. We want to map it to a partition of integers. To do this, observe first that $(\lambda_1, ..., \lambda_k)$ is the class of $[\sigma, \pi]$ if and only if for every $i = 1, ..., k$, there is a block $b_i \in \pi$ such that b_i contains exactly λ_i blocks of σ. In particular, $k = |\pi|$. We now construct a bijection ψ, between $[\sigma, \pi]$ and the lattice products of the $\mathcal{P}([\lambda_i])$'s, as follows:

i)   for $i = 1, ..., k$, write $b_{i,j}$, $j = 1, ..., \lambda_i$, to indicate the blocks of σ contained in b_i;
ii) for every partition $\rho \in [\sigma, \pi]$ and every $i = 1, ..., k$, construct a partition $\zeta(i, \rho)$ of the set of integers $[\lambda_i] = \{1, ..., \lambda_i\}$ by the following rule: for every $j, l \in \{1, ..., \lambda_i\}$, one has

$$j \sim_{\zeta(i,\rho)} l$$

(that is, j and l belong to the same block of $\zeta(i, \rho)$) if and only if

the union $b_{i,j} \cup b_{i,l}$ is contained in a block of ρ;

iii) define the application $\psi : [\sigma, \pi] \to \mathcal{P}([\lambda_1]) \times \cdots \times \mathcal{P}([\lambda_k])$ as

$$\rho \mapsto \psi(\rho) := (\zeta(1, \rho), ..., \zeta(k, \rho)). \tag{2.6.37}$$

It is easily seen that the application ψ in (2.6.37) is indeed an order-preserving bijection, verifying (2.6.36) for $P = [\sigma, \pi]$ and $Q = \mathcal{P}([\lambda_1]) \times \cdots \times \mathcal{P}([\lambda_k])$.   ∎

Example 2.6.7 Let $p = [4] = \{1, 2, 3, 4\}$, and consider the following two partitions in $\mathcal{P}(b)$:

$$\sigma = \{\{1, 2\}, \{3\}, \{4\}\} = \{b_1, b_2, b_3\},$$
$$\pi = \hat{1} = \{1, 2, 3, 4\}.$$

The segment $[\sigma, \pi]$ is such that $[\sigma, \pi] = \{\sigma, \pi_1, \pi_2, \pi_3, \hat{1}\}$, where

$$\pi_1 = \{\{1, 2, 3\}, \{4\}\}$$
$$\pi_2 = \{\{1, 2, 4\}, \{3\}\}$$
$$\pi_3 = \{\{1, 2\}, \{3, 4\}\}.$$

Since the only block of π contains 3 blocks of σ, namely b_1, b_2, b_3, one has $\lambda_1 = 3$ (here, $k = |\pi| = 1$). Consider now the mapping

$$\zeta : [\sigma, \pi] \to \mathcal{P}([3]) = \mathcal{P}(\{1, 2, 3\})$$

which associates to a partition $\rho \in [\sigma, \pi]$, the partition of $[3] = \{1, 2, 3\}$, which puts j and l in the same block if and only if the blocks b_j and b_l are contained in the same block of ρ. Then

$$\zeta(\sigma) = \zeta\{b_1, b_2, b_3\} = \{\{1\}, \{2\}, \{3\}\}$$
$$\zeta(\pi_1) = \zeta\{b_1 \cup b_2, b_3\} = \{\{1, 2\}, \{3\}\}$$
$$\zeta(\pi_2) = \zeta\{b_1 \cup b_3, b_2\} = \{\{1, 2\}, \{3\}\}$$
$$\zeta(\pi_3) = \zeta\{b_1, b_2 \cup b_3\} = \{\{1\}, \{2, 3\}\}$$
$$\zeta(\hat{1}) = \zeta\{b_1 \cup b_2 \cup b_3\} = \{\{1, 2, 3\}\}.$$

Observe that ζ is an order preserving bijection. For instance, $\pi_3 \leq \pi$ and $\zeta(\pi_3) \leq \zeta(\pi)$.

Remark. By inspection of the preceding proof of Proposition 2.6.6, one sees that, if $\sigma \leq \pi$ are two partitions as in the statement of Proposition 2.6.6, then the segment $[\sigma, \pi]$ is a lattice (with respect to the partial order relation $\leq$) with minimal and maximal elements given, respectively, by σ and π. Recall that a *lattice* is a partially ordered set P such that each pair $x, z \in P$ has a least upper bound (denoted by $x \vee z$) and a greatest lower bound (denoted by $x \wedge z$).

We shall now provide a self-contained proof of (2.5.21), which is inspired by the discussion contained in [80, Ch. 10]. Our arguments hinge on the following general lemma.

Lemma 2.6.8 *Let P be a finite lattice, with partial order $\preceq$ and Möbius function μ. Denote by $\hat{0}$ and $\hat{1}$, respectively, the minimal and maximal element of P. Then, for every $z \neq \hat{0}$, one has that*

$$\sum_{x \,:\, x \vee z = \hat{1}} \mu(\hat{0}, x) = 0. \tag{2.6.38}$$

Proof. From (2.6.34) one infers that

$$\delta(x \vee z, \hat{1}) = \sum_{x \vee z \preceq t \preceq \hat{1}} \mu(t, \hat{1}) = \sum_{\hat{0} \preceq t \preceq \hat{1}} \mu(t, \hat{1})\zeta(x, t)\zeta(z, t),$$

where the second equality derives from the fact that $x \vee z \preceq t$ if and only if $x \preceq t$ and $z \preceq t$. Using these relations, we deduce that

$$\sum_{x : x \vee z = \hat{1}} \mu(\hat{0}, x) = \sum_{\hat{0} \preceq x \preceq \hat{1}} \mu(\hat{0}, x)\delta(x \vee z, \hat{1}) = \sum_{\hat{0} \preceq x \preceq \hat{1}} \mu(\hat{0}, x) \sum_{\hat{0} \preceq t \preceq \hat{1}} \mu(t, \hat{1})\zeta(x, t)\zeta(z, t)$$

$$= \sum_{z \preceq t \preceq \hat{1}} \mu(t, \hat{1}) \sum_{\hat{0} \preceq x \preceq t} \mu(\hat{0}, x)\zeta(x, t) = \sum_{z \preceq t \preceq \hat{1}} \mu(t, \hat{1})(\mu * \zeta)(\hat{0}, t)$$

$$= \sum_{z \preceq t \preceq \hat{1}} \mu(t, \hat{1})\delta(\hat{0}, t) = 0,$$

since $\mu * \zeta = \delta$ and $\hat{0} \prec z$ by assumption. ∎

Proof that (2.5.21) defines the Möbius function of $\mathcal{P}(b)$. Our general definiton of the Möbius function led to formula (2.6.35). We now want to show that this implies formula (2.5.21) for the Möbius function of $\mathcal{P}(b)$. Without loss of generality, we can assume that $b = [n]$ for some integer $n \geq 1$. Also, we shall use the notation $\mu_j, j \geq 1$, to indicate the quantity $\mu(\hat{0}, \hat{1})$, whenever μ is the Möbius function associated with the lattice $\mathcal{P}([j])$. We first show that $\mu_1 = 1$, $\mu_2 = -1$ and $\mu_3 = 2$. Indeed, $\mathcal{P}([1])$ contains a single element $x = \{\{1\}\}$ and hence $\mu(x, x) = 1$. On the other hand $\mathcal{P}([2])$ contains only $\hat{0}$ and $\hat{1}$ and hence (2.6.35) implies $0 = \mu(\hat{0}, \hat{0}) + \mu(\hat{0}, \hat{1}) = 1 + \mu_2$, yielding $\mu_2 = -1$. Finally, $\mathcal{P}([3]) = \{\hat{0}, \pi_1, \pi_2, \pi_3, \hat{1}\}$, where $\pi_1 = \{\{1, 2\}, \{3\}\}$, $\pi_2 = \{\{1, 3\}, \{2\}\}$ and $\pi_3 = \{\{1\}, \{2, 3\}\}$. Formula (2.6.35) implies therefore that $0 = \mu(\hat{0}, \hat{0}) + \mu(\hat{0}, \pi_i)$, that is, $\mu(\hat{0}, \pi_i) = -1$, for $i = 1, 2, 3$. To conclude, use the fact that

$$0 = \sum_{\hat{0} \preceq z \preceq \hat{1}} \mu(\hat{0}, z) = 1 - 1 - 1 - 1 + \mu(\hat{0}, \hat{1}),$$

implying $\mu_3 = \mu(\hat{0}, \hat{1}) = 2$. Now consider two partitions $\sigma, \pi \in \mathcal{P}([n])$ such that $\sigma \leq \pi$, and the segment $[\sigma, \pi]$ has class $(\lambda_1, ..., \lambda_k) \vdash |\sigma|$. Using Proposition 2.6.5 and Proposition 2.6.6, one has the decomposition

$$\mu(\sigma, \pi) = \prod_{i=1}^{k} \mu_{\lambda_i}.$$

It follows that, in order to prove (2.5.21), it is sufficient to show that, for every $j \geq 1$,

$$\mu_j = (-1)^{j-1}(j - 1)!. \tag{2.6.39}$$

We already pointed out that (2.6.39) is true for $j = 1, 2, 3$ so that we can assume that $j \geq 4$. According to Lemma 2.6.8, for every partition $\rho \in \mathcal{P}([j])$ such that $\rho > \hat{0}$, one has that

$$\sum_{\pi:\pi\vee\rho=\hat{1}} \mu(\hat{0},\pi) = 0. \tag{2.6.40}$$

We now want to explicitly compute the left-hand side of (2.6.40) when $\rho = \{\{1\}, \{2\}, ..., \{j-2\}, \{j-1, j\}\}$, that is, when ρ is the partition obtained by merging the last two blocks of $\hat{0}$. In this case, the class $\{\pi : \pi \vee \rho = \hat{1}\}$ is given by the following union:

$$\{\pi : \pi \vee \rho = \hat{1}\} = \hat{1} \cup \{\pi_A : A \subseteq [j-2]\},$$

where the π_A are defined as follows:

- $\pi_\emptyset = \{\{1, 2, ..., j-2, j\}, \{j-1\}\}$;
- $\pi_{[j-2]} = \{[j-1], \{j\}\} = \{\{1, ..., j-1\}, \{j\}\}$;
- $\pi_A = \{A \cup \{j-1\}, \{j\} \cup ([j-2]\backslash A)\}$, for every $A \neq \emptyset, [j-2]$.

Since each partition π_A has one block of size $|A| + 1$ and one block of size $1 + (j - 2 - |A|) = j - |A| - 1$, applying again Proposition 2.6.5 and Proposition 2.6.6 one sees that

$$\mu(\hat{0},\pi_A) = \mu_{|A|+1}\,\mu_{j-|A|-1}.$$

Plugging these expressions into (2.6.40) yields

$$0 = \sum_{\pi:\pi\vee\rho=\hat{1}} \mu(\hat{0},\pi) = \mu_j + \sum_{A\subseteq[j-2]} \mu(\hat{0},\pi_A)$$

$$= \mu_j + \sum_{A\subseteq[j-2]} \mu_{|A|+1}\,\mu_{j-|A|-1}$$

$$= \mu_j + \sum_{k=0}^{j-2} \binom{j-2}{k}\mu_{k+1}\,\mu_{j-k-1}.$$

We thus get an expression of μ_j in terms of μ_l, $l \leq j - 1$, valid for $j \geq 4$. We now conclude the proof of (2.6.39) (and therefore of (2.5.21)) by induction. Suppose that $\mu_l = (-1)^{l-1}(l - 1)!$ for every $l \leq j - 1$, where $j \geq 4$: then,

$$\mu_j = -\sum_{k=0}^{j-2} \binom{j-2}{k}[(-1)^k k!]\,[(-1)^{j-k-2}(j - k - 2)!] = (-1)^{j-1}(j - 1)!.$$

This concludes the proof. ∎

3

Combinatorial expressions of cumulants and moments

We recall here the definition of *cumulant*, and we present several of its properties. A classic discussion of cumulants is contained in the book by Shiryaev [141]; see also the papers by Rota and Shen [131], Speed [145] and Surgailis [151], as well as Pitman [119, pp. 20-23] and the references therein. From now on, we assume that every random variable is defined on a common suitable probability space $(\Omega, \mathcal{F}, \mathbb{P})$. The symbol $\mathbb{E}$ indicates the mathematical expectation associated with the probability measure $\mathbb{P}$.

3.1 Cumulants

For $n \geq 1$, we consider a vector of real-valued random variables $\mathbf{X}_{[n]} = (X_1, ..., X_n)$ such that $\mathbb{E}|X_j|^n < \infty$, $\forall j = 1, ..., n$. For every subset $b = \{j_1, ..., j_k\} \subseteq [n] = \{1, ..., n\}$, we write

$$\boxed{\mathbf{X}_b = (X_{j_1}, ..., X_{j_k})} \quad \text{and} \quad \boxed{\mathbf{X}^b = X_{j_1} \times \cdots \times X_{j_k},} \qquad (3.1.1)$$

where $\times$ denotes the usual product. For instance, $\forall m \leq n$,

$$\mathbf{X}_{[m]} = (X_1, ..., X_m) \quad \text{and} \quad \mathbf{X}^{[m]} = X_1 \times \cdots \times X_m.$$

Remark. $\mathbf{X}_b$ is a vector of random variables, whereas $\mathbf{X}^b$ is a random variable.

For every $b = \{j_1, ..., j_k\} \subseteq [n]$ and $(z_1, ..., z_k) \in \mathbb{R}^k$, we let

$$g_{\mathbf{X}_b}(z_1, .., z_k) = \mathbb{E}\left[\exp\left(i \sum_{\ell=1}^k z_\ell X_{j_\ell}\right)\right].$$

The *joint cumulant* of the components of the vector $\mathbf{X}_b$ is defined as

$$\chi(\mathbf{X}_b) = (-i)^k \frac{\partial^k}{\partial z_1 \cdots \partial z_k} \ln g_{\mathbf{X}_b}(z_1, ..., z_k) \mid_{z_1 = ... = z_k = 0}, \qquad (3.1.2)$$

G. Peccati, M.S. Taqqu: Wiener Chaos: Moments, Cumulants and Diagrams –
A survey with computer implementation.
© Springer-Verlag Italia 2011

thus

$$\chi(X_1, ..., X_k) = (-i)^k \frac{\partial^k}{\partial z_1 \cdots \partial z_k} \ln \mathbb{E}\left[\exp\left(i \sum_{\ell=1}^k z_l X_l\right)\right]\Bigg|_{z_1 = ... = z_k = 0}.$$

We recall the following facts.

(i) The application $\mathbf{X}_b \mapsto \chi(\mathbf{X}_b)$ is *homogeneous*, that is, for every $\mathbf{h} = (h_1, ..., h_k) \in \mathbb{R}^k$,

$$\chi(h_1 X_{j_1}, ..., h_k X_{j_k}) = (\Pi_{\ell=1}^k h_\ell) \times \chi(\mathbf{X}_b).$$

(ii) The application $\mathbf{X}_b \mapsto \chi(\mathbf{X}_b)$ is invariant with respect to the permutations of b.

(iii) $\chi(\mathbf{X}_b) = 0$, if the vector $\mathbf{X}_b$ has the form $\mathbf{X}_b = \mathbf{X}_{b'} \cup \mathbf{X}_{b''}$, with $b', b'' \neq \varnothing$, $b' \cap b'' = \varnothing$ and $\mathbf{X}_{b'}$ and $\mathbf{X}_{b''}$ independent.

(iv) If $\mathbf{Y} = \{Y_j : j \in J\}$ is a Gaussian family and if $\mathbf{X}_{[n]}$ is a vector obtained by juxtaposing $n \geq 3$ elements of $\mathbf{Y}$ (with possible repetitions), then $\chi(\mathbf{X}_{[n]}) = 0$.

Properties (i) and (ii) follow immediately from (3.1.2). To see how to deduce (iii) from (3.1.2), just observe that, if $\mathbf{X}_b$ has the structure described in (iii), then

$$\ln g_{\mathbf{X}_b}(z_1, ..., z_k) = \ln g_{\mathbf{X}_{b'}}(z_\ell : j_\ell \in b') + \ln g_{\mathbf{X}_{b''}}(z_\ell : j_\ell \in b'')$$

(by independence), so that

$$\frac{\partial^k}{\partial z_1 \cdots \partial z_k} \ln g_{\mathbf{X}_b}(z_1, ..., z_k)$$
$$= \frac{\partial^k}{\partial z_1 \cdots \partial z_k} \ln g_{\mathbf{X}_{b'}}(z_\ell : j_\ell \in b') + \frac{\partial^k}{\partial z_1 \cdots \partial z_k} \ln g_{\mathbf{X}_{b''}}(z_\ell : j_\ell \in b'') = 0,$$

since we are taking derivatives over non-existent variables.

Finally, property (iv) is proved by using the fact that, if $\mathbf{X}_{[n]}$ is obtained by juxtaposing $n \geq 3$ elements of a Gaussian family (even with repetitions), then $\ln g_{\mathbf{X}_b}(z_1, .., z_k)$ has necessarily the form $\sum_l a(l) z_l + \sum_{i,j} b(i,j) z_i z_j$, where $a(k)$ and $b(i,j)$ are coefficients not depending on the z_l's. All the derivatives of order higher than 2 are then zero.

When $|b| = n$, one says that the cumulant $\chi(\mathbf{X}_b)$, given by (3.1.2), *has order* n. When $\mathbf{X}_{[n]} = (X_1, ..., X_n)$ is such that $X_j = X$, $\forall j = 1, ..., n$, where X is a random variable in $L^n(\mathbb{P})$ (that is, $\mathbb{E}|X|^n < \infty$), we use the notation χ_n and write

$$\left| \; \chi(\mathbf{X}_{[n]}) = \chi(X, ..., X) = \chi_n(X) \; \right| \tag{3.1.3}$$

and we say that $\chi_n(X)$ is the *n*th *cumulant* (or the *cumulant of order n*) of X. Of course, in this case one has that

$$\chi_n(X) = (-i)^n \frac{\partial^n}{\partial z^n} \ln g_X(z)\Bigg|_{z=0},$$

where $g_X(z) = \mathbb{E}[\exp(izX)]$. Note that, if $X, Y \in L^n(\mathbb{P})$ $(n \geq 1)$ are independent random variables, then (3.1.2) implies that

$$\chi_n(X + Y) = \chi_n(X) + \chi_n(Y),$$

since $\chi_n(X + Y)$ involves the derivatives of

$$\ln \mathbb{E}[\exp(iz(X + Y))] = \ln \mathbb{E}[\exp(izX)] + \ln \mathbb{E}[\exp(izY)]$$

with respect to z. In particular, by setting Y to be equal to a constant $c \in \mathbb{R}$, one obtains the remarkable properties that

$$\chi_1(X + c) = \mathbb{E}(X + c) = \mathbb{E}(X) + c = \chi_1(X) + c, \tag{3.1.4}$$
$$\chi_n(X + c) = \chi_n(X), \quad n \geq 2, \tag{3.1.5}$$

since $\chi_n(c) = 0$ for every $n \geq 2$.

Example 3.1.1 Let N be a Poisson random variable with parameter $x > 0$, that is, $\mathbb{P}[N = k] = e^{-x}x^k/k!$, for every integer $k \geq 0$. Then, elementary computations show that $g_N(z) = \exp[x(e^{iz} - 1)]$, thus yielding $\chi_n(N) = x$, for every $n \geq 1$.

3.2 Relations involving moments and cumulants

We want to relate expectations of products of random variables, such as $\mathbb{E}[X_1 X_2 X_3]$, to cumulants of vectors of random variables, such as $\chi(X_1, X_2, X_3)$. Note the dissymmetry: moments involve products, while cumulants involve vectors. We will have, for example,

$$\chi(X_1, X_2) = \mathbb{E}[X_1 X_2] - \mathbb{E}[X_1]\mathbb{E}[X_2],$$

and hence

$$\chi(X_1, X_2) = \text{Cov}(X_1, X_2),$$

the covariance of the vector (X_1, X_2). Conversely, we will have

$$\mathbb{E}[X_1 X_2] = \chi(X_1)\chi(X_2) + \chi(X_1, X_2).$$

Thus, using the notation introduced above, we will establish precise relations between objects of the type

$$\left| \quad \chi(\mathbf{X}_b) = \chi(X_j : j \in b) \quad \text{and} \quad \mathbb{E}[\mathbf{X}^b] = \mathbb{E}[\Pi_{j \in b} X_j]. \quad \right|$$

We can do this also for random variables that are products of other random variables: for instance, to obtain $\chi(Y_1 Y_2, Y_3)$, we apply the previous formula with $X_1 = Y_1 Y_2$ and $X_2 = Y_3$, and get $\chi(Y_1 Y_2, Y_3) = \mathbb{E}[Y_1 Y_2 Y_3] - \mathbb{E}[Y_1 Y_2]\mathbb{E}[Y_3]$. We shall also state a formula, due to Malyshev, which expresses $\chi(Y_1 Y_2, Y_3)$ in terms of other cumulants, namely in this case

$$\chi(Y_1 Y_2, Y_3) = \chi(Y_1, Y_3)\chi(Y_2) + \chi(Y_1)\chi(Y_2, Y_3) + \chi(Y_1, Y_2, Y_3).$$

The next result, first proved in [62] (for Parts 1 and 2) and [67] (for Part 3), contains three crucial relations, linking the cumulants and the moments associated with a random vector $\mathbf{X}_{[n]}$. We use the properties of the lattices of partitions, as introduced in the previous chapter.

Proposition 3.2.1 (Leonov and Shiryaev [62] and Malyshev [67]) *For every* $b \subseteq [n]$,

1.

$$\mathbb{E}\left[\mathbf{X}^b\right] = \sum_{\pi = \{b_1, \ldots, b_k\} \in \mathcal{P}(b)} \chi(\mathbf{X}_{b_1}) \cdots \chi(\mathbf{X}_{b_k}) ; \qquad (3.2.6)$$

2.

$$\chi(\mathbf{X}_b) = \sum_{\sigma = \{a_1, \ldots, a_r\} \in \mathcal{P}(b)} (-1)^{r-1} (r-1)! \mathbb{E}(\mathbf{X}^{a_1}) \cdots \mathbb{E}(\mathbf{X}^{a_r}) ; \qquad (3.2.7)$$

3. $\forall \sigma = \{b_1, \ldots, b_k\} \in \mathcal{P}(b)$,

$$\chi\left(\mathbf{X}^{b_1}, \ldots, \mathbf{X}^{b_k}\right) = \sum_{\substack{\tau = \{t_1, \ldots, t_s\} \in \mathcal{P}(b) \\ \tau \vee \sigma = \hat{1}}} \chi(\mathbf{X}_{t_1}) \cdots \chi(\mathbf{X}_{t_s}). \qquad (3.2.8)$$

Remark. To the best of our knowledge, our forthcoming proof of equation (3.2.8) (which is known as *Malyshev's formula*) is new. As an illustration of (3.2.8), consider the cumulant $\chi(X_1 X_2, X_3)$, in which case one has $\sigma = \{b_1, b_2\}$, with $b_1 = \{1, 2\}$ and $b_2 = \{3\}$. There are three partitions $\tau \in \mathcal{P}([3])$ such that $\tau \vee \sigma = \hat{1} = \{1, 2, 3\}$, namely $\tau_1 = \hat{1}$, $\tau_2 = \{\{1, 3\}, \{2\}\}$ and $\tau_3 = \{\{1\}, \{2, 3\}\}$, from which it follows that

$$\chi(X_1 X_2, X_3) = \chi(X_1, X_2, X_3) + \chi(X_1, X_3)\chi(X_2) + \chi(X_1)\chi(X_2, X_3).$$

Proof of Proposition 3.2.1. The proof of (3.2.6) is obtained by differentiating the characteristic function and its logarithm, and by identifying corresponding terms (for more details on these computations, see [141], [131, Sect. 6] or [145]). We now show how to obtain (3.2.7) and (3.2.8) from (3.2.6).

Relation (3.2.6) implies that, $\forall \sigma = \{a_1, ..., a_r\} \in \mathcal{P}(b)$,

$$\prod_{j=1}^{r} \mathbb{E}[\mathbf{X}^{a_j}] = \sum_{\substack{\pi = \{b_1, ..., b_k\} \le \sigma \\ \pi \in \mathcal{P}(b)}} \chi(\mathbf{X}_{b_1}) \cdots \chi(\mathbf{X}_{b_k}). \tag{3.2.9}$$

We can therefore set

$$G(\sigma) = \prod_{j=1}^{r} \mathbb{E}[\mathbf{X}^{a_j}]$$

and

$$F(\pi) = \chi(\mathbf{X}_{b_1}) \cdots \chi(\mathbf{X}_{b_k})$$

in (2.5.25) and (2.5.26), so as to deduce that, for every $\pi = \{b_1, ..., b_k\} \in \mathcal{P}(b)$,

$$\chi(\mathbf{X}_{b_1}) \cdots \chi(\mathbf{X}_{b_k}) = \sum_{\sigma = \{a_1, ..., a_r\} \le \pi} \mu(\sigma, \pi) \prod_{j=1}^{r} \mathbb{E}[\mathbf{X}^{a_j}]. \tag{3.2.10}$$

Relation (3.2.7) is therefore a particular case of (3.2.10), obtained by setting $\pi = \hat{1}$ and by using the equality $\mu(\sigma, \hat{1}) = (-1)^{|\sigma|-1}(|\sigma| - 1)!$, which is a consequence of (2.5.21).

To deduce Malyshev's formula (3.2.8) from (3.2.7) and (3.2.9), write $\mathbf{X}^{b_j} = Y_j$, $j = 1, ..., k$ (recall that the $\mathbf{X}^{b_j}$ are random variables defined in (3.1.1)), and apply (3.2.7) to the vector $\mathbf{Y} = (Y_1, ..., Y_k)$ to obtain that

$$\chi(\mathbf{X}^{b_1}, ..., \mathbf{X}^{b_k}) = \chi(Y_1, ..., Y_k) = \chi(\mathbf{Y}) \tag{3.2.11}$$

$$= \sum_{\beta = \{p_1, ..., p_r\} \in \mathcal{P}([k])} (-1)^{r-1}(r - 1)! \mathbb{E}(\mathbf{Y}^{p_1}) \cdots \mathbb{E}(\mathbf{Y}^{p_r}).$$

Write $\sigma = \{b_1, ..., b_k\}$, and observe that σ is a partition of the set b, while the partitions β in (3.2.11) are partitions of the first k integers. Now fix

$$\beta \in \{p_1, ..., p_r\} \in \mathcal{P}([k]).$$

For $i = 1, ..., r$, take the union of the blocks $b_j \in \sigma$ having $j \in p_i$, and call this union u_i. One obtains therefore a partition $\pi = \{u_1, ..., u_r\} \in \mathcal{P}(b)$ such that $|\pi| = |\beta| = r$. Thanks to (2.5.21) and (2.5.23),

$$(-1)^{r-1}(r - 1)! = \mu(\beta, \hat{1}) = \mu(\pi, \hat{1}) \tag{3.2.12}$$

(note that the two Möbius functions appearing in (3.2.12) are associated with different lattices: indeed, $\mu(\beta, \hat{1})$ refers to $\mathcal{P}([k])$, whereas $\mu(\pi, \hat{1})$ is associated with $\mathcal{P}(b)$).

With this notation, one has also that $\mathbb{E}\left(\mathbf{Y}^{p_1}\right)\cdots\mathbb{E}\left(\mathbf{Y}^{p_r}\right)=\mathbb{E}\left(\mathbf{X}^{u_1}\right)\cdots\mathbb{E}\left(\mathbf{X}^{u_r}\right)$, so that, by (3.2.9),

$$\mathbb{E}\left(\mathbf{Y}^{p_1}\right)\cdots\mathbb{E}\left(\mathbf{Y}^{p_r}\right)=\mathbb{E}\left(\mathbf{X}^{u_1}\right)\cdots\mathbb{E}\left(\mathbf{X}^{u_r}\right)=\sum_{\substack{\tau=\{t_1,\ldots,t_s\}\leq\pi\\ \tau\in\mathcal{P}(b)}}\chi\left(\mathbf{X}_{t_1}\right)\cdots\chi\left(\mathbf{X}_{t_s}\right).$$

(3.2.13)

By plugging (3.2.12) and (3.2.13) into (3.2.11) we obtain finally that

$$\begin{aligned}\chi\left(\mathbf{X}^{b_1},\ldots,\mathbf{X}^{b_k}\right) &= \sum_{\sigma\leq\pi\leq\hat{1}}\mu\left(\pi,\hat{1}\right)\sum_{\tau=\{t_1,\ldots,t_s\}:\tau\leq\pi}\chi\left(\mathbf{X}_{t_1}\right)\cdots\chi\left(\mathbf{X}_{t_s}\right)\\ &= \sum_{\tau\in\mathcal{P}(b)}\chi\left(\mathbf{X}_{t_1}\right)\cdots\chi\left(\mathbf{X}_{t_s}\right)\sum_{\pi\in[\tau\vee\sigma,\hat{1}]}\mu\left(\pi,\hat{1}\right)\\ &= \sum_{\tau:\tau\vee\sigma=\hat{1}}\chi\left(\mathbf{X}_{t_1}\right)\cdots\chi\left(\mathbf{X}_{t_s}\right),\end{aligned}$$

where the last equality is a consequence of (2.6.33), since

$$\sum_{\pi\in[\tau\vee\sigma,\hat{1}]}\mu\left(\pi,\hat{1}\right)=\begin{cases}1 & \text{if }\tau\vee\sigma=\hat{1}\\ 0 & \text{otherwise.}\end{cases}$$

(3.2.14)

∎

For a single random variable X, one has (3.1.3): hence, Proposition 3.2.1 implies:

Corollary 3.2.2 *Let X be a random variable such that $\mathbb{E}\,|X|^n<\infty$. Then,*

$$\mathbb{E}\left[X^n\right]=\sum_{\pi=\{b_1,\ldots,b_k\}\in\mathcal{P}([n])}\chi_{|b_1|}\left(X\right)\cdots\chi_{|b_k|}\left(X\right)$$

(3.2.15)

$$=\sum_{\lambda=(1^{r_1}2^{r_2}\cdots n^{r_n})\vdash n}\begin{bmatrix}n\\ \lambda\end{bmatrix}\times[\chi_1(X)]^{r_1}\times\cdots\times[\chi_n(X)]^{r_n}$$

(3.2.16)

$$=\sum_{\substack{1r_1+\cdots+r_n=n\\ r_i\geq 0}}\frac{n!}{(1!)^{r_1}\cdots(n!)^{r_n}r_n!}[\chi_1(X)]^{r_1}\times\cdots\times[\chi_n(X)]^{r_n}$$

(3.2.17)

$$=\sum_{s=0}^{n-1}\binom{n-1}{s}\chi_{s+1}(X)\times\mathbb{E}(X^{n-s-1}),$$

(3.2.18)

$$\chi_n\left(X\right)=\sum_{\sigma=\{a_1,\ldots,a_r\}\in\mathcal{P}([n])}(-1)^{r-1}\left(r-1\right)!\mathbb{E}\left(X^{|a_1|}\right)\cdots\mathbb{E}\left(X^{|a_r|}\right),$$

(3.2.19)

where in (3.2.16) we used the notation of formula (2.3.8), with the convention that $[\chi_l(X)]^0=1$.

Proof. Formulae (3.2.15) and (3.2.19) follow immediately from Proposition 3.2.1. Equality (3.2.16) is a consequence of Proposition 2.3.6 in the special case $f(\pi) = \chi_{|b_1|}(X)\cdots\chi_{|b_k|}(X)$, whenever $\pi = (b_1, ..., b_k)$. Equation (3.2.17) is a more explicit expression of (3.2.16). Finally, equation (3.2.18) can be deduced from (3.2.16) by recursion or, alternatively, as a consequence of the following elementary computations:

$$\mathbb{E}(X^n) = (-i)^n \frac{d^n}{dz^n} g_X(z)|_{z=0}$$

$$= (-i)^n \frac{d^{n-1}}{dz^{n-1}} \left[\left(\frac{d}{dz} \ln g_X(z) \right) g_X(z) \right]\bigg|_{z=0}$$

$$= \left[\sum_{s=0}^{n-1} (-i)^{s+1} \binom{n-1}{s} \frac{d^{s+1}}{dz^{s+1}} (\ln g_X(z)) \times (-i)^{n-1-s} \frac{d^{n-1-s}}{dz^{n-1-s}} g_X(z) \right]\bigg|_{z=0},$$

where $g_X(z) = \mathbb{E}[\exp(izX)]$, and $\frac{d^0}{dz^0} g_X(z) = 1$ (by convention). Indeed, for $s = 0, ..., n-1$,

$$(-i)^{s+1} \frac{d^{s+1}}{dz^{s+1}} \ln g_X(z)|_{z=0} = \chi_{s+1}(X) \qquad \text{and}$$

$$(-i)^{n-1-s} \frac{d^{n-1-s}}{dz^{n-1-s}} g_X(z)|_{z=0} = \mathbb{E}(X^{n-s-1}). \qquad \blacksquare$$

Remark. In [88] it is shown that formula (3.2.18) can be used in order to derive explicit expressions for the cumulants of random variables defined on Wiener space, based on infinite-dimensional integration by parts formulae. The expressions derived in [88] represent an actual alternative to the ones obtained in Chapter 7 by means of combinatorial devices. Several generalizations of (3.2.18) are discussed in the paper by Barbour [6].

Example 3.2.3 (i) Formula (3.2.7), applied respectively to $b = \{1\}$ and to $b = \{1, 2\}$, gives immediately the well-known relations

$$\chi(X) = \mathbb{E}(X) \quad \text{and} \quad \chi(X, Y) = \mathbb{E}(XY) - \mathbb{E}(X)\mathbb{E}(Y) = \mathbf{Cov}(X, Y). \tag{3.2.20}$$

$$\chi(X_1, X_2, X_3) = \mathbb{E}(X_1 X_2 X_3) - \mathbb{E}(X_1 X_2)\mathbb{E}(X_3)$$
$$-\mathbb{E}(X_1 X_3)\mathbb{E}(X_2) - \mathbb{E}(X_2 X_3)\mathbb{E}(X_1)$$
$$+2\mathbb{E}(X_1)\mathbb{E}(X_2)\mathbb{E}(X_3),$$

so that, in particular,

$$\chi_3(X) = \mathbb{E}(X^3) - 3\mathbb{E}(X^2)\mathbb{E}(X) + 2\mathbb{E}(X)^3.$$

(ii) Let $\mathbf{G}_{[n]} = (G_1, ..., G_n)$, $n \geq 3$, be a Gaussian vector such that $\mathbb{E}(G_i) = 0$, $i = 1, ..., n$. Then, for every $b \subseteq [n]$ such that $|b| \geq 3$, we know from Section 3.1 that

$$\chi(\mathbf{G}_b) = \chi(G_i : i \in b) = 0.$$

By applying this relation and formulae (3.2.6) and (3.2.20) to $\mathbf{G}_{[n]}$, one therefore obtains the well-known relation (sometimes called *Wick formula*)

$$\mathbb{E}\left[G_1 \times G_2 \times \cdots \times G_n\right] \tag{3.2.21}$$

$$= \begin{cases} \sum_{\pi=\{\{i_1,j_1\},\ldots,\{i_k,j_k\}\}\in\mathcal{P}([n])} \mathbb{E}\left(G_{i_1}G_{j_1}\right)\cdots\mathbb{E}\left(G_{i_k}G_{j_k}\right), & n \text{ even} \\ 0, & n \text{ odd.} \end{cases}$$

Observe that, on the RHS of (3.2.21), the sum is taken over all partitions π of $[n]$ such that each block of π contains exactly two elements. A little inspection shows that, for n even, the number of such partitions is given by the double factorial $(n-1)!!$, which is defined as $1!! = 1$ and, for $n \geq 4$,

$$(n-1)!! = (n-1) \times (n-3) \times \cdots \times 3 \times 1 \tag{3.2.22}$$

(for instance: $3!! = 3$, $5!! = 15$, and so on). Using this fact and taking $\mathbf{G}_{[n]}$ such that $G_i = G$, $i = 1, \ldots, n$, where G is a centered Gaussian random variable with variance σ^2, one deduces from the previous computation the well known fact that

$$\mathbb{E}\left[G^n\right] = \begin{cases} \sigma^n \times (n-1)!!, & n \text{ even} \\ 0, & n \text{ odd.} \end{cases} \tag{3.2.23}$$

Relation (3.2.23) implies in particular that the moments of G satisfy the following recurrence relation: for every $n \geq 2$

$$\mathbb{E}(G^{n+1}) = n \times \sigma^2 \times \mathbb{E}(G^{n-1}). \tag{3.2.24}$$

We conclude this section by showing an interesting consequence of relation (3.2.24), known as *Stein's Lemma*, providing a useful characterization of one-dimensional Gaussian distributions. Note that this result is the basis of the so-called *Stein's method* for normal approximations. The reader is referred to the survey by Chen and Shao [14] for an introduction to the subject. Stein's Lemma should also be compared with the forthcoming Chen-Stein Lemma 3.3.3, providing a characterization of the Poisson distribution.

Lemma 3.2.4 (Stein's Lemma) *Let G be a Gaussian random variable with zero mean and variance $\sigma^2 > 0$, and let W be any real-valued random variable. Then, $W \stackrel{law}{=} G$ if and only if, for every smooth function h verifying $\mathbb{E}|h'(G)| < \infty$, one has that*

$$\mathbb{E}[Wh(W)] = \sigma^2 \times \mathbb{E}[h'(W)]. \tag{3.2.25}$$

Proof. If W is a centered Gaussian random variable with variance σ^2, then integration by parts (or a Fubini argument) shows that relation (3.2.25) is verified. To prove the reverse implication, assume that W verifies (3.2.25) for every test function h as in the statement. By selecting $h(x) = x^n$, $n \geq 0$, we see that relation (3.2.25) implies that $\mathbb{E}(W) = 0$, $\mathbb{E}(W^2) = \sigma^2$, and, for every $n \geq 3$,

$$\mathbb{E}[W^{n+1}] = n \times \sigma^2 \times \mathbb{E}[W^{n-1}].$$

By virtue of (3.2.24), we therefore deduce that $\mathbb{E}(W^n) = \mathbb{E}(G^n)$ for every $n \geq 1$, and the conclusion follows from the fact that the normal distribution is determined by its moments (see the forthcoming Lemma 3.2.5). ∎

For the sake of completeness, we shall provide a self-contained proof of the fact that the Gaussian distribution is determined by its moments. Given a probability measure ν on $\mathbb{R}$, we denote by

$$m_n(\nu) = \int_{\mathbb{R}} x^n \nu(dx), \quad n \geq 0,$$

the sequence of its moments (whenever they are well-defined).

Lemma 3.2.5 *Let $\sigma > 0$ and let g_σ denote the centered Gaussian probability distribution of variance σ^2, that is, $g_\sigma(dx) = (2\pi\sigma^2)^{-1/2}e^{-x^2/2\sigma^2}dx$. Then, g_σ is determined by its moments. This means that, if ν is another probability measure (having moments of all order) verifying $m_n(\nu) = m_n(g_\sigma)$ for every $n \geq 1$, then necessarily $\nu = g_\sigma$.*

Proof. Let ν be as in the statement. We have to show that, for every $t \in \mathbb{R}$, $\int_{\mathbb{R}} e^{itx}\nu(dx) = \int_{\mathbb{R}} e^{itx}g_\sigma(dx)$. Since $m_k(\nu) = m_k(g_\sigma)$ for every k, we can write, thanks to Taylor's expansion and the triangle and Cauchy-Schwarz inequalities,

$$\left| \int_{\mathbb{R}} e^{itx}\nu(dx) - \int_{\mathbb{R}} e^{itx}g_\sigma(x) \right|$$

$$\leq \int_{\mathbb{R}} \left| e^{itx} - \sum_{k=0}^{n} \frac{(itx)^k}{k!} \right| \nu(dx) + \int_{\mathbb{R}} \left| e^{itx} - \sum_{k=0}^{n} \frac{(itx)^k}{k!} \right| g_\sigma(dx)$$

$$\leq \int_{\mathbb{R}} \left| \frac{|tx|^{n+1}}{(n+1)!} \right| \nu(dx) + \int_{\mathbb{R}} \left| \frac{|tx|^{n+1}}{(n+1)!} \right| g_\sigma(dx)$$

$$\leq \left(\int_{\mathbb{R}} \frac{|tx|^{2n+2}}{(n+1)!^2}\nu(dx) \right)^{1/2} + \left(\int_{\mathbb{R}} \frac{|tx|^{2n+2}}{(n+1)!^2}g_\sigma(dx) \right)^{1/2}$$

$$= \sqrt{\frac{|t|^{2n+2}\, m_{2n+2}(\nu)}{(n+1)!^2}} + \sqrt{\frac{|t|^{2n+2}\, m_{2n+2}(g_\sigma)}{(n+1)!^2}} = 2\sqrt{\frac{|t|^{2n+2}m_{2n+2}(g_\sigma)}{(n+1)!^2}},$$

for every $n \geq 1$. The conclusion is obtained by observing that, thanks to Stirling's formula,

$$\lim_{n\to\infty} \frac{|t|^{2n+2}\, m_{2n+2}(g_\sigma)}{(n+1)!^2} = \lim_{n\to\infty} \frac{|t\sigma|^{2n+2} \times (2n+1)!!}{(n+1)!^2} = 0,$$

where we have used the notation (3.2.22). ∎

3.3 Bell and Touchard polynomials and the Poisson distribution

We now want to connect Corollary 3.2.2 to the class of Bell and Touchard polynomials introduced in Definition 2.4.1. We start with a statement reexpressing formulae (3.2.16) and (3.2.19), respectively, in terms of complete and partial Bell polynomials.

Proposition 3.3.1 *Let X be a random variable such that $\mathbb{E}|X|^n < \infty$, and write $m_j = \mathbb{E}(X^j)$ and $q_j = \chi_j(X)$, $j = 1, ..., n$. Then,*

$$\mathbb{E}(X^n) = B_n(q_1, ..., q_n) \tag{3.3.26}$$

$$\chi_n(X) = \sum_{k=1}^{n} (-1)^{k-1}(k-1)! B_{n,k}(m_1, ..., m_{n-k+1}), \tag{3.3.27}$$

where B_n and $B_{n,k}$ are, respectively, the nth complete Bell polynomial (see (2.4.13)) and the partial Bell polynomial of index (n, k) (see (2.4.11)).

Proof. Formula (3.3.26) is a direct consequence of (3.2.16) and (2.4.14). To see that (3.3.27) also holds, rewrite (3.2.19) as

$$\chi_n(X) = \sum_{k=1}^{n} (-1)^{k-1}(k-1)! \left[\sum_{\substack{\lambda=(1^{r_1}2^{r_2}\cdots n^{r_n})\vdash n \\ \lambda \text{ has length } k}} \begin{bmatrix} n \\ \lambda \end{bmatrix} m_1^{r_1} \times \cdots \times m_{n-k+1}^{r_{n-k+1}} \right],$$

$$\tag{3.3.28}$$

so that the conclusion is deduced from (2.4.12).   ∎

In what follows, for every $x > 0$, we shall denote by $N(x)$ a Poisson random variable with parameter x. The following statement provides an elegant connection between the moments of the random variables $N(x)$ and Touchard polynomials.

Proposition 3.3.2 *For every $n \geq 0$, one has that*

$$\mathbb{E}[N(x)^n] = T_n(x), \quad x > 0, \tag{3.3.29}$$

where T_n is the nth Touchard polynomial, as defined in (2.4.15). In particular formulae (2.4.19) and (2.4.20) hold.

Proof. The statement is trivial in the case $n = 0$. Fix $n \geq 1$ and observe that, according to Example 3.1.1, one has that $\chi_k(N(x)) = x$, for every $k \geq 1$. Plugging this expression into (3.3.26) in the case $X = N(x)$, one deduces that

$$\mathbb{E}(N(x)^n) = B_n(x, ..., x) = \sum_{k=1}^{n} B_{n,k}(x, ..., x) = T_n(x),$$

where the last two equalities follow from (2.4.13) and (2.4.17). We shall now use properties of the Poisson distribution to prove Proposition 2.4.4, namely that (2.4.19) and (2.4.20) are verified. To see that (2.4.19) holds, we use (3.2.18) and the fact that $\chi_k(N(x)) = x$ for every k to deduce that

$$T_{n+1}(x) = \mathbb{E}[N(x)^{n+1}] = \sum_{k=0}^{n} \binom{n}{k} x \times \mathbb{E}(N(x)^{n-k}) = x \sum_{k=0}^{n} \binom{n}{k} T_{n-k}(x).$$

To prove (2.4.20), fix $x, y > 0$ and let $N'(y)$ be a Poisson random variable with parameter y, independent of $N(x)$. Since $\mathbb{E}[e^{iz(N(x)+N'(y))}] = \exp[(x+y)(e^{iz}-1)]$, one has that $N(x) + N'(y)$ has the same law as $N(x+y)$. It follows that

$$T_n(x+y) = \mathbb{E}[(N(x) + N'(y))^n] = \sum_{k=0}^{n} \binom{n}{k} \mathbb{E}[N(x)^k N'(y)^{n-k}].$$

By independence, $\mathbb{E}[N(x)^k N'(y)^{n-k}] = \mathbb{E}[N(x)^k] \times \mathbb{E}[N(y)^{n-k}] = T_k(x) T_{n-k}(y)$, and the proof is concluded. ∎

Remarks. (a) By virtue of (2.4.18), and since $\mathbb{P}[N(1) = k] = e^{-1}/k!$, specializing (3.3.29) to the case $x = 1$ yields the following remarkable summation formulae involving the Bell numbers $B_n = T_n(1)$: for every $n \geq 1$,

$$B_n = \frac{1}{e} \sum_{k=1}^{\infty} \frac{k^n}{k!}. \tag{3.3.30}$$

Formula (3.3.30) was first proved by Dobiński [22] in 1877.

(b) Formula (3.3.29) may be also used to deduce an explicit expression for the generating function of Bell numbers. Indeed, for every $x > 0$ one has that

$$\sum_{n=0}^{\infty} T_n(x) \frac{t^n}{n!} = \sum_{n=0}^{\infty} \mathbb{E}[N(x)^n] \frac{t^n}{n!} = \mathbb{E}\left[\sum_{n=0}^{\infty} \frac{N(x)^n t^n}{n!} \right] = \mathbb{E}[e^{tN(x)}] = e^{x(e^t-1)}, \tag{3.3.31}$$

implying, by setting $x = 1$,

$$\sum_{n=0}^{\infty} B_n \frac{t^n}{n!} = e^{e^t-1}. \tag{3.3.32}$$

Reading backwards the chain of equalities (3.3.31), one deduces in particular that, for every $x, t > 0$,

$$\lim_{n\to\infty} T_n(x) \frac{t^n}{n!} = \lim_{n\to\infty} \mathbb{E}[N(x)^n] \frac{t^n}{n!} = 0. \tag{3.3.33}$$

As a further illustration, we will now use Proposition 3.3.2 to deduce a short proof of the so-called *Chen-Stein Lemma*, providing an elegant characterization of the Poisson distribution. The reader is referred, for example, to the survey by Erhardsson [27] for an introduction to the powerful *Chen-Stein method* for Poisson approximations and its many applications. Note that Lemma 3.3.3 is usually proved by solving a finite difference equation. Also, one should compare with the previously discussed Stein's Lemma 3.2.4.

Lemma 3.3.3 (Chen-Stein Lemma) *Let Z be a random variable with values in $\mathbb{N}$. Then, Z has a Poisson distribution with parameter $x > 0$ if and only if, for every bounded function $f : \mathbb{N} \to \mathbb{N}$,*

$$\mathbb{E}[Zf(Z)] = x\mathbb{E}[f(Z+1)]. \tag{3.3.34}$$

Proof. If Z has a Poisson distribution with parameter x, then for every bounded f

$$\mathbb{E}[Zf(Z)] = e^{-x} \sum_{k=1}^{\infty} kf(k)\frac{x^k}{k!} = e^{-x} \sum_{j=0}^{\infty} f(j+1)\frac{x^{j+1}}{j!} = x\mathbb{E}[f(Z+1)],$$

where we used the change of variable $j = k - 1$, thus showing that (3.3.34) is verified. Now suppose that Z is an integer-valued random variable satisfying (3.3.34), and write $N(x)$ to indicate a Poisson random variable with parameter x. Approximating the monomial $a \mapsto a^n$ ($n \geq 1$) by means of the truncated (bounded) functions $a \mapsto a^n 1_{(a \leq m)}$, $m \geq 1$, one deduces by monotone convergence that relation (3.3.34) continues to hold whenever f is a polynomial. Choosing $f(a) = a^n, n \geq 0$, in (3.3.34), one obtains therefore

$$\mathbb{E}[Z^{n+1}] = x\mathbb{E}[(Z+1)^n] = x \sum_{k=0}^{n} \binom{n}{k} \mathbb{E}[Z^{n-k}].$$

Since $\mathbb{E}[Z^0] = 1 = T_0(x)$ by definition, we deduce from the recurrence relation (2.4.19) that $\mathbb{E}[Z^n] = T_n(x) = \mathbb{E}[N(x)^n]$, for every $n \geq 1$. To conclude, we shall prove by a direct argument that the Poisson distribution is characterized by its moments. Indeed, applying the triangle inequality and the standard estimate $\left| e^{itu} - \sum_{k=0}^{n} \frac{(tu)^k}{k!} \right| \leq |tu|^{n+1}/(n+1)!$ we deduce that, for every real t and every integer n,

$$\left| \mathbb{E}[e^{itN(x)}] - \mathbb{E}[e^{itZ}] \right|$$

$$\leq \left| \mathbb{E}[e^{itN(x)}] - \sum_{k=0}^{n} \mathbb{E}[N(x)^k]\frac{t^k}{k!} \right| + \left| \mathbb{E}[e^{itZ}] - \sum_{k=0}^{n} \mathbb{E}[N(x)^k]\frac{t^k}{k!} \right|$$

$$\leq \left| \mathbb{E}[e^{itN(x)}] - \sum_{k=0}^{n} \mathbb{E}[N(x)^k]\frac{t^k}{k!} \right| + \left| \mathbb{E}[e^{itZ}] - \sum_{k=0}^{n} \mathbb{E}[Z^k]\frac{t^k}{k!} \right|$$

$$\leq \frac{\mathbb{E}[N(x)^{n+1}] \times |t|^{n+1}}{(n+1)!} + \frac{\mathbb{E}[Z^{n+1}] \times |t|^{n+1}}{(n+1)!}$$

$$= 2\frac{\mathbb{E}[N(x)^{n+1}] \times |t|^{n+1}}{(n+1)!} \to 0,$$

where the convergence to zero takes place (by virtue of (3.3.33)) as $n \to \infty$.   ∎

Remark. By inspection of the preceding proof, one sees immediately that relation (3.3.34) continues to hold for Z Poisson of parameter x and for real-valued functions f with polynomial growth and not necessarily with values in $\mathbb{N}$.

As above, for every $x > 0$ let $N(x)$ be a Poisson random variable with parameter x. Formula (3.3.29) yields in particular that, for every $n \geq 0$, there exists a polynomial of degree at most n, denoted by $\tilde{T}_n$, such that

$$\tilde{T}_n(x) = \mathbb{E}[(N(x) - x)^n], \quad x > 0. \tag{3.3.35}$$

Of course, $\tilde{T}_0(x) = 1$ and $\tilde{T}_1(x) = 0$. The next statement uses relation (3.3.34) to deduce a complete characterization of the family $\{\tilde{T}_n : n \geq 0\}$.

Proposition 3.3.4 *For every $n \geq 1$,*

$$\tilde{T}_{n+1}(x) = x \sum_{k=0}^{n-1} \binom{n}{k} \tilde{T}_k(x). \tag{3.3.36}$$

Proof. One has that

$$\tilde{T}_{n+1}(x) = \mathbb{E}[(N(x) - x)(N(x) - x)^n] = \mathbb{E}[N(x)(N(x) - x)^n] - x\tilde{T}_n(x).$$

By applying (3.3.34) to the function $f(a) = (a - x)^n$, we deduce that

$$\mathbb{E}[N(x)(N(x) - x)^n] = x\mathbb{E}[(N(x) + 1 - x)^n] = x \sum_{k=0}^{n} \binom{n}{k} \tilde{T}_k(x),$$

whence

$$\tilde{T}_{n+1}(x) = x \sum_{k=0}^{n} \binom{n}{k} \tilde{T}_k(x) - xT_n(x) = x \sum_{k=0}^{n-1} \binom{n}{k} \tilde{T}_k(x). \quad ∎$$

Example 3.3.5 Using (3.3.36), one deduces that

$$\begin{aligned}
\tilde{T}_2(x) &= x\tilde{T}_0(x) = x \\
\tilde{T}_3(x) &= x[\tilde{T}_0(x) + 2\tilde{T}_1(x)] = x \\
\tilde{T}_4(x) &= x[\tilde{T}_0(x) + 3\tilde{T}_1(x) + 3\tilde{T}_2(x)] = x + 3x^2 \\
\tilde{T}_5(x) &= x[\tilde{T}_0(x) + 4\tilde{T}_1(x) + 6\tilde{T}_2(x) + 4\tilde{T}_3(x)] = x + 10x^2.
\end{aligned} \tag{3.3.37}$$

Remark. The polynomials $\tilde{T}_n$ are sometimes called *centered Touchard polynomials*. They provide the centered moments of the Poisson distribution. See Privault [123] for several generalizations of formula (3.3.36).

Another interesting combinatorial property of the family $\{\tilde{T}_n\}$ is given in the next statement.

Proposition 3.3.6 *For every $n \geq 1$ the quantity $\tilde{T}_n(1)$ equals the number of partitions of $[n]$ that have no singletons (that is, the partitions with blocks of size at least 2).*

Proof. Using (3.1.4)–(3.1.5) and Example 3.1.1, one sees that $\chi_1(N(1) - 1) = 0$ and $\chi_n(N(1) - 1) = \chi_n(N(1))$ for every $n \geq 2$. Combining these facts with the identity (3.2.15) yields

$$\tilde{T}_n(1) = \mathbb{E}[(N(1) - 1)^n] = \sum_{\pi \in \mathcal{P}([n])} \mathbf{1}_{\{\pi \text{ has no singletons}\}},$$

from which we deduce the conclusion.   ∎

Example 3.3.7 We can, for example, combine Proposition 3.3.6 with (3.3.37) and deduce the following facts:

(i) there is only $\tilde{T}_2(1) = 1$ partition of $[2]$ and $\tilde{T}_3(1) = 1$ partition of $[3]$ with no singletons (in both cases, it is the trivial partition $\hat{1}$);

(ii) there are exactly $\tilde{T}_4(1) = 4$ partitions of $[4]$ with no singletons, namely: $\hat{1}$, $\{\{1,2\},\{3,4\}\}$, $\{\{1,3\},\{2,4\}\}$ and $\{\{1,4\},\{2,3\}\}$;

(iii) there are exactly $\tilde{T}_5(1) = 11$ partitions of $[5]$ without singletons (list them as an exercise).

4

Diagrams and multigraphs

In this chapter, we translate some of the notions presented in Chapter 2 into the language of diagrams and multigraphs, which are often used in order to compute cumulants and moments of non-linear functionals of random fields (see, for example, [10, 12, 34, 35, 68, 150, 151]). We will have, in particular, the correspondence indicated in Table 4.1, as will be explained in the sequel.

Table 4.1. Correspondence

	Diagram	*Multigraph*
$\pi \vee \sigma = \hat{1}$	connected	connected
$\pi \wedge \sigma = \hat{0}$	non-flat	no loops

We will consider, in addition, Gaussian and circular diagrams.

4.1 Diagrams

Consider a finite set b. A *diagram* is a graphical representation of a pair of partitions $(\pi, \sigma) \subseteq \mathcal{P}(b)$, such that $\pi = \{b_1, ..., b_k\}$ and $\sigma = \{t_1, ..., t_l\}$. It is obtained as follows.

1. Order the elements of each block b_i, for $i = 1, ..., k$.
2. Associate with each block $b_i \in \pi$ a row of $|b_i|$ *vertices* (represented as dots), in such a way that the jth vertex of the ith row corresponds to the jth element of the block b_i.
3. For every $a = 1, ..., l$, draw a closed curve around the vertices corresponding to the elements of the block $t_a \in \sigma$.

We will denote by $\Gamma(\pi, \sigma)$ the diagram of a pair of partitions (π, σ).

G. Peccati, M.S. Taqqu: Wiener Chaos: Moments, Cumulants and Diagrams –
A survey with computer implementation.
© Springer-Verlag Italia 2011

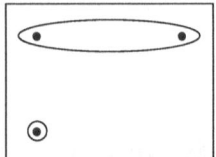

Fig. 4.1. A simple diagram

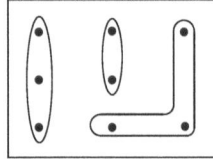

Fig. 4.2. A diagram built from two three-block partitions

Example 4.1.1   (i) If $b = [3] = \{1, 2, 3\}$ and $\pi = \sigma = \{\{1, 2\}, \{3\}\}$, then $\Gamma(\pi, \sigma)$ is represented in Fig. 4.1.

(ii) If $b = [8]$, and $\pi = \{\{1, 2, 3\}, \{4, 5\}, \{6, 7, 8\}\}$ and $\sigma = \{\{1, 4, 6\}, \{2, 5\}, \{3, 7, 8\}\}$, then $\Gamma(\pi, \sigma)$ is represented in Fig. 4.2. Hence, the rows in $\Gamma(\pi, \sigma)$ indicate the sets in π and the curves indicate the sets in σ.

Remarks.

(a) We use the terms "element" and "vertex" interchangeably.

(b) Note that the diagram generated by the pair (π, σ) is different, in general, from the diagram generated by (σ, π).

(c) Each diagram is a finite *hypergraph*. We recall that a finite hypergraph is an object of the type (V, E), where V is a finite set of vertices, and E is a collection of (not necessarily disjoint) nonempty subsets of V. The elements of E are usually called *edges*. In our setting, these are the blocks of σ.

(d) Note that, once a partition π is specified, the diagram $\Gamma(\pi, \sigma)$ encodes all the information on σ.

Now fix a finite set b. In what follows, we will list and describe several type of diagrams. They can be all characterized in terms of the lattice structure of $\mathcal{P}(b)$, namely the partial ordering $\leq$ and the join and meet operations $\vee$ and $\wedge$, as described in Chapter 2. Recall that $\hat{1} = \{b\}$, and $\hat{0}$ is the partition whose elements are the singletons of b.

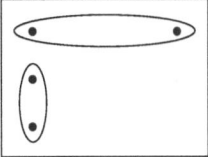

Fig. 4.3. A non-connected diagram

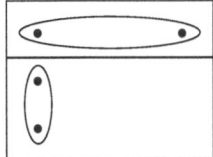

Fig. 4.4. Dividing a non-connected diagram

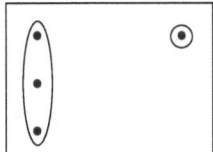

Fig. 4.5. A connected diagram

Connected Diagrams. The diagram $\Gamma(\pi, \sigma)$ associated with two partitions (π, σ) is said to be *connected* if

$$\pi \vee \sigma = \hat{1},$$

that is, if the only partition ρ such that $\pi \leq \rho$ and $\sigma \leq \rho$ is the maximal partition $\hat{1}$.

The diagram appearing in Fig. 4.2 is connected, whereas the one in Fig. 4.1 is not (indeed, in this case $\pi \vee \sigma = \pi \vee \pi = \pi \neq \hat{1}$). Another example of a non-connected diagram (see Fig. 4.3) is obtained by taking $b = [4]$, $\pi = \{\{1,2\},\{3\},\{4\}\}$ and $\sigma = \{\{1,2\},\{3,4\}\}$, so that $\pi \leq \sigma$ (each block of π is contained in a block of σ) and $\pi \vee \sigma = \sigma \neq \hat{1}$.

In other words, $\Gamma(\pi, \sigma)$ *is connected if and only if the rows of the diagram (the blocks of π) cannot be divided into two subsets, each defining a separate diagram.* Fig. 4.4 shows that the diagram in Fig. 4.3 can be so divided, and thus is disconnected (non-connected).

The diagram in Fig. 4.5, which has the same partition π, but $\sigma = \{\{1,3,4\},\{2\}\}$, is connected.

Note that we do not use the term 'connected' as one usually does in graph theory (indeed, the diagrams we consider in this section are always *non-connected hypergraphs*, since their edges are disjoint by construction).

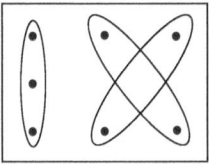

Fig. 4.6. A non-flat diagram

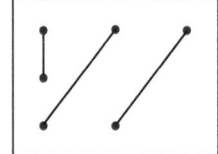

Fig. 4.7. A Gaussian diagram

Non-flat Diagrams. The diagram $\Gamma(\pi, \sigma)$ is *non-flat* if

$$\pi \wedge \sigma = \hat{0},$$

that is, if the only partition ρ such that $\rho \leq \pi$ and $\rho \leq \sigma$ is the minimal partition $\hat{0}$. It is easily seen that $\pi \wedge \sigma = \hat{0}$ if and only if for any two blocks $b_j \in \pi$, $t_a \in \sigma$, the intersection $b_j \cap t_a$ either is empty or contains exactly one element. Graphically, a non-flat graph is such that the closed curves defining the blocks of σ cannot join two vertices in the same row (thus having a 'flat' or 'horizontal' portion). The diagrams in Fig. 4.1-4.3 are all flat (not non-flat). Observe that the diagram in Fig. 4.2 is flat because it has a flat portion: the block $\{3, 7, 8\}$ of σ has $(7, 8)$ as a flat portion. The diagram in Fig. 4.5 is non-flat. Another non-flat diagram is given in Fig. 4.6, and is obtained by taking

$$b = [7], \; \pi = \{\{1, 2, 3\}, \{4\}, \{5, 6, 7\}\} \text{ and } \sigma = \{\{1, 4, 5\}, \{2, 7\}, \{3, 6\}\}.$$

Gaussian Diagrams. We say that the diagram $\Gamma(\pi, \sigma)$ is *Gaussian*, whenever every block of σ contains exactly two elements. Plainly, Gaussian diagrams exists only if there is an even number of vertices. When a diagram is Gaussian, one usually represents the blocks of σ not by closed curves, but by segments connecting two vertices (which are viewed as the edges of the resulting graph). For instance, a Gaussian (non-flat and connected) diagram is obtained in Fig. 4.7, where we have taken

$$b = [6], \; \pi = \{\{1, 2, 3\}, \{4\}, \{5, 6\}\} \text{ and } \sigma = \{\{1, 4\}, \{2, 5\}, \{3, 6\}\}.$$

Whenever a block $a = \{i, j\}$ is such that $i \in b_1$ and $j \in b_2$, where b_1, b_2 are blocks of π, we shall say that *a links* b_1 *and* b_2. For instance, in Fig. 4.7, the blocks $\{1, 2, 3\}$ and $\{5, 6\}$ of π are linked by the two blocks of σ given by $\{2, 5\}$ and $\{3, 6\}$. In the terminology of graph theory, a Gaussian diagram is a *non-connected (non-directed) graph*. Since every vertex is connected with exactly another vertex, one usually says that such a graph is a *perfect matching*.

Circular (Gaussian) Diagrams. Consider two partitions $\pi = \{b_1, ..., b_k\}$ and $\sigma = \{t_1, ..., t_l\}$ such that the blocks of σ have size $|t_a| = 2$ for every $a = 1, ..., l$. Then, the diagram $\Gamma(\pi, \sigma)$ (which is Gaussian) is said to be *circular* if each row of $\Gamma(\pi, \sigma)$ is linked to both the previous and the next row, with no other possible links except for the first and the last row, which should also be linked together. This implies that the diagram is connected.

Formally, the diagram $\Gamma(\pi, \sigma)$, where $\pi = \{b_1, ..., b_k\}$ and $\sigma = \{t_1, ..., t_l\}$, is circular (Gaussian) whenever the following properties hold (recall that $i \sim_\sigma j$ means that i and j belong to the same block of σ):

(a) for every $p = 2, ..., k - 1$ there exist $j_1 \sim_\sigma i_1$ and $j_2 \sim_\sigma i_2$ such that $j_1, j_2 \in b_p$, $i_1 \in b_{p-1}$ and $i_2 \in b_{p+1}$;
(b) for every $p = 2, ..., k - 1$ and every $j \in b_p$, $j \sim_\sigma i$ implies that $i \in b_{p-1} \cup b_{p+1}$;
(c) there exist $j_1 \sim_\sigma i_1$ and $j_2 \sim_\sigma i_2$ such that $j_1, j_2 \in b_k$, $i_1 \in b_{k-1}$ and $i_2 \in b_1$;
(d) for every $j \in b_k$, $j \sim_\sigma i$ implies that $i \in b_1 \cup b_{k-1}$;
(e) there exist $j_1 \sim_\sigma i_1$ and $j_2 \sim_\sigma i_2$ such that $j_1, j_2 \in b_1$, $i_1 \in b_2$ and $i_2 \in b_k$, (vi) for every $j \in b_1$, $j \sim_\sigma i$ implies that $i \in b_k \cup b_2$.

Example 4.1.2

(i) A circular diagram is obtained by taking $b = [10]$ and

$$\pi = \{\{1, 2\}, \{3, 4\}, \{5, 6\} \{7, 8\}, \{9, 10\}\}$$
$$\sigma = \{\{1, 3\}, \{2, 9\}, \{4, 6\}, \{5, 7\}, \{8, 10\}\},$$

which implies that $\Gamma(\pi, \sigma)$ is the diagram in Fig. 4.8.
To illustrate condition (a) above for this example, consider $p = 2$, that is the second block $b_2 = \{3, 4\}$ of π. Then there is $j_1 = 3$, $j_2 = 4$ in b_2 and also $i_1 = 1$ in b_1 and $i_2 = 6$ in b_3 such that $j_1 \sim_\sigma i_1$ and $j_2 \sim_\sigma i_2$.

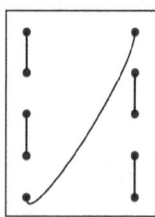

Fig. 4.8. A circular diagram

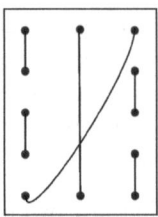

Fig. 4.9. A circular diagram with rows of different size

(ii) Another example of a circular diagram is given in Fig. 4.9.
   It is obtained from $b = [12]$ and

$$\pi = \{\{1,2,3\},\{4,5\},\{6,7\},\{8,9\},\{10,11,12\}\}$$
$$\sigma = \{\{1,4\},\{2,11\},\{3,10\},\{5,7\},\{6,8\},\{9,12\}\}.$$

We can now express various formulas involving moments and cumulants in terms
of diagrams. We shall do this here and in the sequel.

Example 4.1.3

(i) Malyshev's formula (3.2.8) can be expressed in terms of connected diagrams as
   follows:
   For every finite set b and every $\sigma = \{b_1, ..., b_k\} \in \mathcal{P}(b)$,

$$\chi\left(\mathbf{X}^{b_1}, ..., \mathbf{X}^{b_k}\right) = \sum_{\substack{\tau = \{t_1, ..., t_s\} \in \mathcal{P}(b) \\ \Gamma(\sigma, \tau) \text{ is connected}}} \chi\left(\mathbf{X}_{t_1}\right) \cdots \chi\left(\mathbf{X}_{t_s}\right). \qquad (4.1.1)$$

(ii) Suppose that the random variables X_1, X_2, X_3 are such that $\mathbb{E}|X_i|^3 < \infty$,
   $i = 1, 2, 3$. We have already applied formula (4.1.1) in order to compute the
   cumulant $\chi(X_1 X_2, X_3)$. Here, we shall give a graphical demonstration. Recall that, in this case, $b = [3] = \{1, 2, 3\}$, and that the relevant partition is
   $\sigma = \{\{1, 2\}, \{3\}\}$. There are only three partitions $\tau_1, \tau_2, \tau_3 \in \mathcal{P}([3])$ such
   that $\Gamma(\sigma, \tau_1)$, $\Gamma(\sigma, \tau_2)$ and $\Gamma(\sigma, \tau_3)$ are connected, namely $\tau_1 = \hat{1}$, $\tau_2 =$
   $\{\{1, 3\}, \{2\}\}$ and $\tau_3 = \{\{1\}, \{2, 3\}\}$. The diagrams $\Gamma(\sigma, \tau_1)$, $\Gamma(\sigma, \tau_2)$ and
   $\Gamma(\sigma, \tau_3)$ are represented in Fig. 4.10. Relation (4.1.1) thus implies that

$$\chi(X_1 X_2, X_3) = \chi(X_1, X_2, X_3) + \chi(X_1, X_3)\chi(X_2) + \chi(X_1)\chi(X_2, X_3)$$
$$= \chi(X_1, X_2, X_3) + \mathbf{Cov}(X_1, X_3)\mathbb{E}(X_3) + \mathbb{E}(X_1)\mathbf{Cov}(X_2, X_3),$$

where we have used (3.2.20).

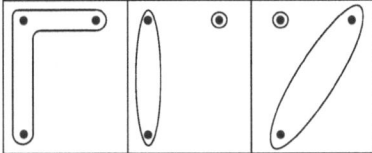

Fig. 4.10. Computing cumulants by connected diagrams

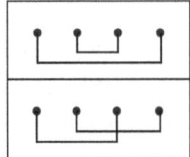

Fig. 4.11. Two partitons: one non-crossing (above) and one crossing (below)

Remark. Another class of partitions admitting a neat graphical representation is the collection of *non-crossing partitions*. Given $n \geq 2$, we say that a partition $\pi \in \mathcal{P}([n])$ is non-crossing if one cannot find integers $1 \leq p_1 < q_1 < p_2 < q_2 \leq n$ such that $p_1 \sim_\pi p_2$, $q_1 \sim_\pi q_2$ and p_1 and p_2 are not in the same block as q_1 and q_2. To understand the terminology, one has to represent a partition $\pi \in \mathcal{P}([n])$ as follows: (i) draw the elements of $[n]$ as dots on the same line, (ii) supply each dot with a vertical line under it, and (iii) join the vertical lines of the elements in the same block with a horizontal line. The partition π is non-crossing if and only if the lines involved in this representation can be drawn in such a way that they do not cross. Consider for instance the two partitions of $[4]$ depicted in Fig. 4.11. The upper partition is $\{\{1,4\},\{2,3\}\}$, and it is non-crossing, whereas the lower partition is $\{\{1,3\},\{2,4\}\}$, which is crossing. One customarily denotes by $NC(n)$ the class of non-crossing partitions of $[n]$. Note that $\mathcal{P}(n) = NC(n)$, for $n = 2, 3$, and that one has $NC(n) \subset \mathcal{P}([n])$, with strict inclusion, for every $n \geq 4$. One remarkable feature of the class $NC(n)$ is that it can be endowed with a lattice structure analogous to the one of $\mathcal{P}(n)$. Non-crossing partitions play a crucial role in free probability, where they are used, for example, to define the concept of *free cumulant*. The reader is referred to the monograph by Nica and Speicher [80] for a detailed introduction to the combinatorics of non-crossing partitions in the context of free probability.

4.2 Solving the equation $\sigma \wedge \pi = \hat{0}$

Let π be a partition of $[n] = \{1, ..., n\}$. One is often asked, as will be the case in Section 6.1, to find all partitions $\sigma \in \mathcal{P}([n])$ such that

$$\sigma \wedge \pi = \hat{0}, \tag{4.2.2}$$

where, as usual, $\hat{0} = \{\{1\}, ..., \{n\}\}$, that is, $\hat{0}$ is the partition made up of singletons. The use of diagrams provides an easy way to solve (4.2.2), since (4.2.2) holds if and only if the diagram $\Gamma(\pi, \sigma)$ is non-flat. Hence, proceed as in Section 4.1, by (1) ordering the blocks of π, (2) associating with each block of π a row of the diagram, the number of points in a row being equal to the number of elements in the block, and (3) drawing non-flat closed curves around the points of the diagram.

Example 4.2.1   (i) Let $n = 2$ and $\pi = \{\{1\}, \{2\}\} = \hat{0}$. Then, $\sigma_1 = \pi = \hat{0}$ and $\sigma_2 = \hat{1}$ (as represented in Fig. 4.12) solve equation (4.2.2). Note that $\mathcal{P}([2]) = \{\sigma_1, \sigma_2\}$.
  (ii) Let $n = 3$ and $\pi = \{\{1, 2\}, \{3\}\}$. Then, $\sigma_1 = \hat{0}$, $\sigma_2 = \{\{1, 3\}, \{2\}\}$ and $\sigma_3 = \{\{1\}, \{2, 3\}\}$ (see Fig. 4.13) are the only elements of $\mathcal{P}([3])$ solving (4.2.2).
  (iii) Let $n = 4$ and $\pi = \{\{1, 2\}, \{3, 4\}\}$. Then, there are exactly seven $\sigma \in \mathcal{P}([4])$ solving (4.2.2). They are all represented in Fig. 4.14.
  (iv) Let $n = 4$ and $\pi = \{\{1, 2\}, \{3\}, \{4\}\}$. Then, there are ten $\sigma \in \mathcal{P}([4])$ that are solutions of (4.2.2). They are all represented in Fig. 4.15.

In what follows (see, for example, Theorem 7.1.3 below), we will sometimes be called to solve jointly the equations $\sigma \wedge \pi = \hat{0}$ and $\sigma \vee \pi = \hat{1}$, that is, given π, to

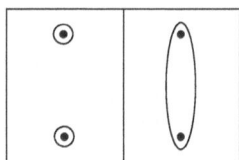

Fig. 4.12. Solving $\sigma \wedge \pi = \hat{0}$ in the simplest case

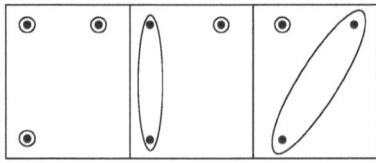

Fig. 4.13. Solving $\sigma \wedge \pi = \hat{0}$ in a three-vertex diagram

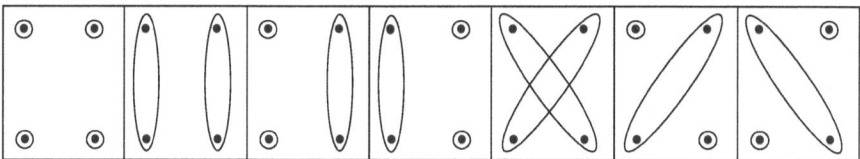

Fig. 4.14. The seven solutions of $\sigma \wedge \pi = \hat{0}$ in a four-vertex diagram

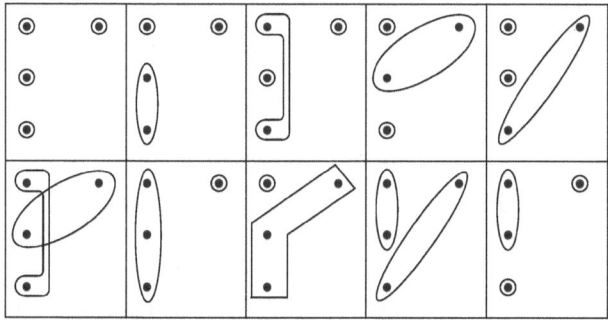

Fig. 4.15. The ten solutions of $\sigma \wedge \pi = \hat{0}$ in a three-row diagram

find all diagrams $\Gamma(\pi, \sigma)$ that are non-flat ($\sigma \wedge \pi = \hat{0}$) and connected ($\sigma \vee \pi = \hat{1}$). Having found, as before, all those that are non-flat, one just has to choose among them those that are connected, that is, the diagrams whose rows cannot be divided into two subset, each defining a separate diagram. These are: the second diagram in Fig. 4.12, the last two in Fig. 4.13, the last six in Fig. 4.14, the sixth to ninth of Fig. 4.15. Again as an example, observe that the second diagram in Fig. 4.15 is not connected: indeed, in this case, $\pi = \{\{1,2\}, \{3\}, \{4\}\}$, $\sigma = \{\{1\}, \{2\}, \{3,4\}\}$, and $\pi \vee \sigma = \{\{1,2\}, \{3,4\}\} \neq \hat{1}$.

4.3 From Gaussian diagrams to multigraphs

A *multigraph* is a graph in which (a) two vertices can be connected by more than one edge, and (b) loops (that is, edges connecting one vertex to itself) are allowed. Such objects are sometimes called "pseudographs", but we will avoid this terminology. In what follows, we show how a multigraph can be derived from a Gaussian diagram. This representation of Gaussian diagrams can be used in the computation of moments and cumulants (see [35] or [68]).

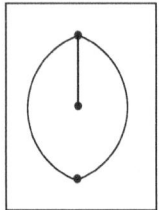

Fig. 4.16. A multigraph with three vertices

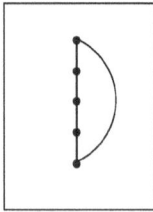

Fig. 4.17. A multigraph built from a circular diagram

Fix a set b and consider partitions $\pi, \sigma \in \mathcal{P}(b)$ such that $\Gamma(\pi, \sigma)$ is Gaussian and $\pi = \{b_1, ..., b_k\}$. Then, the multigraph $\hat{\Gamma}(\pi, \sigma)$, with k vertices and $|b|/2$ edges, is obtained from $\Gamma(\pi, \sigma)$ as follows.

1. Identify each row of $\Gamma(\pi, \sigma)$ with a vertex of $\hat{\Gamma}(\pi, \sigma)$, in such a way that the ith row of $\Gamma(\pi, \sigma)$ corresponds to the ith vertex v_i of $\hat{\Gamma}(\pi, \sigma)$.
2. Draw an edge linking v_i and v_j for every pair (x, y) such that $x \in b_i$, $y \in b_j$ and $x \sim_\sigma y$.

Example 4.3.1 (i) The multigraph obtained from the diagram in Fig. 4.7 is given in Fig. 4.16.
(ii) The multigraph associated with Fig. 4.8 is given in Fig. 4.17 (note that this graph has been obtained from a circular diagram).

The following result is easily verified: it shows how the nature of a Gaussian diagram can be deduced from its graph representation.

Proposition 4.3.2 *Fix a finite set b, as well as a pair of partitions $(\pi, \sigma) \subseteq \mathcal{P}(b)$ such that the diagram $\Gamma(\pi, \sigma)$ is Gaussian and $|\pi| = k$. Then,*

1. *$\Gamma(\pi, \sigma)$ is connected if and only if $\hat{\Gamma}(\pi, \sigma)$ is a connected multigraph.*
2. *$\Gamma(\pi, \sigma)$ is non-flat if and only if $\hat{\Gamma}(\pi, \sigma)$ has no loops.*
3. *$\Gamma(\pi, \sigma)$ is circular if and only if the vertices $v_1, ..., v_k$ of $\hat{\Gamma}(\pi, \sigma)$ are such that: (i) there is an edge linking v_i and v_{i+1} for every $i = 1, ..., k - 1$, and (ii) there is an edge linking v_k and v_1.*

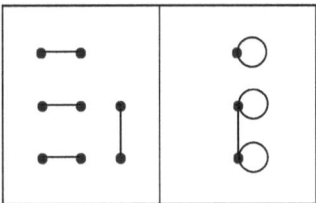

Fig. 4.18. A non-connected flat diagram and its multigraph

As an illustration, in Fig. 4.18 we present the picture of a flat and non-connected diagram (on the left), whose multigraph (on the right) is non-connected and displays three loops.

This situation corresponds to the case $b = [8]$,

$$\pi = \{\{1,2\}, \{3,4,5\}, \{6,7,8\}\}, \text{ and}$$
$$\sigma = \{\{1,2\}, \{3,4\}, \{5,8\}, \{6,7\}\}.$$

Section 4.4 provides a summary, as well as intuitive explanations.

4.4 Summarizing diagrams and multigraphs

The diagram $\Gamma(\pi, \sigma)$ is:

Table 4.2. Diagrams

connected	$\pi \vee \sigma = \hat{1}$	if the rows of π cannot be divided into two subsets, each defining a separate diagram
non-flat	$\pi \wedge \sigma = \hat{0}$	if the closed curves defining the blocks of σ do not have a horizontal portion
Gaussian		if each block of σ contains exactly two elements
circular		if each row is linked to both the previous and next row (with no other links)

The multigraph corresponding to a Gaussian diagram $\Gamma(\pi, \sigma)$ is:

connected	$\pi \vee \sigma = \hat{1}$	if the multigraph is connected
non-flat	$\pi \wedge \sigma = \hat{0}$	if the multigraph has no loops
circular		if the edges connecting the vertices form a cycle

Intuitive explanation for diagrams

1. *Connected diagrams: $\sigma \vee \pi = \hat{1}$.* Suppose that the rows of π can be divided into two subsets, each defining a separate subdiagram. Then there is no way of joining together the elements of these two subdiagrams in order to get $\sigma \vee \pi = \hat{1}$. This is the case, for example, in Fig. 4.1 where $\pi = \sigma = \{\{1, 2\}, \{3\}\}$, so that their join equals $\{\{1, 2\}, \{3\}\}$ and not $\hat{1} = \{1, 2, 3\}$. The diagram is therefore not connected. On the other hand, there are no such problems with the diagram of Fig. 4.5 which is connected.
2. *Non-flat diagrams: $\sigma \wedge \pi = \hat{0}$.* Since the blocks of π are depicted horizontally, if σ had some of its blocks partially horizontal as well, then π and σ would have that part in common, and thus one would not have $\sigma \wedge \pi = \hat{0}$. In Fig. 4.2, the points 7 and 8 lie in a partly horizontal block of σ and hence $\sigma \wedge \pi$ will not contain them as singletons.
3. *Gaussian diagrams.* As it will turn out, partitions σ with blocks of size strictly greater than 2 are often not present in a Gaussian context, for instance in the context of moments and cumulants (see, for example, Corollary 7.3.1). This is why diagrams $\Gamma(\pi, \sigma)$, where every block of σ has exactly two elements are called Gaussian.
4. *Circular diagrams.* Circular diagrams are easily recognizable because the blocks of σ form a cycle (this is readily apparent in the multigraph representation). They appear in the formulas for moments and cumulants of a Gaussian chaos of order $q = 2$. For instance, the diagram in Fig. 4.8 is circular. See also the multigraph representation in Fig. 4.17.

5

Wiener-Itô integrals and Wiener chaos

We will now introduce the notion of a *completely random measure* on a Polish space $(Z, \mathcal{Z})$, as well as those of a *stochastic measure of order* $n \geq 2$, a *diagonal measure* and a *multiple (stochastic) Wiener-Itô integral*. All these concepts can be unified by means of the formalism introduced in Chapters 2–4. We shall merely remark that the domain of the multiple stochastic integrals defined in this chapter can be extended to more general (and possibly random) classes of integrands. We refer the interested reader to the paper by Kallenberg ans Szulga [51], as well as to the monographs by Kussmaul [59], Kwapień and Woyczyński [61, Ch. 10] and Linde [64], for several results in this direction.

5.1 Completely random measures

Diagonals and subdiagonals play an important role in the context of multiple integrals. The following definitions provide a convenient way to specify them. In what follows, we will denote by $(Z, \mathcal{Z})$ a Polish space, where $\mathcal{Z}$ is the associated Borel σ-field. A Polish space is a complete separable metric space. The fact that we endow Z with the Borel σ-field $\mathcal{Z}$ makes $(Z, \mathcal{Z})$ a so-called *Borel space*. Typical examples include $\mathbb{R}^d$ with the Euclidean metric, the space $C[0, 1]$ of continuous functions on $[0, 1]$ with the sup-norm metric and the space $C[0, \infty)$ of continuous functions on $[0, \infty)$ with the topology of uniform convergence on bounded sets (see, for example, [32]).

Definition 5.1.1 *For every $n \geq 1$, we write $(Z^n, \mathcal{Z}^n) = (Z^{\otimes n}, \mathcal{Z}^{\otimes n})$, with $Z^1 = Z$. For every partition $\pi \in \mathcal{P}([n])$ and every $B \in \mathcal{Z}^n$, we set*

$$Z_\pi^n \triangleq \{(z_1, ..., z_n) \in Z^n : z_i = z_j \ \text{if and only if} \ i \sim_\pi j\} \quad \text{and}$$

$$B_\pi \triangleq B \cap Z_\pi^n. \tag{5.1.1}$$

G. Peccati, M.S. Taqqu: Wiener Chaos: Moments, Cumulants and Diagrams –
A survey with computer implementation.
© Springer-Verlag Italia 2011

Recall that $i \sim_\pi j$ means that the elements i and j belong to the same block of the partition π. Relation (5.1.1) states that the variables z_i and z_j should be equated if and only if i and j belong to the same block of π.

Example 5.1.2 (i) Since $\hat{0} = \{\{1\}, ..., \{n\}\}$, no two elements can belong to the same block, and therefore $B_{\hat{0}}$ coincides with the collection of all vectors $(z_1, ..., z_n) \in B$ such that $z_i \neq z_j$, $\forall i \neq j$.

(ii) Since $\hat{1} = \{\{1, ..., n\}\}$, all elements belong to the same block and therefore

$$B_{\hat{1}} = \{(z_1, ..., z_n) \in B : z_1 = z_2 = ... = z_n\}.$$

A set such as $B_{\hat{1}}$ is said to be *purely diagonal*.

(iii) Suppose $n = 3$ and $\pi = \{\{1\}, \{2, 3\}\}$. Then, $B_\pi = \{(z_1, z_2, z_3) \in B : z_2 = z_3, z_1 \neq z_2\}$.

The following decomposition lemma (whose proof is immediate and left to the reader) will be used a number of times.

Lemma 5.1.3 *For every set $B \in \mathscr{Z}^n$,*

$$B = \cup_{\sigma \in \mathcal{P}([n])} B_\sigma = \cup_{\sigma \geq \hat{0}} B_\sigma.$$

Moreover $B_\pi \cap B_\sigma = \varnothing$ if $\pi \neq \sigma$.

One has also that

$$(A_1 \times \cdots \times A_n)_{\hat{1}} = \underbrace{((\cap_{i=1}^n A_i) \times \cdots \times (\cap_{i=1}^n A_i))}_{n \text{ times}}_{\hat{1}}; \tag{5.1.2}$$

indeed, since all coordinates are equal in the LHS of (5.1.2), their common value must be contained in the intersection of the sets.

Example 5.1.4 As an illustration of (5.1.2), let $A_1 = [0, 1]$ and $A_2 = [0, 2]$ be intervals in $\mathbb{R}^1$, and draw the rectangle $A_1 \times A_2 = [0, 1] \times [0, 2] \in \mathbb{R}^2$. The set $(A_1 \times A_2)_{\hat{1}}$ (that is, the subset of $A_1 \times A_2$ composed of vectors whose coordinates are equal) is therefore identical to the diagonal of the square $(A_1 \cap A_2) \times (A_1 \cap A_2) = [0, 1]^2$. The set $(A_1 \times A_2)_{\hat{1}}$ can be visualized as the thick diagonal segment in Fig. 5.1.

We shall now define a "completely random measure" φ, often called an "independently scattered random measure". It has two characteristics: it is a measure and it takes values in a space of random variables. It will be denoted with its arguments as $\varphi(B, \omega)$, where B is a Borel set and ω is a point in the underlying sample space Ω.

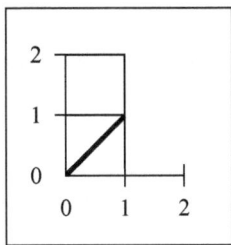

Fig. 5.1. A purely diagonal set

The "size" of φ will be controlled by a non-random, σ-finite and non-atomic measure ν, where

$$\nu(B) = \mathbb{E}\varphi(B)^2.$$

The measure ν is called the *control measure* of φ. The fact that ν is non-atomic means that $\nu(\{z\}) = 0$ for every $z \in Z$.

The measure φ will be used to define multiple integrals, where one integrates either over a whole subset of Z^p, $p \geq 1$, or over a subset "without diagonals". In the first case we will need to suppose $\mathbb{E}|\varphi(B)|^p < \infty$; in the second case, one may suppose as little as $\mathbb{E}\varphi(B)^2 < \infty$, which is a good reason for working with multiple integrals where one excludes diagonals. In the first case, since $p \geq 1$ will be arbitrary, we shall suppose that $\mathbb{E}|\varphi(B)|^p < \infty$, $\forall p \geq 1$, that is, $\varphi \in \cap_{p \geq 1} L^p(\mathbb{P})$. We now present the formal definition of φ.

Definition 5.1.5 (1) *Let ν be a positive, σ-finite and non-atomic measure on $(Z, \mathcal{Z})$, and let*

$$\mathcal{Z}_\nu = \{B \in \mathcal{Z} : \nu(B) < \infty\}. \tag{5.1.3}$$

*A centered **completely random measure** (in $\cap_{p \geq 1} L^p(\mathbb{P})$) on $(Z, \mathcal{Z})$ with **control measure** ν, is a function $\varphi(\cdot, \cdot)$, from $\mathcal{Z}_\nu \times \Omega$ to $\mathbb{R}$, such that*

(i) *for every fixed $B \in \mathcal{Z}_\nu$, the application $\omega \mapsto \varphi(B, \omega)$ is a random variable;*
(ii) *for every fixed $B \in \mathcal{Z}_\nu$, $\varphi(B) \in \cap_{p \geq 1} L^p(\mathbb{P})$;*
(iii) *for every fixed $B \in \mathcal{Z}_\nu$, $\mathbb{E}[\varphi(B)] = 0$;*
(iv) *for every collection of pairwise disjoint elements of $\mathcal{Z}_\nu$, $B_1, ..., B_n$, the variables $\varphi(B_1), ..., \varphi(B_n)$ are independent;*
(v) *for every $B, C \in \mathcal{Z}_\nu$,*

$$\boxed{\mathbb{E}[\varphi(B)\varphi(C)] = \nu(B \cap C).} \tag{5.1.4}$$

(Note that this implies that $\mathbb{E}[\varphi(\varnothing)^2] = \nu(\varnothing) = 0$, and therefore that $\varphi(\varnothing) = 0$ a.s.-$\mathbb{P}$.)

(2) *When $\varphi(\cdot)$ verifies the properties (i) and (iii)–(v) above and $\varphi(B) \in L^2(\mathbb{P})$, $\forall B \in \mathcal{Z}_\nu$ (so that it is not necessarily true that $\varphi(B) \in L^p(\mathbb{P})$, $p > 2$), we say that φ is a **completely random measure in** $L^2(\mathbb{P})$.*

Remark. In what follows, we will systematically assume that $\nu \neq 0$, that is, that there exists $C \in \mathcal{Z}$ such that $\nu(C) > 0$ and also that ν is non-atomic. A Borel space Z can be either finite, countable or have the cardinality of the continuum (see, for example, Dudley [24, Th. 13.1.1] for more details). Since we assume that ν is non-atomic, we conclude that Z has the cardinality of the continuum, and hence that $(Z, \mathcal{Z})$ is Borel-isomorphic to $([0, 1], \mathcal{B}([0, 1]))$, where $\mathcal{B}([0, 1])$ stands for the Borel σ-field of $[0, 1]$. Recall that $(Z, \mathcal{Z})$ is Borel-isomorphic to $([0, 1], \mathcal{B}([0, 1]))$ if and only if there exists a bijection f from Z onto $[0, 1]$ such that f and f^{-1} are both measurable.

Two crucial remarks on additivity. (a) Let $B_1, ..., B_n, ...$ be a sequence of *disjoint* elements of $\mathcal{Z}_\nu$, and let φ be a completely random measure on $(Z, \mathcal{Z})$ with control ν. Then, for every finite $N \geq 2$, one has that $\cup_{n=1}^N B_n \in \mathcal{Z}_\nu$, and, by using Properties (iii), (iv) and (v) in Definition 5.1.5, one has that

$$\mathbb{E}\left[\left(\varphi\left(\cup_{n=1}^N B_i\right) - \sum_{n=1}^N \varphi(B_n)\right)^2\right] = \nu\left(\cup_{n=1}^N B_n\right) - \sum_{n=1}^N \nu(B_n) = 0, \quad (5.1.5)$$

because ν is a measure, and therefore it is finitely additive. Relation (5.1.5) implies in particular that

$$\varphi\left(\cup_{n=1}^N B_n\right) = \sum_{n=1}^N \varphi(B_n), \quad \text{a.s.-}\mathbb{P}. \quad (5.1.6)$$

Now suppose that $\cup_{n=1}^\infty B_n \in \mathcal{Z}_\nu$. Then, by (5.1.6) and again by virtue of Properties (iii), (v) and (vi) in Definition 5.1.5,

$$\mathbb{E}\left[\left(\varphi\left(\cup_{n=1}^\infty B_n\right) - \sum_{n=1}^N \varphi(B_n)\right)^2\right] = \mathbb{E}\left[\left(\varphi\left(\cup_{n=1}^\infty B_n\right) - \varphi\left(\cup_{n=1}^N B_i\right)\right)^2\right]$$

$$= \nu\left(\cup_{n=N+1}^\infty B_n\right) \underset{N\to\infty}{\to} 0,$$

because ν is σ-additive. This entails in turn that

$$\left| \varphi\left(\cup_{n=1}^\infty B_n\right) = \sum_{n=1}^\infty \varphi(B_n), \quad \text{a.s.-}\mathbb{P}, \right| \quad (5.1.7)$$

where the series on the RHS converges in $L^2(\mathbb{P})$, and hence a.s. because the summands are independent (this is a consequence of Lévy's famous "Equivalence Theorem" – see, for example, [24, Th. 9.7.1, p. 320]). Relation (5.1.7) simply means that the application

$$\mathcal{Z}_\nu \to L^2(\mathbb{P}) : B \mapsto \varphi(B),$$

is σ-additive, and therefore that *every completely random measure acts like a σ-additive measure with values in the Hilbert space $L^2(\mathbb{P})$*. See, for example, Engel [26], Kussmaul [59] or Linde [64] for further discussions on vector-valued measures.

(b) In general, it is *not* true that, for a completely random measure φ and for a fixed $\omega \in \Omega$, the application

$$\mathcal{Z}_\nu \to \mathbb{R} : B \mapsto \varphi(B, \omega)$$

acts like a σ-additive real-valued (signed) measure. The most remarkable example of this phenomenon is given by Gaussian completely random measures. This point is related to the content of Section 5.11.

Notation. We consider the spaces $(Z, \mathcal{Z})$ and $(Z^n, \mathcal{Z}^n) = (Z^{\otimes n}, \mathcal{Z}^{\otimes n})$. Do not confuse the subset Z_π^n in (5.1.1), where π denotes a partition, with the class $\mathcal{Z}_\nu^n$ in the forthcoming formula (5.1.8) (see also (5.1.3)), where ν denotes a control measure.

Now fix a completely random measure φ on $(Z, \mathcal{Z})$. We want now to define a corresponding measure $\varphi^{[n]}$ on $(Z^n, \mathcal{Z}^n)$ and we want that measure to have the nice properties stated in the next definition. If $\varphi^{[n]}$ has these nice properties, then we will say that the underlying measure φ is a "good" measure. We start by defining $\varphi^{[n]}$ on rectangles. For every $n \geq 2$ and every rectangle

$$C = C_1 \times \cdots \times C_n, \ C_j \in \mathcal{Z}_\nu,$$

we define

$$\varphi^{[n]}(C) \triangleq \varphi(C_1) \times \cdots \times \varphi(C_n),$$

so that the application $C \mapsto \varphi^{[n]}(C)$ can be extended to a finitely additive application on the ring of finite linear combinations of rectangular sets in Z^n with finite ν^n measure, with values in the set of $\sigma(\varphi)$-measurable random variables. In the next definition we focus on those completely random measures such that the application $\varphi^{[n]}$ admits a unique infinitely additive (and square-integrable) extension on $\mathcal{Z}^n$ (more precisely, on the class of those elements of $\mathcal{Z}^n$ having finite ν^n measure). Here, the infinite additivity (also called σ-additivity) is in the sense of the $L^1(\mathbb{P})$ convergence. Note that we write $\varphi^{[n]}$ to emphasize the dependence of $\varphi^{[n]}$ not only on n, but also on the set $[n] = \{1, ..., n\}$, whose lattice of partitions will be considered later. The basic idea is as follows: the completely random measure φ is defined on Z; it will be "good" if it induces a measure on Z^n with the nice properties listed in the following definition.

Definition 5.1.6 *For $n \geq 2$, we write*

$$\mathcal{Z}_\nu^n = \{C \in \mathcal{Z}^n : \nu^n(C) < \infty\}. \tag{5.1.8}$$

*A completely random measure φ, verifying points (i)–(v) of Definition 5.1.5, is said to be **good** if, for every fixed $n \geq 2$, there exists a (unique) collection of random variables*

$$\varphi^{[n]} = \left\{\varphi^{[n]}(C) : C \in \mathcal{Z}^n\right\}$$

such that

(i) $\left\{ \varphi^{[n]}(C) : C \in \mathcal{Z}_\nu^n \right\} \subseteq L^2(\mathbb{P})$;
(ii) *for every rectangle* $C = C_1 \times \cdots \times C_n$, $C_j \in \mathcal{Z}_\nu$,

$$\varphi^{[n]}(C) = \varphi(C_1) \cdots \varphi(C_n); \tag{5.1.9}$$

(iii) $\varphi^{[n]}$ *is a σ-additive random measure in the following sense: if $C \in \mathcal{Z}_\nu^n$ is such that $C = \cup_{j=1}^\infty C_j$, with the $\{C_j\}$ disjoints, then*

$$\varphi^{[n]}(C) = \sum_{j=1}^\infty \varphi^{[n]}(C_j), \quad \text{with convergence (at least) in } L^1(\mathbb{P}). \tag{5.1.10}$$

Remarks.

(a) The notion of a "completely random measure" can be traced back to Kingman's seminal paper [55]. For further references on completely random measures, see also the two surveys by Surgailis [151] and [152] (note that, in such references, completely random measures are called "independently scattered measures"). The use of the term "good", to indicate completely random measures satisfying the requirements of Definition 5.1.6, is taken from Rota and Wallstrom [132]. Existence of good measures is discussed in Engel [26] and Kwapień and Woyczyński [61, Ch. 10]. For further generalizations of Engel's results the reader is referred, for example, to [51] and [133].

(b) We will sometimes write

$$\varphi^{[1]} = \varphi.$$

Note that, in the case $n = 1$, the σ-additivity (5.1.7) of $\varphi = \varphi^{[1]}$ in the a.s. sense follows from point (v) of Definition 5.1.5. In the case $n \geq 2$, the assumption that the measure φ is good implies σ-additivity in the sense of (5.1.10).

(c) The σ-additivity of the measures $\varphi^{[n]}$ immediately implies the following *continuity property*. Let $\{C, C_k : k \geq 1\}$ be a collection of elements of $\mathcal{Z}_\nu^n$ such that, (i) either $C_1 \subseteq C_2 \subseteq C_3 \subseteq \dots$ and $C = \cup_k C_k$, or (ii) $C_1 \supseteq C_2 \supseteq C_3 \supseteq \dots$ and $C = \cap_k C_k$. Then, as $k \to \infty$, the sequence $\varphi^{[n]}(C_k)$ converges to $\varphi(C)$ in $L^1(\mathbb{P})$. The proof of this fact follows from (5.1.10) and an argument analogous to the non-random case.

Example 5.1.7 The following two examples of good completely random measures will play a crucial role in the subsequent sections. Note that, in this book, we are not going to provide a complete proof of the fact that a given measure is good (the reader is referred to [26] and [61, Ch. 10] for a full treatment of this point). However, in Section

5.8 we shall provide some hints about the arguments one must use in order to prove that a measure is good, in the special case of measures generated by Lévy processes on the real line. Note that, by a Borel isomorphism argument similar to the ones already used in this section, one can always represent a completely random measure (CRM) on a Polish space with non-atomic control in terms of a CRM on the real line.

(i) A *centered Gaussian random measure* with control ν is a collection $G = \{G(B) : B \in \mathcal{Z}_\nu\}$ of jointly Gaussian random variables, centered and such that, for every $B, C \in \mathcal{Z}_\nu$, $\mathbb{E}[G(C)G(B)] = \nu(C \cap B)$. The family G is clearly a completely random measure. The fact that G is also good is classic, and can be seen as a special case of the main results in [26]. Some details can be found in Section 5.8.

(ii) A *compensated Poisson measure with control* ν is a completely random measure $\widehat{N} = \{\widehat{N}(B) : B \in \mathcal{Z}_\nu\}$, as in Definition 5.1.5, such that,

$$\forall B \in \mathcal{Z}_\nu, \ \widehat{N}(B) \overset{law}{=} N(B) - \nu(B),$$

where $N(B)$ is a Poisson random variable with parameter

$$\nu(B) = \mathbb{E}N(B) = \mathbb{E}N(B)^2.$$

The fact that $\widehat{N}$ is also good derives once again from the main findings of [26]. A more direct proof of this last fact can be obtained by observing that, for almost every ω, $\widehat{N}^{[n]}(\cdot, \omega)$ must necessarily coincide with the canonical product (signed) measure (on $(Z^n, \mathcal{Z}^n)$) associated with the signed measure on $(Z, \mathcal{Z})$ given by $\widehat{N}(\cdot, \omega) = N(\cdot, \omega) - \nu(\cdot)$ (indeed, such a canonical product measure satisfies necessarily (5.1.9)).

(iii) A ω-based proof as at Point (ii) cannot be obtained in the Gaussian case. Indeed, if G is a Gaussian measure as in Point (i), one has that, in general, for almost every ω, the mapping $B \mapsto G(B, \omega)$ *does not define* a signed measure. As an example, consider the case where G is generated by a standard Brownian motion $\{W_t : t \in [0, 1]\}$ on $[0, 1]$, in such a way that $G([0, t]) = W_t$ (note that the paths of W are assumed to be almost surely continuous). Then, the fact that G is a signed measure with positive probability, would imply that the mapping $t \mapsto W_t$ has finite variation with positive probability (this is a standard property of signed measures). However, this is inconsistent with the well-known fact that

$$\lim_{n \to \infty} \sum_{k=1}^{n} \left(W_{\frac{k}{n}} - W_{\frac{k-1}{n}}\right)^2 = 1,$$

where the limit is in the sense of the convergence in probability. Note that this last relation is a very special case of Theorem 5.11.1 below.

5.2 Single Wiener-Itô integrals

Let φ be a completely random measure in the sense of Definition 5.1.5, with control measure ν. Our aim in this section is to define (single) Wiener-Itô integrals with respect to φ.

Proposition 5.2.1 *Let φ be a completely random measure in $L^2(\mathbb{P})$, with σ-finite control measure ν. Then, there exists a unique continuous linear operator $h \mapsto \varphi(h)$, from $L^2(\nu)$ into $L^2(\mathbb{P})$, such that*

$$\varphi(h) = \sum_{j=1}^{m} c_j \varphi(B_j) \tag{5.2.11}$$

for every elementary function of the type

$$h(z) = \sum_{j=1}^{m} c_j \mathbf{1}_{B_j}(z), \tag{5.2.12}$$

where $c_j \in \mathbb{R}$ and the sets B_j are in $\mathcal{Z}_\nu$ and pairwise disjoint.

Proof. In what follows, we call *simple kernel* a kernel h as in (5.2.12). For every simple kernel h, set $\varphi(h)$ to be equal to (5.2.11). Then, by using Properties (iii), (v) and (vi) in Definition 5.1.5, one has that, for every pair of simple kernels h, h',

$$\mathbb{E}\left[\varphi(h)\varphi(h')\right] = \int_Z h(z) h'(z) \nu(dz). \tag{5.2.13}$$

Since simple kernels are dense in $L^2(\nu)$, the proof is completed by the following (standard) approximation argument. If $h \in L^2(\nu)$ and $\{h_n\}$ is a sequence of simple kernels converging to h, then (5.2.13) implies that $\{\varphi(h_n)\}$ is a Cauchy sequence in $L^2(\mathbb{P})$, and one defines $\varphi(h)$ to be the $L^2(\mathbb{P})$ limit of $\varphi(h_n)$. One easily verifies that the definition of $\varphi(h)$ does not depend on the chosen approximating sequence $\{h_n\}$. The application $h \mapsto \varphi(h)$ is therefore well-defined, and (by virtue of (5.2.13)) it is an isomorphism from $L^2(\nu)$ into $L^2(\mathbb{P})$. ∎

The random variable $\varphi(h)$ is usually written as

$$\int_Z h(z)\varphi(dz), \quad \int_Z h\,d\varphi \quad \text{or} \quad I_1^\varphi(h), \tag{5.2.14}$$

and it is called the *Wiener-Itô stochastic integral* of h with respect to φ. By inspection of the previous proof, one sees that Wiener-Itô integrals verify the isometric relation

$$\left| \; \mathbb{E}\left[\varphi(h)\varphi(g)\right] = \int_Z h(z) g(z) \nu(dz) = (g, h)_{L^2(\nu)}, \quad \forall g, h \in L^2(\nu). \; \right| \tag{5.2.15}$$

Observe also that

$$\mathbb{E}\varphi(h) = 0.$$

If $B \in \mathcal{Z}_\nu$, we write interchangeably $\varphi(B)$ or $\varphi(1_B)$ (the two objects coincide, thanks to (5.2.11)). For every $h \in L^2(\nu)$, the law of the random variable $\varphi(h)$ is infinitely divisible (see Section 5.3).

5.3 Infinite divisibility of single integrals

Let φ be a completely random measure in the sense of Definition 5.1.5, with control measure ν. Our aim in this section is to give a characterization of the single Wiener-Itô integrals, as appearing in Proposition 5.2.1, as infinitely divisible random variables. Since these integrals have mean zero and finite variance, we do not use the most general representations of infinitely divisible distributions but only those corresponding to random variables with mean zero and finite variance. Infinitely divisible laws are introduced in many textbooks, see, for example, Billingsley [8], Sato [136] or Steutel and van Ham [148], to which we refer the reader for further details. We recall the definition.

Definition 5.3.1 *A random variable X is said to be **infinitely divisible** if, for any $n \geq 1$, there are i.i.d. random variables $X_{j,n}$, $j = 1, ..., n$, such that the following equality in law holds:*

$$X \stackrel{law}{=} \sum_{j=1}^{n} X_{j,n}. \tag{5.3.16}$$

Proposition 5.3.2 *Let φ be a completely random measure on the Polish space $(Z, \mathcal{Z})$, with non-atomic control ν. Then, for every $B \in \mathcal{Z}_\nu$ the random variable $\varphi(B)$ is infinitely divisible.*

Proof. We can assume that $\nu \neq 0$. As already observed in a remark following Definition 5.1.5, since ν is non-atomic and Z is a Polish space whose Borel σ-field is given by $\mathcal{Z}$, there exists a one-to-one mapping $f : (Z, \mathcal{Z}) \rightarrow ([0, 1], \mathcal{B}([0, 1]))$ such that f and f^{-1} are both measurable. For every $A \in \mathcal{B}([0, 1])$, write $\nu_f(A) = \nu(f^{-1}(A))$. It is clear that ν_f is a Borel non-atomic measure on $([0, 1], \mathcal{B}([0, 1]))$, and also that the collection of random variables

$$\varphi_f = \{\varphi_f(A) : A \in \mathcal{B}([0, 1]), \ \nu_f(A) < \infty\},$$

where $\varphi_f(A) = \varphi(f^{-1}(A))$, is a completely random measure on $([0, 1], \mathcal{B}([0, 1]))$ with control ν_f (as defined in Definition 5.1.5). Since, for every $B \in \mathcal{Z}_\nu$, one has that $\nu_f(f(B)) = \nu(B) < \infty$ and $\varphi(B) = \varphi_f(f(B))$, it is therefore sufficient to prove that, for every $A \in \mathcal{B}([0, 1])$ such that $\nu_f(A) < \infty$, the variable $\varphi_f(A)$ has an infinitely divisible law. To show this, fix such a set A and, for every integer $n \geq 1$, write $A_{0,n} = A \cap [0, \frac{1}{n}]$, and $A_{i,n} = A \cap (\frac{i}{n}, \frac{i+1}{n}]$, $i = 1, ..., n-1$. By construction, one has that, for a fixed n, the random variables $\varphi(A_{i,n})$ are independent. Moreover, by additivity, for every n,

$$\varphi_f(A) = \sum_{i=0}^{n-1} \varphi_f(A_{i,n}). \tag{5.3.17}$$

If the $\varphi_f(A_{i,n})$, $i = 0, 1, ..., n-1$, were identically distributed, then (5.3.16) would be satisfied and we would conclude that $\varphi_f(A)$ is infinitely divisible. To reach this conclusion in our situation, note that, since ν_f is non-atomic, the mapping $t \mapsto \nu_f(A \cap [0, t])$ is continuous and hence uniformly continuous on $[0, 1]$, since $[0, 1]$ is a compact interval. Therefore, for any $\epsilon > 0$ one has that (by Chebyshev's inequality),

$$\max_{i=0,...,n-1} \mathbb{P}[|\varphi_f(A_{i,n})| > \epsilon]$$
$$\leq \frac{\max_{i=0,...,n-1} \nu_f(A_{i,n})}{\epsilon^2}$$
$$= \frac{\max_{i=0,...,n-1}[\nu_f(A \cap [0, (i+1)/n]) - \nu_f(A \cap [0, i/n])]}{\epsilon^2}$$
$$\longrightarrow 0,$$

as $n \to \infty$. In addition, one has trivially that $\sum_{i=0}^{n-1} \varphi_f(A_{i,n})$ converges in distribution to $\varphi_f(A)$. We can now use Khintchine's Theorem (see below) in the special case $X_{i,n} = \varphi_f(A_{i,n})$ and $S = \varphi_f(A)$, and deduce immediately that $\varphi_f(A)$ is infinitely divisible and so is therefore $\varphi(B)$ for every $B \in \mathcal{Z}_\nu$. ∎

In the previous proof, we have used the so-called Khintchine's Theorem, that we state (for the sake of completeness) in a form that is adapted to our framework. For a proof, see for example, Sato [136, Th. 9.3, p. 47].

Theorem 5.3.3 (Khintchine's Theorem) *Let $\{X_{i,n} : n \geq 1, \ i = 0, ..., n-1\}$ be an array of real valued random variables such that, for a fixed n, $X_{0,n}, X_{1,n}, ..., X_{n-1,n}$ are independent. Assume that, for every $\epsilon > 0$,*

$$\lim_{n \to \infty} \max_{i=0,1,...,n-1} \mathbb{P}(|X_{i,n}| > \epsilon) = 0,$$

and that, as $n \to \infty$, the sums $S_n = \sum_{i=0}^{n-1} X_{i,n}$, $n \geq 1$, converge in law to some random variable S. Then, S has an infinitely divisible distribution.

Having established that $\varphi(B)$ is infinitely divisible, we shall now determine its so-called *Lévy-Khintchine representation*. Using the one for infinitely divisible distributions with finite variance (see, for example, Billingsley [8, Th. 28.1]), we have that, for every $B \in \mathcal{Z}_\nu$, there exists a unique pair $(c^2(B), \alpha_B)$ such that $c^2(B) \in [0, \infty)$

and α_B is a measure on $\mathbb{R}$ satisfying

$$\alpha_B(\{0\}) = 0 \quad \text{and} \quad \int_{\mathbb{R}} x^2 \alpha_B(dx) < \infty, \qquad (5.3.18)$$

and, for every $\theta \in \mathbb{R}$,

$$\left| \mathbb{E}\left[\exp\left(i\theta\varphi(B)\right)\right] = \exp\left[-\frac{c^2(B)\theta^2}{2} + \int_{\mathbb{R}}\left(\exp\left(i\theta x\right) - 1 - i\theta x\right)\alpha_B(dx)\right]. \right|$$

$$(5.3.19)$$

The measure α_B is called a *Lévy measure*, and the components of the pair $(c^2(B), \alpha_B)$ are called the *Lévy-Khintchine characteristics* associated with $\varphi(B)$. Also, the exponent

$$-\frac{c^2(B)\theta^2}{2} + \int_{\mathbb{R}}\left(\exp\left(i\theta x\right) - 1 - i\theta x\right)\alpha_B(dx)$$

is known as the *Lévy-Khintchine exponent* associated with $\varphi(B)$. Plainly, if φ is Gaussian, then $\alpha_B = 0$ for every $B \in \mathcal{Z}_\nu$ (the reader is referred, for example, to [136] for an exhaustive discussion of infinitely divisible laws). The representation (5.3.19) is valid for infinitely divisible distributions with mean zero and finite variance.

The following result relates higher moments of the random variable $\varphi(B)$ to the "moments" of the measure α_B.

Lemma 5.3.4 *For any $k \geq 3$*

$$\mathbb{E}|\varphi(B)|^k < \infty \quad \text{if and only if} \quad \int_{\mathbb{R}} |x|^k \alpha_B(dx) < \infty. \qquad (5.3.20)$$

Proof. This a consequence of a result by Wolfe [163, Th. 2]. ∎

Proposition 5.3.2 states that $\varphi(B)$ is infinitely divisible. The next proposition states that $\varphi(h)$, $h \in L^2(\nu)$, is also infinitely divisible and provides a description of the Lévy-Khintchine exponent of $\varphi(h)$. The proof is taken from [108] and uses arguments and techniques developed in [127] (see also [60, Sect. 5]). Following the proof, we present an interpretation of the result.

Proposition 5.3.5 *For every $B \in \mathcal{Z}_\nu$, let $\left(c^2(B), \alpha_B\right)$ denote the pair such that $c^2(B) \in [0, \infty)$, α_B verifies (5.3.18) and*

$$\mathbb{E}\left[\exp\left(i\theta\varphi(B)\right)\right] = \exp\left[-\frac{c^2(B)\theta^2}{2} + \int_{\mathbb{R}}\left(\exp\left(i\theta x\right) - 1 - i\theta x\right)\alpha_B(dx)\right].$$

$$(5.3.21)$$

Then, the following holds:

1. *The application $B \mapsto c^2(B)$, from $\mathcal{Z}_\nu$ to $[0, \infty)$, extends to a unique σ-finite measure $c^2(dz)$ on $(Z, \mathcal{Z})$, such that*

$$c^2(dz) \ll \nu(dz),$$

   where ν is the control measure of φ.
2. *There exists a unique measure α on $(Z \times \mathbb{R}, \mathcal{Z} \times \mathcal{B}(\mathbb{R}))$ such that*

$$\alpha(B \times C) = \alpha_B(C),$$

   for every $B \in \mathcal{Z}_\nu$ and $C \in \mathcal{B}(\mathbb{R})$.
3. *There exists a function*

$$\rho_\nu : Z \times \mathcal{B}(\mathbb{R}) \mapsto [0, \infty]$$

   such that

   (i) *for every $z \in Z$, $\rho_\nu(z, \cdot)$ is a Lévy measure[1] on $(\mathbb{R}, \mathcal{B}(\mathbb{R}))$ satisfying*

$$\int_{\mathbb{R}} x^2 \rho_\nu(z, dx) < \infty;$$

   (ii) *for every $C \in \mathcal{B}(\mathbb{R})$, $\rho_\nu(\cdot, C)$ is a Borel measurable function;*
   (iii) *for every positive function $g(z, x) \in \mathcal{Z} \otimes \mathcal{B}(\mathbb{R})$,*

$$\int_Z \int_{\mathbb{R}} g(z, x) \rho_\nu(z, dx) \nu(dz) = \int_Z \int_{\mathbb{R}} g(z, x) \alpha(dz, dx). \qquad (5.3.22)$$

4. *For every $(\theta, z) \in \mathbb{R} \times Z$, define*

$$K_\nu(\theta, z) = -\frac{\theta^2}{2}\sigma_\nu^2(z) + \int_{\mathbb{R}}\left(e^{i\theta x} - 1 - i\theta x\right)\rho_\nu(z, dx), \qquad (5.3.23)$$

   where

$$\sigma_\nu^2(z) = \frac{dc^2}{d\nu}(z);$$

   then, for every $h \in L^2(\nu)$,

$$\int_Z |K_\nu(\theta h(z), z)| \nu(dz) < \infty,$$

   and

$$\mathbb{E}\left[\exp(i\theta\varphi(h))\right] \qquad (5.3.24)$$

$$= \exp\left[\int_Z K_\nu(\theta h(z), z)\nu(dz)\right]$$

$$= \exp\left[-\frac{\theta^2}{2}\int_Z h^2(z)\sigma_\nu^2(z)\nu(dz)\right.$$

$$\left. + \int_Z \int_{\mathbb{R}}\left(e^{i\theta h(z)x} - 1 - i\theta h(z)x\right)\rho_\nu(z, dx)\nu(dz)\right].$$

---

[1] That is, $\rho_\nu(z, \{0\}) = 0$ and $\int_{\mathbb{R}} \min(1, x^2) \rho_\nu(z, dx) < \infty$.

Sketch of the proof. The proof follows from results contained in [127, Sect. II]. Point 1 is indeed a direct consequence of [127, Proposition 2.1 (a)]. In particular, whenever $B \in \mathcal{Z}$ is such that $\nu(B) = 0$, then $\mathbb{E}[\varphi(B)^2] = 0$ (due to Point (vi) of Definition 5.1.5) and therefore $c^2(B) = 0$, thus implying

$$c^2 \ll \nu.$$

Point 2 follows from the first part of the statement of [127, Lemma 2.3]. To establish Point 3 define, as in [127, p. 456],

$$\gamma(B) = c^2(B) + \int_{\mathbb{R}} \min\left(1, x^2\right) \alpha_B(dx) = c^2(B) + \int_{\mathbb{R}} \min\left(1, x^2\right) \alpha(B, dx),$$
$$(5.3.25)$$

whenever $B \in \mathcal{Z}_\nu$, and observe (see [127, Definition 2.2]) that $\gamma(\cdot)$ can be canonically extended to a σ-finite and positive measure on $(Z, \mathcal{Z})$. Moreover, since $\nu(B) = 0$ implies $\varphi(B) = 0$ a.s.-$\mathbb{P}$, the uniqueness of the Lévy-Khinchine characteristics implies as before $\gamma(B) = 0$, and therefore

$$\gamma(dz) \ll \nu(dz).$$

Observe also that, by standard arguments, one can select a version of the density $(d\gamma/d\nu)(z)$ such that

$$(d\gamma/d\nu)(z) < \infty$$

for every $z \in Z$. According to [127, Lemma 2.3], there exists a function $\rho : Z \times \mathcal{B}(\mathbb{R}) \mapsto [0, \infty]$, such that:

(a) $\rho(z, \cdot)$ is a Lévy measure on $\mathcal{B}(\mathbb{R})$ for every $z \in Z$;
(b) $\rho(\cdot, C)$ is a Borel measurable function for every $C \in \mathcal{B}(\mathbb{R})$;
(c) for every positive function $g(z, x) \in \mathcal{Z} \otimes \mathcal{B}(\mathbb{R})$,

$$\int_Z \int_{\mathbb{R}} g(z, x) \rho(z, dx) \gamma(dz) = \int_Z \int_{\mathbb{R}} g(z, x) \alpha(dz, dx).$$
$$(5.3.26)$$

In particular, by using (5.3.26) in the case $g(z, x) = 1_B(z) x^2$ for $B \in \mathcal{Z}_\mu$,

$$\int_B \int_{\mathbb{R}} x^2 \rho(z, dx) \gamma(dz) = \int_{\mathbb{R}} x^2 \alpha_B(dx) < \infty,$$

since $\varphi(B) \in L^2(\mathbb{P})$, and we deduce that ρ can be chosen in such a way that, for every $z \in Z$,

$$\int_{\mathbb{R}} x^2 \rho(z, dx) < \infty.$$

Now define, for every $z \in Z$ and $C \in \mathcal{B}(\mathbb{R})$,

$$\rho_\nu(z, C) = \left(\frac{d\gamma}{d\nu}(z)\right) \rho(z, C),$$

so that

$$\rho_\nu (z, dx)\, \nu (dz) = \rho (z, dx) \frac{d\gamma}{d\nu} (z)\, \nu(dz) = \rho (z, dx)\, \gamma(dz), \qquad (5.3.27)$$

and observe that, due to the previous discussion, the application $\rho_\nu : Z \times \mathcal{B}(\mathbb{R}) \mapsto [0, \infty]$ trivially satisfies properties (i)-(iii) in the statement of Point 3, which is therefore proved.

To prove Point 4, first define (as before) a function $h \in L^2(\nu)$ to be *simple* if

$$h(z) = \sum_{i=1}^{n} a_i \mathbf{1}_{B_i}(z),$$

where $a_i \in \mathbb{R}$, and $(B_1, ..., B_n)$ is a finite collection of disjoint elements of $\mathcal{Z}_\nu$. Because of (5.3.21), the relations (5.3.24) hold for such a simple kernel $h(z)$.

Of course, the class of simple functions (which is a linear space) is dense in $L^2(\nu)$, and therefore for every $L^2(\nu)$ there exists a sequence h_n, $n \geq 1$, of simple functions such that $\int_Z (h_n(z) - h(z))^2 \nu(dz) \to 0$. As a consequence, since ν is σ-finite there exists a subsequence n_k such that $h_{n_k}(z) \to h(z)$ for ν-a.e. $z \in Z$ (and therefore for γ-a.e. $z \in Z$) and moreover, for every $B \in \mathcal{Z}$, the random sequence $\varphi(\mathbf{1}_B h_n)$ is a Cauchy sequence in $L^2(\mathbb{P})$, and hence it converges in probability. In the terminology of [127, p. 460], this implies that every $h \in L^2(\nu)$ is φ-integrable, and that, for every $B \in \mathcal{Z}$, the random variable $\varphi(h \mathbf{1}_B)$, defined according to Proposition 5.2.1, coincides with $\int_B h(z)\, \varphi(dz)$, i.e. the integral of h with respect to the restriction of $\varphi(\cdot)$ to B, as defined in [127, p. 460]. As a consequence, by using a slight modification of [127, Proposition 2.6][2] (which basically exploits the fact that (5.3.24) is trivially satisfied by simple kernels, and then proceeds by approximation), the function K_0 on $\mathbb{R} \times Z$ given by

$$K_0(\theta, z) = -\frac{\theta^2}{2}\sigma_0^2(z) + \int_\mathbb{R} \left(e^{i\theta x} - 1 - i\theta x \right) \rho(z, dx),$$

where $\sigma_0^2(z) = (dc^2/d\gamma)(z)$, is such that $\int_Z |K_0(\theta h(z), z)|\, \gamma(dz) < \infty$ for every $h \in L^2(\nu)$, and also

$$\mathbb{E}\left[\exp(i\theta\varphi(h)) \right]$$

$$= \mathbb{E}\left[\exp\left(i\theta \int_Z h(z)\, \varphi(dz) \right) \right]$$

$$= \int_Z \left[-\frac{(\theta h(z))^2}{2}\sigma_0^2(z) + \int_\mathbb{R} \left(e^{i\theta h(z)x} - 1 - i\theta h(z)x \right) \rho(z, dx) \right] \gamma(dz)$$

$$= \int_Z K_0(\theta h(z), z)\, \gamma(dz). \qquad (5.3.28)$$

The fact that, by definition, K_ν in (5.3.23) verifies

$$K_\nu(\theta h(z), z) = K_0(\theta h(z), z) \frac{d\gamma}{d\nu}(z), \quad \forall z \in Z, \forall h \in L^2(\nu), \forall \theta \in \mathbb{R},$$

yields (5.3.24).   ∎

---

[2] The difference lies in the choice of the truncation.

Interpretation of Proposition 5.3.5 The random variable $\varphi(h)$ can be viewed as a "sum" of random variables $h(z)\varphi(dz)$, each with its own Lévy measure. To make this statement precise, it is best to look at $\varphi(1_B)$, where B be a given set in $\mathcal{Z}_\nu$. The characteristic function of the random variable $\varphi(B)$ or $\varphi(1_B)$ involves the Lévy characteristic

$$(c^2(B), \alpha_B),$$

where $c^2(B)$ is a non-negative constant and $\alpha_B(dx)$ is a Lévy measure on $\mathbb{R}$. We want now to view $B \in \mathcal{Z}_\nu$ as a "variable" and thus to extend $c^2(B)$ to a measure $c^2(dz)$ on $(Z, \mathcal{Z})$, and $\alpha_B(dx)$ to a measure $\alpha(dz, dx)$ on $\mathcal{Z} \otimes \mathcal{B}(\mathbb{R})$. Consider first α_B. According to Proposition 5.3.5, it is possible to extend $\alpha_B(dx)$ to a measure $\alpha(dz, dx)$ on $\mathcal{Z} \otimes \mathcal{B}(\mathbb{R})$, which can be expressed as

$$\alpha(dz, dx) = \rho_\nu(z, dx)\nu(dz) = \rho(z, dx)\gamma(dz), \tag{5.3.29}$$

by (5.3.27), where both ρ_ν and ρ are functions on $Z \times \mathcal{B}(\mathbb{R})$. The functions ρ_ν and ρ are both Radon-Nykodim derivatives: $\rho_\nu(z, dx)$ is the Radon-Nykodim derivative of $\alpha(dz, dx)$ with respect to the control measure of $\nu(dz)$ of $\varphi(dz)$, and $\rho(z, dx)$ is the Radon-Nykodim derivative of $\alpha(dz, dx)$ with respect to the Lévy measure defined in (5.3.25). Observe, however, that ρ enters only in the proof, and not in the statement of Proposition 5.3.5. Thus, in view of (5.3.29), the measure $\alpha(dz, dx)$ is obtained as a "mixture" of the Lévy measures $\rho_\nu(z, \cdot)$ and $\rho(z, \cdot)$ over the variable z, using respectively either the control measure ν or the Lévy measure γ as a mixing measure.

A similar approach is applied to the Gaussian part of the exponent in (5.3.21), involving $c^2(B)$. The coefficient $c^2(B)$ can be extended to a measure $c^2(dz)$, and this measure can be moreover expressed as

$$c^2(dz) = \sigma_\nu^2(z)\nu(dz), \tag{5.3.30}$$

where σ_ν^2 is the density of c^2 with respect to the control measure ν of φ. This allows us to represent the characteristic function of the Wiener-Itô integral $\varphi(h)$ as in (5.3.24). In that expression, the function $h(z)$ appears explicitly in the Lévy-Khintchine exponent as a factor to the argument θ of the characteristic function. One would obtain the Lévy-Khintchine characteristics by making a change of variables, setting for example, $u = h(z)x$ and $v = x$.

Example 5.3.6 (i) If $\varphi = G$ is a centered Gaussian measure with control measure ν, then $\alpha = 0$ and $c^2 = \nu$ (therefore $\sigma_\nu^2 = 1$) and, for $h \in L^2(\nu)$,

$$\mathbb{E}\left[\exp\left(i\theta G(h)\right)\right] = \exp\left[-\frac{\theta^2}{2}\int_Z h^2(z)\nu(dz)\right].$$

(ii) If $\varphi = \widehat{N}$ is a compensated Poisson measure with control measure ν, then $c^2(\cdot) = 0$ and

$$\rho_\nu(z, dx) = \delta_1(dx)$$

for all $z \in Z$, where δ_1 is the Dirac mass at $x = 1$. It follows that, for $h \in L^2(\nu)$,

$$\mathbb{E}\left[\exp\left(i\theta\widehat{N}(h)\right)\right] = \exp\left[\int_Z \left(e^{i\theta h(z)} - 1 - i\theta h(z)\right)\nu(dz)\right]. \quad (5.3.31)$$

(iii) We are interested in the random variable

$$\varphi(B) = \int_{\mathbb{R}}\int_Z u\mathbf{1}_B(z)\,\widehat{N}(du, dz),$$

where $(Z, \mathcal{Z})$ is a Polish space and $\widehat{N}$ is a centered Poisson random measure on

$$\boxed{\mathbb{R} \times Z}$$

(endowed with the usual product σ-field) with σ-finite control measure $\nu(du, dz)$. The random variable $\varphi(B)$ describes the sum of Poisson jumps of size u during the "time" interval B (if $Z \subset \mathbb{R}_+$).

Define the measure μ on $(Z, \mathcal{Z})$ by

$$\mu(B) = \int_{\mathbb{R}}\int_Z u^2\mathbf{1}_B(z)\,\nu(du, dz).$$

Then, the mapping

$$B \mapsto \varphi(B) = \int_{\mathbb{R}}\int_Z u\mathbf{1}_B(z)\,\widehat{N}(du, dz), \quad (5.3.32)$$

where $B \in \mathcal{Z}_\mu = \{B \in \mathcal{Z} : \mu(B) < \infty\}$, is a completely random measure on $(Z, \mathcal{Z})$, with control measure μ. In particular, by setting

$$k_B(u, z) = u\mathbf{1}_B(z),$$

one has that the characteristic function of $\widehat{N}(k_B)$ is given by (5.3.31), with $\nu(dz)$ replaced by $\nu(du, dz)$. Thus

$$\mathbb{E}\left[\exp\left(i\theta\varphi(B)\right)\right] = \mathbb{E}\left[\exp\left(i\theta\widehat{N}(k_B)\right)\right]$$

$$= \exp\left[\int_{\mathbb{R}}\int_Z \left(e^{i\theta k_B(u,z)} - 1 - i\theta k_B(u, z)\right)\nu(du, dz)\right]$$

$$= \exp\left[\int_{\mathbb{R}}\int_Z \left(e^{i\theta u\mathbf{1}_B(z)} - 1 - i\theta u\mathbf{1}_B(z)\right)\nu(du, dz)\right]$$

$$= \exp\left[\int_{\mathbb{R}}\int_Z \left(e^{i\theta u} - 1 - i\theta u\right)\mathbf{1}_B(z)\,\nu(du, dz)\right] \quad (5.3.33)$$

$$= \exp\left[\int_{\mathbb{R}} \left(e^{i\theta u} - 1 - i\theta u\right)\alpha_B(du)\right], \quad (5.3.34)$$

where

$$\alpha_B (du) = \int_Z 1_B (z) \, \nu (du, dz) \qquad (5.3.35)$$

(compare with (5.3.19)). A key step in the argument is (5.3.33).

(iv) Keep the framework of Point (iii). When the measure ν is a product measure of the type $\nu (du, dx) = \rho (du) \beta (dx)$, where β is σ-finite and $\rho (du)$ verifies $\rho (\{0\}) = 0$ and $\int_{\mathbb{R}} u^2 \rho (du) < \infty$ (and therefore $\alpha_B (du) = \beta (B) \rho (du)$), one says that the completely random measure φ in (5.3.32) is *homogeneous* (see, for example, [105]). In particular, for a homogeneous measure φ, relation (5.3.34) gives

$$\mathbb{E} \left[\exp (i\theta\varphi (B)) \right] = \exp \left[\beta (B) \int_{\mathbb{R}} \left(e^{i\theta u} - 1 - i\theta u \right) \rho (du) \right]. \qquad (5.3.36)$$

(v) From (5.3.36) and the classic results on infinitely divisible random variables (see, for example, [136]), one deduces that a centered and square-integrable random variable Y is infinitely divisible if and only if the following holds: there exists a homogeneous completely random measure φ on some space $(Z, \mathcal{Z})$, as well as an independent centered standard Gaussian random variable G, such that

$$Y \overset{law}{=} aG + \varphi (B), \text{ for some } a \geq 0 \text{ and } B \in \mathcal{Z}.$$

(vi) Let the framework and notation of the previous Point (iii) prevail, and assume moreover that:

(1) $(Z, \mathcal{Z}) = ([0, \infty), \mathcal{B} ([0, \infty)))$, and

(2) $\nu (du, dx) = \rho (du) \, dx$, where dx stands for the restriction of the Lebesgue measure on $[0, \infty)$, and ρ verifies $\rho (\{0\}) = 0$ and $\int_{\mathbb{R}} u^2 \rho (du) < \infty$.

Then, the process

$$\left| \quad t \mapsto \varphi ([0, t]) = \int_{\mathbb{R}} \int_{[0,t]} u \widehat{N} (du, dz), \quad t \geq 0, \quad \right| \qquad (5.3.37)$$

is a *centered and square-integrable Lévy process* (with no Gaussian component) started from zero: in particular, the stochastic process $t \mapsto \varphi ([0, t])$ has independent and stationary increments. Let

$$X(t) = \varphi([0, t]), \ t \geq 0,$$

denote this process. It has the following interpretation. In view of (5.3.37), if the Poisson measure $\widehat{N}$ has a point in the box (du, dt), then the process $X(t)$ has a jump of size u in the time interval $(t, t + dt)$.

(vii) Conversely, every centered and square-integrable Lévy process Z_t with no Gaussian component is such that $Z_t \overset{law}{=} \varphi([0,t])$ (in the sense of stochastic processes) for some $\varphi([0,t])$ defined as in (5.3.37). To see this, just use the fact that, for every t,

$$\mathbb{E}\left[\exp\left(i\theta Z_t\right)\right] = \exp\left[t\int_{\mathbb{R}} \left(e^{i\theta u} - 1 - i\theta u\right)\rho\left(du\right)\right], \tag{5.3.38}$$

where the Lévy measure verifies $\rho(\{0\}) = 0$ and $\int_{\mathbb{R}} u^2 \rho(du) < \infty$, and observe that this last relation implies that $\varphi([0,t])$ and Z_t have the same finite-dimensional distributions. This fact is the starting point of the paper by Farré *et al.* [28], concerning Hu-Meyer formulae for Lévy processes.

Remark. Let $(Z, \mathcal{Z})$ be a Polish space. Point 4 in Proposition 5.3.5 implies that every centered completely random measure φ on $(Z, \mathcal{Z})$ has the same law as a random mapping of the type

$$\left| \varphi : B \mapsto G(B) + \int_{\mathbb{R}} \int_{Z} u 1_B(z)\, \widehat{N}(du, dz), \right|$$

where G and $\widehat{N}$ are, respectively, a Gaussian measure on Z and an independent compensated Poisson measure on $\mathbb{R} \times Z$.

Example 5.3.7 (i) (*Gamma random measures*) Let $(Z, \mathcal{Z})$ be a Polish space, let $\lambda > 0$ (interpret λ as a "rate") and let $\widehat{N}$ be a centered Poisson random measure on $\mathbb{R} \times Z$ with σ-finite control measure

$$\nu(du, dz) = \frac{\exp(-\lambda u)}{u} 1_{u>0}\, du\beta(dz),$$

where $\beta(dz)$ is a σ-finite measure on $(Z, \mathcal{Z})$. Now define the completely random measure φ according to (5.3.32). By using (5.3.35) and the fact that

$$\alpha_B(du) = \beta(B)\frac{\exp(-\lambda u)}{u} 1_{u>0}\, du,$$

one infers that, for every $B \in \mathcal{Z}$ such that $\beta(B) < \infty$ and every real θ,

$$\mathbb{E}\left[\exp\left(i\theta\varphi(B)\right)\right]$$
$$= \exp\left[\beta(B)\int_0^\infty \left(e^{i\theta u} - 1 - i\theta u\right)\frac{\exp(-\lambda u)}{u}\,du\right]$$
$$= \exp\left[\beta(B)\int_0^\infty \left(e^{i\lambda^{-1}\theta u} - 1\right)\frac{\exp(-u)}{u}\,du\right]\exp\left(-i\theta\lambda^{-1}\beta(B)\right)$$
$$= \frac{1}{(1 - i\lambda^{-1}\theta)^{\beta(B)}}\exp\left(-i\lambda^{-1}\theta\beta(B)\right), \tag{5.3.39}$$

thus yielding that $\varphi(B)$ is a centered Gamma random variable, with rate λ (mean $1/\lambda$) and shape parameter $\beta(B)$. To verify (5.3.39), observe that, if a is a suitable constant, then

$$\int_0^\infty \frac{e^{iau} - 1}{u} e^{-u} du = i \int_0^\infty \left(\int_0^a e^{iux} dx \right) e^{-u} du$$

$$= i \int_0^a \int_0^\infty e^{-(1-ix)u} du dx = i \int_0^a \frac{1}{1 - ix} dx = -\ln(1 - ia).$$

The completely random measure $\varphi(B)$ has control measure β, and it is called a (centered) *Gamma random measure*. Note that $\varphi(B) + \beta(B) > 0$, a.s.-$\mathbb{P}$, whenever $0 < \beta(B) < \infty$ because it gives us back the usual non-centered Gamma measure. See, for example, [37, 38, 45, 157], and the references therein, for recent results on (non-centered) Gamma random measures.

(ii) (*Dirichlet processes*) Let the notation and assumptions of the previous example prevail, and assume that $0 < \beta(Z) < \infty$ (that is, β is non-zero and finite). Then, $\varphi(Z) + \beta(Z) > 0$ and the mapping

$$B \longmapsto \frac{\varphi(B) + \beta(B)}{\varphi(Z) + \beta(Z)} \tag{5.3.40}$$

defines a random probability measure on $(Z, \mathcal{Z})$, known as *Dirichlet process with parameter* β. Since the groundbreaking paper by Ferguson [30], Dirichlet processes play a fundamental role in Bayesian non-parametric statistics: see, for example, [44, 75] and the references therein. Note that (5.3.40) *does not* define a completely random measure (the independence over disjoint sets fails): however, as shown in [104], one can develop a theory of (multiple) stochastic integration with respect to general Dirichlet processes, by using some appropriate approximations in terms of orthogonal U-statistics. See [119] for a state of the art overview of Dirichlet processes in modern probability.

(iii) (*Compound Poisson processes, I*) Keep the framework and assumptions of formula (5.3.38), and assume that $\rho(\cdot) = \lambda \times \pi(\cdot)$, where $\lambda > 0$ and π is a probability measure on $\mathbb{R}$, not charging zero and such that $\int_{\mathbb{R}} x\pi(dx) = 0$. Then, Z_t has the same law (as a stochastic process) as the following *compound Poisson process*

$$t \mapsto P_t = \sum_{j=1}^{N_t} Y_j, \quad t \geq 0, \tag{5.3.41}$$

where $N = \{N_t : t \geq 0\}$ is a standard Poisson process with parameter λ, and $\{Y_j : j \geq 1\}$ is an i.i.d. sequence with common law equal to π, independent of N. To see this, observe that P_t is a process with independent increments, define

$$\hat{\pi}(\theta) = \int_{\mathbb{R}} e^{i\theta x} \pi(dx)$$

to be the characteristic function of π, and finally exploit independence to write

$$
\mathbb{E}\left[e^{i\theta \sum_{j=1}^{N_t} Y_j}\right]
$$

$$
= \mathbb{E}\left[\hat{\pi}(\theta)^{N_t}\right]
$$

$$
= \exp\left[\lambda t(\hat{\pi}(\theta) - 1)\right]
$$

$$
= \exp\left[\lambda t \int_{\mathbb{R}} (e^{i\theta x} - 1 - \theta x)\pi(dx)\right]
$$

$$
= \mathbb{E}\left[e^{i\theta Z_t}\right]. \tag{5.3.42}
$$

(iv) (*Compound Poisson processes, II*) This is an exercise for you: consider a compound Poisson process P_t as in (5.3.41), except that this time we assume that $\pi(\{0\}) > 0$, that is, the "weights" Y_i can be zero with positive probability. Show that P_t has the same law as a process Z_t as in (5.3.38), with ρ given by:

$$
\rho(A) = \lambda \times \pi(A \cap \{x : x > 0\}).
$$

5.4 Multiple stochastic integrals of elementary functions

We now fix a *good* completely random measure φ, in the sense of Definition 5.1.6 of Section 5.1. We defined there the corresponding random measure $\varphi^{[n]}$ on $\mathcal{Z}_\nu^n$.

We now consider what happens when $\varphi^{[n]}$ is applied not to a set $C \in \mathcal{Z}_\nu^n$ but to its restriction C_π, where π is a partition of $[n] = \{1, ..., n\}$. The set C_π is defined according to (5.1.1), namely C_π is the subset of points $z = (z_1, ..., z_n) \in Z^n$ such that $z_i = z_j$ if and only if $i \sim_\pi j$. We shall also apply $\varphi^{[n]}$ to the union $\cup_{\sigma \geq \pi} C_\sigma$.[3]

It will be convenient to express the result in terms of C, and thus to view $\varphi^{[n]}(C_\pi)$, for example, not as the map $\varphi^{[n]}$ applied to C_π, but as a suitably *restricted* map applied to C.

This restricted map will be denoted $\mathrm{St}_\pi^{\varphi,[n]}$

where "St" stands for "*Stochastic*". In this way, the restriction is embodied in the map, that is, in the measure, rather than in the set.

Thus, fix a good completely random measure φ, as well as an integer $n \geq 2$.

Definition 5.4.1 *For every $\pi \in \mathcal{P}([n])$, we define the two random measures:*[4]

$$
\mathrm{St}_\pi^{\varphi,[n]}(C) \triangleq \varphi^{[n]}(C_\pi), \quad C \in \mathcal{Z}_\nu^n, \tag{5.4.43}
$$

---

[3] From here on (for instance, in formula (5.4.44)), the expressions "$\sigma \geq \pi$" and "$\pi \leq \sigma$" are used interchangeably.

[4] Here, we use a slight variation of the notation introduced by Rota and Wallstrom in [132]. They write $\mathrm{St}_\pi^{[n]}$ and $\varphi_\pi^{[n]}$, instead of $\mathrm{St}_\pi^{\varphi,[n]}$ and $\mathrm{St}_{\geq \pi}^{\varphi,[n]}$ respectively.

and

$$\mathrm{St}_{\geq\pi}^{\varphi,[n]}(C) \triangleq \varphi^{[n]}\left(\cup_{\sigma\geq\pi}C_\sigma\right) = \sum_{\sigma\geq\pi}\mathrm{St}_\sigma^{\varphi,[n]}(C), \quad C \in \mathcal{Z}_\nu^n, \qquad (5.4.44)$$

that are the restrictions of $\varphi^{[n]}$, respectively to the sets Z_π^n and $\cup_{\sigma\geq\pi}Z_\sigma^n$, as defined in (5.1.1).

The notation $\mathrm{St}_\pi^{\varphi,[n]}$ (together with $\mathrm{St}_{\geq\pi}^{\varphi,[n]}(C)$) will be used extensively. It lists:

(i) φ = the random measure on $\mathcal{Z}$;
(ii) n = the dimension of the space $\mathcal{Z}^n$;
(iii) $[n] = \{1, \cdots, n\}$ = the set of indices involved;
(iv) π = the partition which describes the restriction of the map to $\mathcal{Z}_\pi^n$.

In particular, one has the following relations:

- $\mathrm{St}_{\geq\hat{0}}^{\varphi,[n]} = \varphi^{[n]}$, because the subscript " $\geq \hat{0}$ " involves no restriction. Hence, $\mathrm{St}_{\geq\hat{0}}^{\varphi,[n]}$ charges the whole space, and therefore coincides with $\varphi^{[n]}$ (see also Lemma 5.1.3);
- $\mathrm{St}_{\hat{0}}^{\varphi,[n]}$ does not charge diagonals (it charges the whole space minus its diagonals);
- $\mathrm{St}_{\hat{1}}^{\varphi,[n]}$ charges only the full diagonal set $Z_{\hat{1}}^n$, namely $\{z_1 = z_2 = \cdots = z_n\}$;
- for every $\sigma \in \mathcal{P}([n])$ and every $C \in \mathcal{Z}_\nu^n$, $\mathrm{St}_{\geq\sigma}^{\varphi,[n]}(C) = \mathrm{St}_{\geq\hat{0}}^{\varphi,[n]}(C \cap Z_\sigma^n)$.

When $n = 1$, we write

$$\mathrm{St}_{\hat{1}}^{\varphi,[1]}(C) = \mathrm{St}_{\hat{0}}^{\varphi,[1]}(C) = \varphi^{[1]}(C) = \varphi(C), \quad C \in \mathcal{Z}_\nu, \qquad (5.4.45)$$

and, more generally,

$$\mathrm{St}_{\hat{1}}^{\varphi,[1]}(f) = \mathrm{St}_{\hat{0}}^{\varphi,[1]}(f) = \varphi^{[1]}(f) = \varphi(f), \quad f \in L^2(\nu). \qquad (5.4.46)$$

Observe that (5.4.45 and 5.4.46) are consistent with the trivial fact that the class $\mathcal{P}([1])$ contains uniquely the trivial partition $\{\{1\}\}$, so that, in this case, $\hat{1} = \hat{0} = \{\{1\}\}$.

We now define the class $\mathcal{E}(\nu^n)$ of *elementary functions* on Z^n. This is the collection of all functions of the type

$$f(\mathbf{z}_n) = \sum_{j=1}^m k_j \mathbf{1}_{C_j}(\mathbf{z}_n), \qquad (5.4.47)$$

where $k_j \in \mathbb{R}$ and every $C_j \in \mathcal{Z}_\nu^n$ has the form $C_j = C_j^1 \times \cdots \times C_j^n$, $C_j^\ell \in \mathcal{Z}_\nu$ $(\ell = 1, ..., n)$. For every $f \in \mathcal{E}(\nu^n)$ as above, we set

$$\mathrm{St}_\pi^{\varphi,[n]}(f) = \int_{Z^n} f \, d\mathrm{St}_\pi^{\varphi,[n]} = \sum_{j=1}^m k_j \mathrm{St}_\pi^{\varphi,[n]}(C_j) \tag{5.4.48}$$

$$\mathrm{St}_{\geq\pi}^{\varphi,[n]}(f) = \int_{Z^n} f \, d\mathrm{St}_{\geq\pi}^{\varphi,[n]} = \sum_{j=1}^m k_j \mathrm{St}_{\geq\pi}^{\varphi,[n]}(C_j), \tag{5.4.49}$$

and we say that $\mathrm{St}_\pi^{\varphi,[n]}(f)$ (resp. $\mathrm{St}_{\geq\pi}^{\varphi,[n]}(f)$) is the *stochastic integral* of f with respect to $\mathrm{St}_\pi^{\varphi,[n]}$ (resp. $\mathrm{St}_{\geq\pi}^{\varphi,[n]}(f)$). For $C \in \mathcal{Z}_\nu^n$, we write interchangeably $\mathrm{St}_\pi^{\varphi,[n]}(C)$ and $\mathrm{St}_\pi^{\varphi,[n]}(1_C)$ (resp. $\mathrm{St}_{\geq\pi}^{\varphi,[n]}(C)$ and $\mathrm{St}_{\geq\pi}^{\varphi,[n]}(1_C)$). Note that (5.4.44) yields the relation

$$\mathrm{St}_{\geq\pi}^{\varphi,[n]}(f) = \sum_{\sigma \geq \pi} \mathrm{St}_\sigma^{\varphi,[n]}(f).$$

We can therefore apply the Möbius formula (2.5.26) in order to deduce the inverse relation

$$\left| \; \mathrm{St}_\pi^{\varphi,[n]}(f) = \sum_{\sigma \geq \pi} \mu(\pi, \sigma) \, \mathrm{St}_{\geq\sigma}^{\varphi,[n]}(f), \; \right| \tag{5.4.50}$$

(see also [132, Proposition 1]).

Remark. The random variables $\mathrm{St}_\pi^{\varphi,[n]}(f)$ and $\mathrm{St}_{\geq\pi}^{\varphi,[n]}(f)$ are elements of $L^2(\mathbb{P})$ for every $f \in \mathcal{E}(\nu^n)$. While here f is an elementary function, it is neither supposed that f is symmetric nor that it vanishes on the diagonals.

5.5 Wiener-Itô stochastic integrals

We consider the extension of the integrals $\mathrm{St}_\pi^{\varphi,[n]}(f)$ to *non-elementary functions* f in the case $\pi = \hat{0} = \{\{1\}, ..., \{n\}\}$. In view of (5.1.1) and (5.4.43), the random measure $\mathrm{St}_{\hat{0}}^{\varphi,[n]}$ does not charge diagonals (see Definition 5.1.1, as well as the subsequent examples). The fact that one does not integrate on diagonals yields an important simplification.

We start with a heuristic presentation. While relation (5.2.14) involves a simple integral over Z, our goal here is to define integrals over Z^n with respect to $\mathrm{St}_{\hat{0}}^{\varphi,[n]}$. In this case, instead of writing

$$\mathrm{St}_\pi^{\varphi,[n]}(f) = \int_{Z^n} f \, d\mathrm{St}_\pi^{\varphi,[n]}$$

as in (5.4.48), we often write

$$I_n^\varphi(f) = \int_{Z_{\hat{0}}^n} f(z_1, ..., z_n) \, \varphi(dz_1) \cdots \varphi(dz_n). \tag{5.5.51}$$

Thus our goal is to define multiple integrals of functions $f : Z^n \mapsto \mathbb{R}$, of the form (5.5.51).

Since the integration is over $Z_{\hat{0}}^n$ we are excluding diagonals, that is, we are asking that the support of the integrator is restricted to the set of those $(z_1, ..., z_n)$ such that $z_i \neq z_j$ for every $i \neq j$, $1 \leq i, j \leq n$. To define the multiple integral, we approximate the restriction of f to $Z_{\hat{0}}$ by special elementary functions, namely by finite linear combinations of indicator functions $\mathbf{1}_{C_1 \times \cdots \times C_n}$, where the C_j's are pairwise disjoint sets in $\mathcal{Z}_\nu$. This will allow us to define the extension by using isometry, that is, relations of the type

$$\mathbb{E}\left[I_n^\varphi (f)^2\right] = n! \int_{Z^n} f(z_1, ..., z_n)^2\, \nu(dx_1) \cdots \nu(dx_n)$$
$$= n! \int_{Z_{\hat{0}}^n} f(z_1, ..., z_n)^2\, \nu(dx_1) \cdots \nu(dx_n). \qquad (5.5.52)$$

Note that the equality (5.5.52) is due to the fact that the control measure ν is non-atomic, and therefore the associated deterministic product measure never charges diagonals. It is enough, moreover, to suppose that f is symmetric, because if

$$\widetilde{f}(z_1, ..., z_n) = \frac{1}{n!} \sum_{w \in \mathfrak{S}_n} f\left(z_{w(1)}, ..., z_{w(n)}\right) \qquad (5.5.53)$$

is the canonical symmetrization of f ($\mathfrak{S}_n$ is the group of the permutations of $[n]$), then

$$\boxed{I_n^\varphi (f) = I_n^\varphi(\widetilde{f}).} \qquad (5.5.54)$$

This last equality is just a "stochastic equivalent" of the well-known fact that integrals with respect to deterministic symmetric measures are invariant with respect to symmetrizations of the integrands. Indeed, an intuitive explanation of (5.5.54) can be obtained by writing

$$I_n^\varphi (f) = \int_{Z_{\hat{0}}^n} f d\varphi^{[n]} = \int_{Z^n} f\left[\mathbf{1}_{Z_{\hat{0}}^n} d\varphi^{[n]}\right]$$

and by observing that the set $Z_{\hat{0}}$ is symmetric[5], so that $I_n^\varphi (f)$ appears as an integral with respect to the symmetric stochastic measure $\mathbf{1}_{Z_{\hat{0}}} d\varphi^{[n]}$.

From now on, we will denote by $\mathcal{Z}_{s,\nu}^n = \mathcal{Z}_s^n$ (the dependence on ν is dropped, whenever there is no risk of confusion) the *symmetric σ-field* generated by the elements of $\mathcal{Z}_\nu^n$ of the type

$$\widetilde{C} = \bigcup_{w \in \mathfrak{S}_n} C_{w(1)} \times C_{w(2)} \times \cdots \times C_{w(n)}, \qquad (5.5.55)$$

---

[5] That is: $(z_1, ..., z_n) \in Z_{\hat{0}}^n$ implies that $(z_{w(1)}, ..., z_{w(n)}) \in Z_{\hat{0}}^n$ for every $w \in \mathfrak{S}_n$.

where $\{C_j : j = 1, ..., n\} \subset \mathcal{Z}_\nu$ are *pairwise* disjoint and $\mathfrak{S}_n$ is the group of the permutations of $[n]$. The label "s" stands for "symmetric". We stress that, by definition, the class $\mathcal{Z}^n_{s,\nu}$ is a σ-field, whereas $\mathcal{Z}^n_\nu$ is not.

Remarks.

(a) We use the symbol "~" differently for sets and functions. On sets, it corresponds to union over all permutations, see (5.5.55), whereas, for functions it involves a division by the number of permutations (see (5.5.53)).

(b) One can easily show that $\mathcal{Z}^n_s$ is the σ-field generated by the symmetric functions on Z^n that are square-integrable with respect to ν^n, vanishing on every set Z^n_π such that $\pi \neq \hat{0}$, that is, on all diagonals of Z^n of the type $\{z_{i_1} = \cdots = z_{i_j}\}$, $1 \leq i_1 \leq \cdots \leq i_j \leq n$.

By specializing (5.4.43)-(5.4.49) to the case $\pi = \hat{0}$, where

$$\mathrm{St}^{\varphi,[n]}_{\hat{0}}(f) = I^\varphi_n(f),$$

we obtain an intrinsic characterization of *Wiener-Itô multiple stochastic integrals*, as well as of the concept of *stochastic measure of order* $n \geq 2$. The key is the following result, proved in [132, p. 1268].

Proposition 5.5.1 *Let φ be a good completely random measure.*
   (A) *For every $f \in \mathcal{E}(\nu^n)$,*

$$\mathrm{St}^{\varphi,[n]}_{\hat{0}}(f) = \mathrm{St}^{\varphi,[n]}_{\hat{0}}\left(\widetilde{f}\right), \tag{5.5.56}$$

where $\widetilde{f}$ is given in (5.5.53). In other words, the measure $\mathrm{St}^{\varphi,[n]}_{\hat{0}}$ is symmetric.
   (B) *The collection*

$$\left\{ \mathrm{St}^{\varphi,[n]}_{\hat{0}}(C) : C \in \mathcal{Z}^n_\nu \right\} \tag{5.5.57}$$

is the unique symmetric random measure on $\mathcal{Z}^n_\nu$ verifying the two following properties:

   (i) *$\mathrm{St}^{\varphi,[n]}_{\hat{0}}(C) = 0$ for every $C \in \mathcal{Z}^n_\nu$ such that $C \subset Z^n_\pi$ for some $\pi \neq \hat{0}$;*
   (ii) *for every set $\widetilde{C}$ as in (5.5.55), one has*

$$\mathrm{St}^{\varphi,[n]}_{\hat{0}}\left(\widetilde{C}\right) = \mathrm{St}^{\varphi,[n]}_{\hat{0}}(1_{\widetilde{C}}) = n!\varphi(C_1) \times \varphi(C_2) \times \cdots \times \varphi(C_n). \tag{5.5.58}$$

Remark. Note that $\mathrm{St}^{\varphi,[n]}_{\hat{0}}$ is defined on the class $\mathcal{Z}^n_\nu$, which also contains non-symmetric sets. The measure $\mathrm{St}^{\varphi,[n]}_{\hat{0}}$ is "symmetric" in the sense that, for every set $C \in \mathcal{Z}^n_\nu$, the following equality holds:

$$\mathrm{St}^{\varphi,[n]}_{\hat{0}}(C) = \mathrm{St}^{\varphi,[n]}_{\hat{0}}(C_w), \quad \text{a.s.-}\mathbb{P},$$

where w is a permutation of the set $[n]$, and

$$C_w = \left\{ (z_1, ..., z_n) \in Z^n : \left(z_{w(1)}, ..., z_{w(n)} \right) \in C \right\}.$$

Thus, the result does not depend on which permutation one considers.

Notation. For $n \geq 2$, we denote by $L_s^2 (\nu^n)$ the Hilbert space of symmetric and square integrable functions on Z^n (with respect to ν^n). We sometimes use the notational convention: $L_s^2 (\nu^1) = L^2 (\nu^1) = L^2(\nu)$. We write

$$\mathcal{E}_{s,0} (\nu^n) \tag{5.5.59}$$

to indicate the subset of $L_s^2 (\nu^n)$ composed of *elementary <u>symmetric</u> functions vanishing on diagonals*. We also write

$$\mathcal{E}_0 (\nu^n)$$

to indicate the subset of $L^2 (\nu^n)$ composed of *elementary functions vanishing on diagonals*. This means that:

(i) $\mathcal{E}_{s,0} (\nu^n)$ is the class of functions of the type $f = \sum_{j=1}^m k_j 1_{\widetilde{C}_j}$, where $k_j \in \mathbb{R}$ and every $\widetilde{C}_j \subset Z_{\hat{0}}^n$ has the form (5.5.55), and

(ii) $\mathcal{E}_0 (\nu^n)$ is the class of functions of the type $f = \sum_{j=1}^m k_j 1_{C_j}$, where $k_j \in \mathbb{R}$ and every C_j is the Cartesian product of n disjoint subsets of Z with finite ν measure.

The index 0 in $\mathcal{E}_{s,0} (\nu^n)$ refers to the fact that it is a set of functions which equals 0 on the diagonals. Since ν is non-atomic, and ν^n does not charge diagonals, one deduces that $\mathcal{E}_{s,0} (\nu^n)$ is dense in $L_s^2 (\nu^n)$. More precisely, we have the following:

Lemma 5.5.2 *Fix $n \geq 2$. Then, $\mathcal{E}_0 (\nu^n)$ is dense in $L^2 (\nu^n)$, and $\mathcal{E}_{s,0} (\nu^n)$ is dense in $L_s^2 (\nu^n)$.*

Proof. Since, for every $h \in L_s^2 (\nu)$ and every $f \in \mathcal{E}_0 (\nu^n)$, by symmetry,

$$\langle h, f \rangle_{L^2(\nu^n)} = \langle h, \widetilde{f} \rangle_{L^2(\nu^n)},$$

it is enough to prove that $\mathcal{E}_0 (\nu^n)$ is dense in $L^2 (\nu^n)$. We shall only provide a detailed proof for $n = 2$ (the general case is absolutely analogous, albeit it requires a more involved notation). To prove the desired claim, it is therefore sufficient to show that every function of the type

$$h (z_1, z_2) = 1_A (z_1) 1_B (z_2),$$

with $A, B \in \mathcal{Z}_\nu$, is the limit in $L^2 (\nu^2)$ of linear combinations of products of the type $1_{D_1} (z_1) 1_{D_2} (z_2)$, with $D_1, D_2 \in \mathcal{Z}_\nu$ and $D_1 \cap D_2 = \emptyset$. To do this, define

$$C_1 = A \backslash B, \quad C_2 = B \backslash A \text{ and } C_3 = A \cap B,$$

so that
$$h = 1_{C_1} 1_{C_2} + 1_{C_1} 1_{C_3} + 1_{C_3} 1_{C_2} + 1_{C_3} 1_{C_3}.$$

If $\nu(C_3) = 0$, there is nothing to prove. If $\nu(C_3) > 0$, since ν is non-atomic, for every $N \geq 2$ we can find disjoint sets $C_3(i, N) \subset C_3$, $i = 1, ..., N$, such that $\nu(C_3(i, N)) = \nu(C_3)/N$ and $\cup_{i=1}^{N} C_3(i, N) = C_3$. It follows that

$$1_{C_3}(z_1) 1_{C_3}(z_2) = \sum_{1 \leq i \neq j \leq N} 1_{C_3(i,N)}(z_1) 1_{C_3(j,N)}(z_2)$$

$$+ \sum_{i=1}^{N} 1_{C_3(i,N)}(z_1) 1_{C_3(i,N)}(z_2)$$

$$= h_1(z_1, z_2) + h_2(z_1, z_2).$$

Plainly, $h_1 \in \mathcal{E}_0(\nu^n)$, and

$$\|h_2\|_{L^2(\nu^n)}^2 = \sum_{i=1}^{N} \nu(C_3(i, N))^2 = \frac{\nu(C_3)^2}{N}.$$

Since N is arbitrary, we deduce the desired conclusion.   ∎

The proof of the following result can be deduced from (5.5.58) is left to the reader.

Proposition 5.5.3 *Let the above assumptions and notations prevail.*

1. *One has that, $\forall n, m \geq 1$,*

$$\mathbb{E}\left[\text{St}_{\hat{0}}^{\varphi,[m]}(f) \text{St}_{\hat{0}}^{\varphi,[n]}(g) \right] = \delta_{n,m} \times n! \int_{Z^n} f(\mathbf{z}_n) g(\mathbf{z}_n) \nu^n(d\mathbf{z}_n), \quad (5.5.60)$$

   $\forall f \in \mathcal{E}_{s,0}(\nu^m)$ and $\forall g \in \mathcal{E}_{s,0}(\nu^n)$, where $\delta_{n,m} = 1$ if $n = m$, and $= 0$ otherwise (if, for example, $n = 1$, then the result holds for $f \in L^2(\nu)$, by using definition (5.4.46)).

2. *It follows from (5.5.60) that, for every $n \geq 2$, the linear operator $f \mapsto \text{St}_{\hat{0}}^{\varphi,[n]}(f)$, from $\mathcal{E}_{s,0}(\nu^n)$ into $L^2(\mathbb{P})$, can be uniquely extended to a continuous operator (in fact, an isometry) from $L_s^2(\nu^n)$ into $L^2(\mathbb{P})$. This extension enjoys the orthogonality and isometry properties given by (5.5.60).*

We can now present the crucial definition of this section.

Definition 5.5.4 (Wiener chaos) *For every $f \in L_s^2(\nu^n)$, the random variable $\text{St}_{\hat{0}}^{\varphi,[n]}(f)$ is the **multiple stochastic Wiener-Itô integral** (of order n) of f with respect to φ. We also use the classic notation*

$$\left| \quad \text{St}_{\hat{0}}^{\varphi,[n]}(f) = I_n^{\varphi}(f), \quad f \in L_s^2(\nu^n). \quad \right| \qquad (5.5.61)$$

Note that $\forall f \in L_s^2(\nu^m)$ and $\forall g \in L_s^2(\nu^n)$,

$$\left| \quad \mathbb{E}\left[I_m^\varphi(f) I_n^\varphi(g)\right] = \delta_{n,m} \times n! \int_{Z^n} f(\mathbf{z}_n) g(\mathbf{z}_n) \nu^n(d\mathbf{z}_n). \quad \right| \qquad (5.5.62)$$

For $n \geq 2$, the random measure

$$\left\{ \mathrm{St}_{\hat{0}}^{\varphi,[n]}(C) : C \in \mathcal{Z}_\nu^n \right\}$$

*is called the **stochastic measure of order** n associated with φ. When $f \in L^2(\nu^n)$ (not necessarily symmetric), we set*

$$I_n^\varphi(f) = I_n^\varphi(\tilde{f}), \qquad (5.5.63)$$

where $\tilde{f}$ is the symmetrization of f given by (5.5.53). For a completely random measure φ, the Hilbert space

$$C_n^\varphi = \left\{ I_n^\varphi(f) : f \in L_s^2(\nu^n) \right\}, \quad n \geq 1,$$

*is called the nth **Wiener chaos** associated with φ.*

We have supposed so far that $\varphi(C) \in L^p(\mathbb{P})$, $\forall p \geq 1$, for every $C \in \mathcal{Z}_\nu$ (see Definition 5.1.5). We shall now suppose that $\varphi(C) \in L^2(\mathbb{P})$, $C \in \mathcal{Z}_\nu$. In this case, the notion of "good measure" introduced in Definition 5.1.6 and Proposition 5.5.1 does not apply since, in this case, $\varphi^{[n]}$ may not exist. Indeed, consider (5.1.9) for example with $C_1 = \ldots = C_n \in \mathcal{Z}_\nu$. Then, the quantity

$$\mathbb{E}\left| \varphi^{[n]}(C) \right|^2 = \mathbb{E}\left| \varphi(C_1) \right|^{2n}$$

may be infinite because we are not assuming anymore that moments higher than 2 exist (see also [26]). It follows that, for $n \geq 2$, the multiple Wiener-Itô integral cannot be defined as a multiple integral with respect to the restriction to $Z_{\hat{0}}^n$ of the "full stochastic measure" $\varphi^{[n]}$. Nonetheless, one can always proceed as follows, by focusing on sets that are pairwise disjoint, by using isometry and considering integrals where one does not include the diagonals.

Definition 5.5.5 *Let φ be a completely random measure in $L^2(\mathbb{P})$ (and not necessarily in $L^p(\mathbb{P})$, $p > 2$), with non-atomic control measure ν. For $n \geq 2$, let*

$$I_n^\varphi(f) = n! \sum_{k=1}^m \gamma_k \times \left\{ \varphi\left(C_1^{(k)}\right) \varphi\left(C_2^{(k)}\right) \cdots \varphi\left(C_n^{(k)}\right) \right\}, \qquad (5.5.64)$$

for every simple function

$$f \in \sum_{k=1}^p \gamma_k \mathbf{1}_{\tilde{C}^{(k)}} \in \mathcal{E}_{s,0}(\nu^n),$$

where every $\widetilde{C}^{(k)}$ is as in (5.5.55) (in particular, the $\widetilde{C}^{(k)}$ are pairwise disjoint). It is easily seen that the integrals $I_n^\varphi(f)$ defined in (5.5.64) still verify the $L^2(\mathbb{P})$ isometry property (5.5.62). Since the sets of the type $\widetilde{C}$ generate $\mathcal{Z}_s^n$, and ν is non-atomic, the operator $I_n^\varphi(\cdot)$ can be extended to a continuous linear operator from $L_s^2(\nu^n)$ into $L^2(\mathbb{P})$, such that (5.5.62) is verified. When $f \in L^2(\nu^n)$ (not necessarily symmetric), we set

$$\boxed{I_n^\varphi(f) = I_n^\varphi(\widetilde{f}),} \tag{5.5.65}$$

where $\widetilde{f}$ is given by (5.5.53).

Remark. Of course, if $\varphi \in \cap_{p \geq 1} L^p(\mathbb{P})$ (for example, when φ is a Gaussian measure or a compensated Poisson measure), then the definition of I_n^φ obtained from (5.5.64) coincides with the one given in (5.5.61).

5.6 Integral notation

We have already introduced in (5.5.51) the notation I_n^φ which is used when integration over the diagonals is excluded. In the general case, one can use the following integral notation which is quite suggestive: for every $n \geq 2$, every $\sigma \in \mathcal{P}([n])$ and every elementary function $f \in \mathcal{E}(\nu^n)$,

$$\left| \operatorname{St}_\sigma^{\varphi,[n]}(f) = \int_{Z_\sigma^n} f(z_1, ..., z_n)\, \varphi(dz_1) \cdots \varphi(dz_n), \right|$$

and

$$\left| \operatorname{St}_{\geq \pi}^{\varphi,[n]}(f) = \int_{\cup_{\sigma \geq \pi} Z_\sigma^n} f(z_1, ..., z_n)\, \varphi(dz_1) \cdots \varphi(dz_n). \right|$$

For instance:

- if $n = 2$, $f(z_1, z_2) = f_1(z_1) f_2(z_2)$ and $\sigma = \hat{0} = \{\{1\}, \{2\}\}$, then

$$I_2(f) = \operatorname{St}_\sigma^{\varphi,[2]}(f) = \int_{z_1 \neq z_2} f_1(z_1) f_2(z_2)\, \varphi(dz_1)\, \varphi(dz_2)\,;$$

- if $n = 2$, $f(z_1, z_2) = f_1(z_1) f_2(z_2)$ and $\sigma = \hat{1} = \{1, 2\}$, then

$$\operatorname{St}_{\hat{1}}^{\varphi,[2]}(f) = \int_{z_1 = z_2} f_1(z_1) f_2(z_2)\, \varphi(dz_1)\, \varphi(dz_2)\,;$$

- if $n = 2$,

$$\operatorname{St}_{\geq \hat{0}}^{\varphi,[2]}(f) = \operatorname{St}_{\{\{1\},\{2\}\}}^{\varphi,[2]}(f) + \operatorname{St}_{\{\{1,2\}\}}^{\varphi,[2]}(f)$$

$$= \int_{z_1 \neq z_2} f(z_1, z_2)\, \varphi(dz_1)\, \varphi(dz_2) + \int_{z_1 = z_2} f(z_1, z_2)\, \varphi(dz_1)\, \varphi(dz_2)$$

$$= \int_{Z^2} f(z_1, z_2)\, \varphi(dz_1)\, \varphi(dz_2)\,;$$

- if $n = 3$, $f(z_1, z_2, z_3) = f_1(z_1, z_2) f_2(z_3)$ and $\sigma = \{\{1,2\}, \{3\}\}$, then

$$\mathrm{St}_\sigma^{\varphi,[3]}(f) = \int_{\substack{z_3 \neq z_1 \\ z_1 = z_2}} f_1(z_1, z_2) f_2(z_3) \, \varphi(dz_1) \, \varphi(dz_2) \, \varphi(dz_3);$$

- if $n = 3$ and $f(z_1, z_2, z_3) = f_1(z_1, z_2) f_2(z_3)$ and $\sigma = \hat{1} = \{\{1,2,3\}\}$, then

$$\mathrm{St}_\sigma^{\varphi,[3]}(f) = \int_{z_1 = z_2 = z_3} f_1(z_1, z_2) f_2(z_3) \, \varphi(dz_1) \, \varphi(dz_2) \, \varphi(dz_3).$$

5.7 The role of diagonal measures

We now provide the definition of the *diagonal measures* associated with a good measure φ.

Definition 5.7.1 *Let φ be a good random measure on the Borel space $(Z, \mathcal{Z})$, with non-atomic control measure ν. For every $n \geq 1$, we define the mapping*

$$C \mapsto \Delta_n^\varphi(C) \triangleq \mathrm{St}_{\hat{1}}^{\varphi,[n]}(\underbrace{C \times \cdots \times C}_{n \text{ times}}), \quad C \in \mathcal{Z}_\nu, \tag{5.7.66}$$

*to be the nth **diagonal measure** associated with φ.*

Observe that Δ_n^φ is a measure on $(Z, \mathcal{Z})$ which is defined through the measure $\mathrm{St}_{\hat{1}}^{\varphi,[n]}$ on $(Z^n, \mathcal{Z}^n)$. Recall also that $\mathrm{St}_{\hat{1}}^{\varphi,[n]}$ "equates the diagonals", so that, for example, if $C = [0, T]$, $T > 0$, then

$$\Delta_2^\varphi(C) \triangleq \mathrm{St}_{\hat{1}}^{\varphi,[2]}(C \times C) = \int_{\substack{[0,T]^2 \\ z_1 = z_2}} \varphi(dz_1)\varphi(dz_2).$$

The measure Δ_n^φ may be random or not, depending on the φ, as we will see in the sequel. The following statement (corresponding to Proposition 2 in [132]) is one of the keys of the subsequent discussion. It shows that a stochastic measure associated with a partition can be represented as the product of the diagonal measures attached to the blocks of that partition.

Proposition 5.7.2 *Let φ be a good completely random measure. Then, for every $n \geq 2$, every $C_1, ..., C_n \subset \mathcal{Z}_\nu$, and every partition $\pi \in \mathcal{P}([n])$,*

$$\mathrm{St}_{\geq \pi}^{\varphi,[n]}(C_1 \times \cdots \times C_n)$$

$$= \prod_{b = \{i_1, ..., i_{|b|}\} \in \pi} \mathrm{St}_{\hat{1}}^{\varphi,[|b|]}(C_{i_1} \times \cdots \times C_{i_{|b|}}) \tag{5.7.67}$$

$$= \prod_{b = \{i_1, ..., i_{|b|}\} \in \pi} \mathrm{St}_{\hat{1}}^{\varphi,[|b|]}(\underbrace{(\cap_{k=1}^{|b|} C_{i_k}) \times \cdots \times (\cap_{k=1}^{|b|} C_{i_k})}_{|b| \text{ times}}) \tag{5.7.68}$$

$$= \prod_{b = \{i_1, ..., i_{|b|}\} \in \pi} \Delta_{|b|}^\varphi \left(\cap_{k=1}^{|b|} C_{i_k} \right). \tag{5.7.69}$$

Proof. Recall that $\text{St}_{\geq\pi}^{\varphi,[n]} = \sum_{\sigma\geq\pi} \text{St}_{\sigma}^{\varphi,[n]}$, by (5.4.44). To prove the first equality, just observe that both random measures on the LHS and the RHS of (5.7.67) are the restriction of the product measure $\text{St}_{\geq\hat{0}}^{\varphi,[n]}$ to the union of the sets Z_{σ}^n such that $\sigma \geq \pi$. Equality (5.7.68) is an application of (5.1.2).   ∎

The next result, which can be seen as an abstract version of a famous formula by Kailath and Segall [50] (see also the subsequent Section 5.8), provides a neat relation between diagonal and non-diagonal measures. It corresponds to Theorem 2 in [132]. It also relates completely non-diagonal measures $\text{St}_{\hat{0}}^{\varphi,[n]}$ of order n to those of lower orders.

Theorem 5.7.3 (Abstract Kailath-Segall formula) *Let φ be a good completely random measure. Then, for every $n \geq 2$ and every $C_1, ..., C_n \in \mathcal{Z}_{\nu}$,*

$$\text{St}_{\hat{0}}^{\varphi,[n]} (C_1 \times \cdots \times C_n) \tag{5.7.70}$$

$$= \sum (-1)^{|b|-1} (|b|-1)! \, \Delta_{|b|}^{\varphi} \left(\cap_{k=1}^{|b|} C_{i_k}\right) \text{St}_{\hat{0}}^{\varphi,[n-|b|]} \left(C_{j_1} \times \cdots \times C_{j_{n-|b|}}\right),$$

where: (i) the sum runs over all subset $b = \{i_1, ..., i_k\}$ of $[n]$ containing 1, and (ii) we used the convention $\text{St}_{\hat{0}}^{\varphi,[0]} \equiv 1$.

Proof. According to (5.4.50) in the special case $\pi = \hat{0}$, one has the relation

$$\text{St}_{\hat{0}}^{\varphi,[n]} = \sum_{\sigma\in\mathcal{P}([n])} \mu\left(\hat{0}, \sigma\right) \text{St}_{\geq\sigma}^{\varphi,[n]}.$$

Combining this identity with Proposition 5.7.2 (relation (5.7.69)), yields therefore

$$\text{St}_{\hat{0}}^{\varphi,[n]}(C_1 \times \cdots C_n) = \sum_{\sigma\in\mathcal{P}([n])} \mu\left(\hat{0}, \sigma\right) \prod_{a=\{i_1,...,i_{|a|}\}\in\sigma} \Delta_{|a|}^{\varphi} \left(\cap_{k=1}^{|a|} C_{i_k}\right).$$

Now fix a subset b of $[n]$ containing 1, as well as a partition $\sigma \in \mathcal{P}([n])$ such that $b \in \sigma$. We denote by $b^c = [n]\backslash b = \{j_1, ..., j_{n-|b|}\}$ the complement of b, and by σ' the partition of b^c composed of the blocks of σ that are different from b. According to equation (2.5.24) (writing μ' for the Möbius function associated with the lattice $\mathcal{P}(b^c)$) one has that $\mu\left(\hat{0}, \sigma\right) = (-1)^{|b|-1}(|b|-1)!\mu'(\hat{0}, \sigma')$ (if $b = [n]$, one sets by convention $\mu'(\hat{0}, \sigma') = 1$). As a consequence, one can write

$$\text{St}_{\hat{0}}^{\varphi,[n]}(C_1 \times \cdots C_n)$$

$$= \sum_{b\subseteq[n]:1\in b} (-1)^{|b|-1}(|b|-1)!\Delta_{|b|}^{\varphi} \left(\cap_{k=1}^{|b|} C_{i_k}\right)$$

$$\sum_{\sigma'\in\mathcal{P}(b^c)} \mu'(\hat{0}, \sigma') \prod_{a=\{t_1,...,t_{|a|}\}\in\sigma'} \Delta_{|a|}^{\varphi} \left(\cap_{l=1}^{|a|} C_{t_l}\right).$$

We now use once again Proposition 5.7.2 to infer that

$$\sum_{\sigma' \in \mathcal{P}(b^c)} \mu'(\hat{0}, \sigma') \prod_{a=\{t_1,\dots,t_{|a|}\} \in \sigma'} \Delta^\varphi_{|a|} \left(\cap_{l=1}^{|a|} C_{t_l} \right) = \mathrm{St}^{\varphi,[n-|b|]}_{\hat{0}} \left(C_{j_1} \times \cdots \times C_{j_{n-|b|}} \right),$$

thus obtaining the desired conclusion.   ■

Corollary 5.7.4 *Let $C \in \mathcal{Z}_\nu$, fix $n \geq 2$, and, for $j \geq 2$ write $C^{\otimes j}$ for the jth Cartesian product of C (with $C^{\otimes 1} = C$). Then,*

$$\mathrm{St}^{\varphi,[n]}_{\hat{0}} \left(C^{\otimes n} \right) = \sum_{k=1}^{n} \frac{(-1)^{k-1} (n-1)!}{(n-k)!} \Delta^\varphi_k (C) \, \mathrm{St}^{\varphi,[n-k]}_{\hat{0}} \left(C^{\otimes(n-k)} \right), \quad (5.7.71)$$

with the convention $\mathrm{St}^{\varphi,[0]}_{\hat{0}} \equiv 1$.

Proof. Observe that, for every function g on $\mathbb{N}$,

$$\sum_{b \subseteq [n]: 1 \in b} g(|b|) = \sum_{k=1}^{n} \binom{n-1}{k-1} g(k).$$

Since (5.7.70) and (5.1.2) imply that

$$\mathrm{St}^{\varphi,[n]}_{\hat{0}} \left(C^{\otimes n} \right) = \sum_{b \subseteq [n]: 1 \in b} g(|b|),$$

with

$$g(|b|) = (-1)^{|b|-1} (|b| - 1)! \, \Delta^\varphi_{|b|} (C) \, \mathrm{St}^{\varphi,[n-|b|]}_{\hat{0}} \left(C^{\otimes(n-|b|)} \right),$$

the desired conclusion is deduced after elementary simplifications.   ■

5.8 Example: completely random measures on the real line

As usual, our reference for Lévy processes is the book by Sato [136]. Consider a centered square integrable Lévy process on the positive half-line, written $X = \{X_t : t \geq 0\}$. We shall assume that X has finite moments of all orders, that is, $\mathbb{E}|X_t|^n < \infty$, for every $n \geq 1$. According to the discussion developed in the Example 5.3.6, we can assume, without loss of generality, that

$$X_t = \sigma G([0,t]) + \varphi([0,t]), \quad (5.8.72)$$

where $\sigma > 0$, G is a Gaussian measure over $\mathbb{R}_+$ with control given by the Lebesgue measure, and φ is an independent completely random measure on $\mathbb{R}_+$ of the type (5.3.37). In particular, the Poisson measure $\hat{N}$ on $\mathbb{R} \times \mathbb{R}_+$ has homogeneous control

$\rho(du)dx$, and the measure ρ verifies $\rho(\{0\}) = 0$ and (since X has finite moments of all orders) $\int_{\mathbb{R}} |x|^n \rho(dx) < \infty$, for all $n \geq 1$. Note that the application

$$t \mapsto W_t = G\left([0,t]\right), \ t \geq 0,$$

defines a *standard Brownian motion* started from zero. Recall that this means that W_t is a centered Gaussian process with covariance function given by $\mathbb{E}(W_t W_s) = \min(s,t)$. In particular, one has that $W_0 = 0$, and (due, for example, of a standard application of the Kolmogorov criterion – see, for example, [128]) W_t admits a continuous modification. We write $\chi_n(X(t))$, $n \geq 1$, to indicate the sequence of the cumulants of X_t (so that $\chi_1(X(t)) = \mathbb{E}[X_t] = 0$) and introduce the notation

$$K_2 = \sigma^2 + \int_{\mathbb{R}} x^2 \rho(dx), \quad K_n = \int_{\mathbb{R}} x^n \rho(dx), \ n \geq 3. \tag{5.8.73}$$

By using the relation

$$\mathbb{E}\left[\exp\left(i\theta X_t\right)\right] = \exp\left[-\frac{\theta^2 \sigma^2}{2}t + t \int_{\mathbb{R}} \left(e^{i\theta u} - 1 - i\theta u\right) \rho\left(du\right)\right], \tag{5.8.74}$$

as well as the fact that ρ has finite moments of all orders, it is easy to show, by successive differentiations of the argument of the exponential, that

$$\chi_n(X(t)) = K_n t, \ n \geq 2.$$

We now introduce the notation

$$\varphi_X(A) = \sigma G(A) + \varphi(A),$$

for every set $A \subset \mathbb{R}_+$ of finite Lebesgue measure, and we observe that the mapping $A \mapsto \varphi_X(A)$ defines a completely random measure with control

$$\alpha(dx) = \left\{\sigma^2 + \int_{\mathbb{R}} u^2 \rho(du)\right\} dx = K_2 dx,$$

that is, a multiple of Lebesgue measure. The measure φ_X is *good*, in the sense of Definition 5.1.6. A complete proof of this fact can be found, for example, in [26] or [61, Ch. 10]. Here, we merely note the following key step in showing that φ_X is good (see, for example, [61, Lemma 10.1]), namely that for every rectangle of the type $C = C_1 \times \cdots \times C_n$, $n \geq 2$, and where the $C_i \subset \mathbb{R}$ have finite Lebesgue measure, one has that

$$\mathbb{E}|\varphi_X(C_1) \cdots \varphi(C_n)| \leq \alpha^n(C).$$

The existence of the product measures $\varphi_X^{[n]}$ follows from some (quite technical) continuity argument. In what follows, for $n \geq 2$, we shall provide a characterization of the two stochastic measures $\mathrm{St}_0^{\varphi_X,[n]}$ ("completely non-diagonal") and $\Delta_n^{\varphi_X}$ ("diagonal").

To do this, introduce recursively the family of processes

$$P_t^{(0)} \triangleq 1, \ P_t^{(1)} \triangleq X_t, \ P_t^{(n)} \triangleq \int_0^t P_{u-} dP_u, \ n \geq 2, \qquad (5.8.75)$$

where $P_{t-} = \lim_{h \uparrow 0} P_{t+h}$ and the integrals have to be regarded in the usual Itô sense, so that each $P^{(n)}$ is a square-integrable martingale with respect to the natural filtration of X. Write, as usual, $\Delta X_s = X_s - X_{s-}$, $s > 0$. Classic results on Lévy processes (see [99] or [136]) also imply that, for every $n \geq 1$, the *variation processes*

$$X_t^{(1)} \triangleq X_t, \ X_t^{(2)} \triangleq \sigma^2 t + \sum_{0 < s \leq t} (\Delta X_s)^2, \ X_t^{(n)} \triangleq \sum_{0 < s \leq t} (\Delta X_s)^n, \ n \geq 3, \ (5.8.76)$$

are square-integrable (non-centered) Lévy processes, such as the mapping

$$t \mapsto Y_t^{(n)} \triangleq X_t^{(n)} - K_n t$$

defines a square-integrable martingale (with respect to the natural filtration of X) with independent increments. The martingales $Y^{(n)}$, $n \geq 1$, are known as *Teugels martingales* , and play a fundamental role in Nualart and Schoutens' theory of Lévy chaos (see [99]), as well as in the Hu-Meyer formulae proved by Farré *et al.* in [28].

The following statement contains the announced characterization of diagonal and non-diagonal measures using iterated integrals, defined in the Itô sense, that is, as integrals with a random integrand. The arguments used in the proof, in particular the connection with a formula by Kailath and Segall [50], are strongly inspired by [28] (see, in particular, the first ArXiv version of [28]).

Theorem 5.8.1 *Fix $n \geq 2$, and let the above notation and assumptions prevail. Then, for every finite $T > 0$,*

1. $\mathrm{St}_{\hat{0}}^{\varphi X, [n]}([0, T]^n) = n! \, P_T^{(n)}$, and

2. $\Delta_n^\varphi([0, T]) = X_T^{(n)}$.

Proof. By construction of the Itô integral, it is immediately seen that $P_T^{(n)}$ can be obtained as the $L^2(\mathbb{P})$ limit of linear combinations of random variables of the type

$$X(t_1) \times (X(t_2) - X(t_1-)) \times \cdots (X(t_n) - X(t_{n-1}-)),$$

where $0 \leq t_1 < t_2 < \cdots < t_n \leq T$, so that the proof of the first point in the statement is a consequence of Proposition 5.5.1 and of the fact that intervals of the type $(t_i, t_{i+1}]$

generate the Borel σ-field of $\mathbb{R}_+$. To prove Point 2, we use a famous identity proved by Kailath and Segall in [50], stating that

$$P_T^{(n)} = \frac{1}{n} \left(P_T^{(n-1)} X_T^{(1)} - P_T^{(n-2)} X_T^{(2)} + \ldots + (-1)^{n-1} P_T^{(0)} X_T^{(n)} \right)$$

(this relation can be proved by an iterate use of Itô formula). By virtue of Point 1, this formula can be rewritten as

$$\mathrm{St}_{\hat{0}}^{\varphi x, [n]} \left([0, T]^n \right) = \sum_{k=1}^{n} \frac{(-1)^{k-1} (n-1)!}{(n-k)!} X_T^{(k)} \, \mathrm{St}_{\hat{0}}^{\varphi x, [n-k]} \left([0, T]^{(n-k)} \right),$$

so that the desired conclusion is obtained (for example, by recursion on n) as an application of formula (5.7.71). ∎

The previous results can be extended in order to deduce characterizations of more general objects of the type

$$I_n^{\varphi x} (f) = \mathrm{St}_{\hat{0}}^{\varphi x, [n]} (f),$$

where f is symmetric on $\mathbb{R}_+^n$ and square-integrable with respect to the Lebesgue measure. In particular, in this framework, one has that

$$I_n^{\varphi x} (f) = n! \int_0^\infty \left[\int_0^{t_1-} \int_0^{t_2-} \cdots \int_0^{t_{n-1}-} f(t_1, \ldots, t_n) \, dX_{t_n} \cdots dX_{t_2} \right] dX_{t_1},$$

$$(5.8.77)$$

where the RHS of (5.8.77) stands for a usual Itô-type stochastic integral, with respect to X, of the stochastic process

$$t \mapsto v(t) = n! \int_0^{t-} \int_0^{t_2-} \cdots \int_0^{t_{n-1}-} f(t_1, \ldots, t_n) \, dX_{t_n} \cdots dX_{t_2}, \quad t \geq 0.$$

Equality (5.8.77) can be easily proved for elementary functions that are constant on intervals, and the general results are obtained by standard density arguments. Note in particular that $v(t)$ is predictable with respect to the natural filtration of X and also that

$$\mathbb{E} \left[\int_0^\infty v^2(t) \, dt \right] < \infty.$$

5.9 Chaotic representation

When φ is a Gaussian measure or a compensated Poisson measure, multiple stochastic Wiener-Itô integrals play a crucial role, because of the *chaotic representation property* enjoyed by φ. Indeed, when φ is Gaussian or compensated Poisson, one can show that every functional $F(\varphi) \in L^2(\mathbb{P})$ of φ, admits a unique *chaotic (Wiener-Itô) decomposition*

$$F(\varphi) = \mathbb{E}[F(\varphi)] + \sum_{n \geq 1} I_n^\varphi(f_n), \quad f_n \in L_s^2(\nu^n), \quad (5.9.78)$$

where the series converges in $L^2(\mathbb{P})$ (see for instance [20], [46] or [66]), and the kernels $\{f_n\}$ are uniquely determined up to negligible sets. For a proof of the chaotic expansion (5.9.78), see Theorem 8.2.1 below for the case where φ is Gaussian, and Corollary 10.0.5 below for the Poisson case.

Remark. The unicity of the kernels $\{f_n\}$ appearing in (5.9.78) is a consequence of the isometry formula (5.5.62). Indeed, if $\{f_n'\}$ is another sequence of kernels verifying

$$F = \mathbb{E}[F] + \sum_{n \geq 1} I_n^\varphi(f_n')$$

then one has

$$0 = \mathbb{E}[(F - F)^2] = \sum_{n \geq 1} n! \|f_n - f_n'\|_{L^2(\nu^n)}^2,$$

thus implying $f_n = f_n'$, ν^n-almost everywhere, for every n.

Formula (5.9.78) implies that random variables of the type $I_n^\varphi(f_n)$ are the basic "building blocks" of the space of square-integrable functionals of φ. As already pointed out, for a completely random measure φ, the space $C_n^\varphi = \{I_n^\varphi(f) : f \in L_s^2(\nu^n)\}$, $n \geq 1$, is the nth Wiener chaos associated with φ. We set by definition

$$C_0^\varphi = \mathbb{R}$$

(that is, C_0^φ is the collection of all non-random constants) so that, in (5.9.78),

$$\mathbb{E}[F] \in C_0^\varphi.$$

Observe that relation (5.9.78) can be reformulated in terms of Hilbert spaces as follows:

$$L^2(\mathbb{P}, \sigma(\varphi)) = \bigoplus_{n=0}^\infty C_n^\varphi,$$

where $\oplus$ indicates an orthogonal sum in the Hilbert space $L^2(\mathbb{P}, \sigma(\varphi))$.

Remarks

(i) If $\varphi = G$ is a Gaussian measure with non-atomic control ν, for every $p > 2$ and every $n \geq 2$, there exists a universal constant $c_{p,n} > 0$, such that the following *hypercontractivity* property takes place:

$$\mathbb{E}\left[\left|I_n^G(f)\right|^p\right]^{1/p} \leq c_{n,p}\mathbb{E}\left[I_n^G(f)^2\right]^{1/2}, \tag{5.9.79}$$

$\forall f \in L_s^2(\nu^n)$ (see [46, Ch. V and Ch. VI]). Moreover, on every finite sum of Wiener chaoses $\oplus_{j=0}^m C_j^G$ and for every $p \geq 1$, the topology induced by $L^p(\mathbb{P})$ convergence is equivalent to the L^0-topology induced by convergence in probability, that is, convergence in probability is equivalent to convergence in L^p, for every $p \geq 1$ (see, for example, [138]). We refer the reader to [20], [46] or [94, Ch. 1] for an exhaustive analysis of the properties of multiple stochastic Wiener-Itô integrals with respect to a Gaussian measure G.

(ii) The "chaotic representation property" is enjoyed by other processes. One of the most well-known examples is given by the class of *normal martingales*, that is, real-valued martingales on $\mathbb{R}_+$ having a predictable quadratic variation equal to t. See [20] and [65] for a complete discussion of this point. See [99] for a chaotic representation, based on Teugels martingales, associated with general Lévy processes on the real line.

5.10 Computational rules

We are now going to apply our setup to the Gaussian case ($\varphi = G$) and to the Poisson case ($\varphi = \widehat{N} = N - \nu$). We always suppose that the control measure ν of either G or $\widehat{N}$ is non-atomic. Many of the subsequent formulae can be understood intuitively, by applying the following computational rules:

Gaussian case:

$$G(dx)^2 = \nu(dx) \text{ and } G(dx)^n = 0, \text{ for every } n \geq 3. \tag{5.10.80}$$

Poisson case:

$$(\widehat{N}(dx))^n = (N(dx))^n = N(dx), \text{ for every } n \geq 2. \tag{5.10.81}$$

5.11 Multiple Gaussian integrals of elementary functions

Suppose φ is Gaussian. The next result appears in [132, Example G, p. 1272, and Proposition 2, 6 and 12].

Theorem 5.11.1 *Let $\varphi = G$ be a centered Gaussian completely random measure with non-atomic control measure ν. For every $n \geq 2$ and every $A \in \mathcal{Z}_\nu$ (see (5.1.3)):*

$$\mathrm{St}_{\hat{1}}^{G,[n]}(\underbrace{A \times \cdots \times A}_{n \text{ times}}) \triangleq \Delta_n^G(A) = \begin{cases} 0 & n \geq 3 \\ \nu(A) & n = 2, \end{cases} \qquad (5.11.82)$$

(as before, the measure $\Delta_n^G(\cdot)$ is called the diagonal measure of order n associated with G). More generally, for every $n \geq 2$, $\sigma \in \mathcal{P}([n])$ and $A_1, ..., A_n \in \mathcal{Z}_\nu$,

$$\mathrm{St}_{\geq \sigma}^{G,[n]}(A_1 \times \cdots \times A_n)$$
$$= \begin{cases} 0, & \text{if } \exists b \in \sigma : |b| \geq 3 \\ \prod_{b=\{i,j\} \in \sigma} \nu(A_i \cap A_j) \prod_{\ell=1}^{k} G(A_{j_\ell}), & \text{otherwise,} \end{cases} \qquad (5.11.83)$$

and

$$\mathrm{St}_{\sigma}^{G,[n]}(A_1 \times \cdots \times A_n)$$
$$= \begin{cases} 0, & \text{if } \exists b \in \sigma : |b| \geq 3 \\ \prod_{b=\{i,j\} \in \sigma} \nu(A_i \cap A_j) \mathrm{St}_{\hat{0}}^{G,[k]}(A_{j_1} \times \cdots \times A_{j_k}), & \text{otherwise,} \end{cases}$$
$$(5.11.84)$$

where $j_1, ..., j_k$ are the singletons contained in $\sigma \backslash \{b \in \sigma : |b| = 2\}$.

Proof. Although relation (5.11.82) is classic (see, for example, [132, Proposition 6]), we shall give a direct proof. First of all, thanks to a Borel isomorphism argument similar to the one used in the proof of Proposition 5.3.2 (see also the remarks following Definition 5.1.5), we can assume without loss of generality that $Z = [0, 1]$, and that ν is a non-atomic Borel measure on Z. Now fix a set $A \in \mathcal{Z}_\nu$ (that is, $\nu(A) < \infty$) as in (5.11.82). For every integer $k \geq 1$, write $A_{0,k} = A \cap [0, \frac{1}{k}]$, and $A_{i,k} = A \cap (\frac{i}{k}, \frac{i+1}{k}]$, $i = 1, ..., k-1$, so that $\sum_{i=0}^{k-1} \nu(A_{i,k}) = \nu(A)$. By construction, one has that, for a fixed k, the random variables $\varphi(A_{i,k}) \sim N(0, \nu(A_{i,k}))$ are Gaussian and independent. Now define

$$A^{(2,k)} \triangleq \bigcup_{i=0}^{k-1}(A_{i,k} \times A_{i,k}), \quad k \geq 1.$$

Observe that $A^{(2,k)}$ is a subset of $Z \times Z$ and that $A_{\hat{1}} = \cap_{k \geq 1} A^{(2,k)}$. Since the random measure $\varphi^{[2]}$ is continuous on monotone sequences (see Point (c) in the remarks following Definition 5.1.6) one has that $\Delta_2^G(A)$ is the limit in $L^1(\mathbb{P})$ of the sequence

$$\varphi^{[2]}(A^{(2,k)}) = \sum_{i=0}^{k-1} \varphi^{[2]}(A_{i,k} \times A_{i,k}) = \sum_{i=0}^{k-1} \varphi(A_{i,k})^2, \quad k \geq 1,$$

where we have used (5.1.9), as well as the fact that the rectangles $A_{i,k} \times A_{i,k}$ are disjoint for every fixed k. Since $\mathbb{E}[\varphi(A_{i,k})^4] = 3\nu(A_{i,k})^2$, we have

$$\mathbb{E}\left[\left(\varphi^{[2]}(A^{(2,k)}) - \nu(A)\right)^2\right] = \mathbb{E}\left[\left(\sum_{i=0}^{k-1}\left(\varphi(A_{i,k})^2 - \nu(A_{i,k})\right)\right)^2\right]$$

$$= 2\sum_{i=0}^{n}\nu(A_{i,k})^2 \leq 2\nu(A)\max_{i=0,\dots,k-1}\{\nu(A_{i,k})\} \to 0,$$

as $k \to \infty$, since the mapping $t \mapsto \nu(A \cap [0,t])$ is uniformly continuous on $[0,1]$ (because the measure ν is non-atomic). This proves (5.11.82) in the case $n = 2$. To prove (5.11.82) in the general case, fix an integer $n \geq 3$, and observe that, by an argument similar to the one above, the quantity $\Delta_n^G(A)$ is the limit in $L^1(\mathbb{P})$ of the sequence

$$\sum_{i=0}^{k-1}\varphi(A_{i,k})^n, \quad n \geq 1.$$

Then, (5.11.82) follows from the fact that

$$\mathbb{E}\left|\sum_{i=0}^{k-1}\varphi(A_{i,k})^n\right| \leq \sum_{i=0}^{k-1}\mathbb{E}\,|\varphi(A_{i,k})|^n \leq C\sum_{i=0}^{k-1}\nu(A_{i,k})^{\frac{n}{2}}$$

$$\leq C \times \nu(A) \times \left[\max_{i=0,\dots,k-1}\{\nu(A_{i,k})\}\right]^{\frac{n}{2}-1} \to 0,$$

as $k \to \infty$.

Formula (5.11.83) is obtained by combining (5.11.82) and (5.7.67). To prove (5.11.84), suppose first that $\exists b \in \sigma$ such that $|b| \geq 3$. Then, by using Möbius inversion (5.4.50),

$$\mathrm{St}_\sigma^{G,[n]}(A_1 \times \cdots \times A_n) = \sum_{\sigma \leq \rho}\mu(\sigma,\rho)\,\mathrm{St}_{\geq\rho}^{G,[n]}(A_1 \times \cdots \times A_n) = 0,$$

where the last equality is due to (5.11.83) and to the fact that, if $\rho \geq \sigma$ and σ contains a block with more than two elements, then ρ must also contain a block with more than two elements. This proves the first line of (5.11.84). Now suppose that all the blocks of σ have at most two elements, and observe that, by Definition 5.4.1 and (5.1.9),

$$\prod_{\ell=1}^{k}G(A_{j_\ell}) = \mathrm{St}_{\geq\hat{0}}^{G,[k]}(A_{j_1} \times \cdots \times A_{j_k}).$$

The proof is concluded by using the following relations:

$$\text{St}_\sigma^{G,[n]}(A_1 \times \cdots \times A_n)$$
$$= \sum_{\sigma \leq \rho} \mu(\sigma, \rho)\text{St}_{\geq \rho}^{G,[n]}(A_1 \times \cdots \times A_n)$$
$$= \prod_{b=\{i,j\}\in\sigma} \nu(A_i \cap A_j) \sum_{\sigma \leq \rho} \mu(\sigma, \rho) \prod_{b=\{r,l\}\in\rho\backslash\sigma} \nu(A_r \cap A_l) \times$$
$$\times \text{St}_{\geq \hat{0}}^{G,[m]}(A_{i_1} \times \cdots \times A_{i_m})\mathbf{1}_{\{\{i_1\},\ldots,\{i_m\}\text{ are the singletons of }\rho\}},$$

where we write $b = \{r,l\} \in \rho\backslash\sigma$ to indicate that the block b is in ρ and not in σ (equivalently, b is obtained by merging two singletons of σ). Indeed, by the previous discussion, one has that the partitions ρ involved in the previous sums have uniquely blocks of size one or two, and moreover, by Möbius inversion,

$$\sum_{\sigma \leq \rho} \mu(\sigma, \rho) \prod_{b=\{r,l\}\in\rho\backslash\sigma} \nu(A_r \cap A_l) \times$$
$$\times \text{St}_{\geq \hat{0}}^{G,[m]}(A_{i_1} \times \cdots \times A_{i_m})\mathbf{1}_{\{\{i_1\},\ldots,\{i_m\}\text{ are the singletons of }\rho\}}$$
$$= \sum_{\sigma^* \leq \rho^*} \mu(\sigma^*, \rho^*) \prod_{b=\{r,l\}\in\rho^*} \nu(A_r \cap A_l) \times$$
$$\times \text{St}_{\geq \hat{0}}^{G,[m]}(A_{i_1} \times \cdots \times A_{i_m})\mathbf{1}_{\{\{i_1\},\ldots,\{i_m\}\text{ are the singletons of }\rho^*\}}$$
$$= \sum_{\hat{0} \leq \rho^*} \mu(\hat{0}, \rho^*)\text{St}_{\geq \rho^*}^{G,[k]}(A_{j_1} \times \cdots \times A_{j_k})$$
$$= \text{St}_{\hat{0}}^{G,[k]}(A_{j_1} \times \cdots \times A_{j_k}),$$

where σ^* and ρ^* indicate, respectively, the restriction of σ and ρ to $\{j_1, \ldots, j_k\}$, where $\{j_1\}, \ldots, \{j_k\}$ are the singletons of σ (in particular, $\sigma^* = \hat{0}$). Note that the fact that

$$\mu(\sigma, \rho) = \mu(\sigma^*, \rho^*) = \mu(\hat{0}, \rho^*)$$

is a consequence of (2.5.21) and of the fact that ρ has uniquely blocks of size one or two.[6]

$\blacksquare$

Example 5.11.2   (i) One has $\text{St}_{\geq \hat{0}}^{G,[n]}(A_1 \times \cdots \times A_n) = G(A_1)\cdots G(A_n)$, which follows from (5.11.83), since $\hat{0} = \{\{1\}, \ldots, \{n\}\}$, but also directly since the symbol " $\geq \hat{0}$ " entails no restriction on the partition. In integral notation (f is always supposed to be elementary)

$$\text{St}_{\geq \hat{0}}^{G,[n]}(f) = \int_{Z^n} f(z_1, \ldots, z_n) G(dz_1) \cdots G(dz_n).$$

---

[6] Thanks to F. Benaych-Georges for pointing out this argument.

On the other hand, there is no way to "simplify" an object such as $\text{St}_{\hat{0}}^{G,[n]}(A_1 \times \cdots \times A_n)$, which is expressed, in integral notation, as

$$\text{St}_{\hat{0}}^{G,[n]}(f) = I_n^G(f) = \int_{z_1 \neq \cdots \neq z_n} f(z_1, ..., z_n)\, G(dz_1) \cdots G(dz_n).$$

(ii) For $n \geq 3$, one has

$$\text{St}_{\geq \hat{1}}^{G,[n]}(A_1 \times \cdots \times A_n) = \text{St}_{\hat{1}}^{G,[n]}(A_1 \times \cdots \times A_n) = 0,$$

since the partition $\hat{1}$ contains a single block of size ≥ 2. In integral notation,

$$\text{St}_{\hat{1}}^{G,[n]}(f) = \int_{z_1 = \cdots = z_n} f(z_1, ..., z_n)\, G(dz_1) \cdots G(dz_n) = 0.$$

When $n = 1$, however, one has that

$$\text{St}_{\hat{1}}^{G,[1]}(f) = \int_Z f(z)\, G(dz) \sim N\left(0, \int_Z f^2 d\nu\right).$$

When $n = 2$,

$$\text{St}_{\hat{1}}^{G,[2]}(f) = \int_Z f(z, z)\, \nu(dz).$$

(iii) Let $n = 3$ and $\sigma = \{\{1\}, \{2, 3\}\}$, then $\text{St}_{\hat{1}}^{G,[3]}(A_1 \times A_2 \times A_n) = 0$, and therefore

$$\text{St}_{\geq \sigma}^{G,[3]}(A_1 \times A_2 \times A_3)$$
$$= \text{St}_{\sigma}^{G,[3]}(A_1 \times A_2 \times A_3) + \text{St}_{\hat{1}}^{G,[n]}(A_1 \times A_2 \times A_3)$$
$$= \text{St}_{\sigma}^{G,[3]}(A_1 \times A_2 \times A_3) = G(A_1)\, \nu(A_2 \times A_3).$$

In integral notation:

$$\text{St}_{\geq \sigma}^{G,[3]}(f) = \text{St}_{\sigma}^{G,[3]}(f) = \int_Z \int_Z f(z, z, x)\, \nu(dz)\, G(dx).$$

(iv) Let $n = 6$, and $\sigma = \{\{1, 2\}, \{3\}, \{4\}, \{5, 6\}\}$. Then,

$$\text{St}_{\geq \sigma}^{G,[6]}(A_1 \times ... \times A_6) = \nu(A_1 \cap A_2)\, \nu(A_5 \cap A_6)\, G(A_3)\, G(A_4),$$

whereas

$$\text{St}_{\sigma}^{G,[6]}(A_1 \times ... \times A_6) = \nu(A_1 \cap A_2)\, \nu(A_5 \cap A_6)\, \text{St}_{\hat{0}}^{G,[2]}(A_3 \times A_4).$$

These relations can be reformulated in integral notation as

$$\text{St}_{\geq \sigma}^{G,[6]}(f) = \int_Z \int_Z \left\{\int_Z \int_Z f(x, x, y, y, w, z)\, \nu(dx)\, \nu(dy)\right\} G(dw)\, G(dz)$$

$$\text{St}_{\sigma}^{G,[6]}(f) = \int_{w \neq z} \left\{\int_Z \int_Z f(x, x, y, y, w, z)\, \nu(dx)\, \nu(dy)\right\} G(dw)\, G(dz).$$

(v) Let $n = 6$, and $\sigma = \{\{1,2\},\{3\},\{4\},\{5\},\{6\}\}$. Then,

$$\mathrm{St}_{\geq\sigma}^{G,[6]}(A_1 \times \ldots \times A_6) = \nu\,(A_1 \cap A_2)\,G\,(A_3)\,G\,(A_4)\,G\,(A_5)\,G(A_6)$$

and also

$$\mathrm{St}_{\sigma}^{G,[6]}(A_1 \times \ldots \times A_6) = \nu\,(A_1 \cap A_2)\,\mathrm{St}_{0}^{G,[4]}(A_3 \times A_4 \times A_5 \times A_6)\,.$$

5.12 Multiple Poisson integrals of elementary functions

A result analogous to Theorem 5.11.1 holds in the Poisson case. To state this result in a proper way, we shall introduce the following notation. Given $n \geq 2$ and $\sigma \in \mathcal{P}\,[n]$, we write

$$\mathbf{B}_1\,(\sigma) = \{b \in \sigma : |b| = 1\}\,, \tag{5.12.85}$$

to denote the collection of singleton blocks of σ, and

$$\mathbf{B}_2\,(\sigma) = \{b = \{i_1, \ldots, i_\ell\} \in \sigma : \ell = |b| \geq 2\} \tag{5.12.86}$$

to denote the collection of the blocks of σ containing *two or more* elements. We shall also let

$$\mathbf{PB}_2(\sigma) = \text{ the set of all 2-partitions of } \mathbf{B}_2(\sigma), \tag{5.12.87}$$

that is, $\mathbf{PB}_2\,(\sigma)$ is the collection of all ordered pairs $(R_1; R_2)$ of *non-empty* subsets of $\mathbf{B}_2\,(\sigma)$ such that $R_1, R_2 \subset \mathbf{B}_2\,(\sigma)$, $R_1 \cap R_2 = \varnothing$, and $R_1 \cup R_2 = \mathbf{B}_2\,(\sigma)$; whenever $\mathbf{B}_2\,(\sigma) = \varnothing$, one sets $\mathbf{PB}_2\,(\sigma) = \varnothing$.

We stress that $\mathbf{PB}_2\,(\sigma)$ is a partition of $\mathbf{B}_2\,(\sigma)$; the fact that $\mathbf{B}_2\,(\sigma)$ is also a subset of the partition σ should not create confusion.

Example 5.12.1   (i) Let $n = 7$, and $\sigma = \{\{1,2\},\{3,4\},\{5,6\},\{7\}\}$. Then,

$$\mathbf{B}_2\,(\sigma) = \{\{1,2\},\{3,4\},\{5,6\}\}$$

and $\mathbf{PB}_2\,(\sigma)$ contains the six ordered pairs:

$$(\{\{1,2\},\{3,4\}\};\{\{5,6\}\}) \;\; ; \;\; (\{\{5,6\}\};\{\{1,2\},\{3,4\}\})$$
$$(\{\{1,2\},\{5,6\}\};\{\{3,4\}\}) \;\; ; \;\; (\{\{3,4\}\};\{\{1,2\},\{5,6\}\})$$
$$(\{\{3,4\},\{5,6\}\};\{\{1,2\}\}) \;\; ; \;\; (\{\{1,2\}\};\{\{3,4\},\{5,6\}\})\,.$$

For instance, the first ordered pair is made up of $R_1 = \{\{1,2\},\{3,4\}\}$ and $R_2 = \{\{5,6\}\}$, whose union is $\mathbf{B}_2\,(\sigma)$.

(ii) If $n = 5$ and $\sigma = \{\{1,2,3\},\{4\},\{5\}\}$, then $\mathbf{B}_1\,(\sigma) = \{\{4\},\{5\}\}$, $\mathbf{B}_2\,(\sigma) = \{\{1,2,3\}\}$ and $\mathbf{PB}_2\,(\sigma) = \varnothing$.

(iii) If $n = 7$ and $\sigma = \{\{1,2,3\}, \{4,5\}, \{6\}, \{7\}\}$, then $\mathbf{B}_1(\sigma) = \{\{6\}, \{7\}\}$ and $\mathbf{B}_2(\sigma) = \{\{1,2,3\}, \{4,5\}\}$. Also, the set $\mathbf{PB}_2(\sigma)$ contains the two ordered pairs

$$(\{1,2,3\}; \{4,5\}) \quad \text{and} \quad (\{4,5\}, \{1,2,3\}).$$

We shall now suppose that φ is a compensated Poisson measure.

Theorem 5.12.2 *Let* $\varphi = \widehat{N}$ *be a compensated Poisson measure with non-atomic control measure* ν, *and let* $N(\cdot) \triangleq \widehat{N}(\cdot) + \nu(\cdot)$. *For every* $n \geq 2$ *and every* $A \in \mathcal{Z}_\nu$,

$$\mathrm{St}_{\widehat{1}}^{\widehat{N},[n]}(\underbrace{A \times \cdots \times A}_{n \text{ times}}) \triangleq \Delta_n^{\widehat{N}}(A) = N(A) \tag{5.12.88}$$

$(\Delta_n^{\widehat{N}}(\cdot)$ *is the diagonal measure of order* n *associated with* $\widehat{N}$). *Moreover, for every* $A_1, ..., A_n \in \mathcal{Z}_\nu$,

$$\mathrm{St}_{\widehat{1}}^{\widehat{N},[n]}(A_1 \times \cdots \times A_n) = N(A_1 \cap \cdots \cap A_n). \tag{5.12.89}$$

More generally, for every $n \geq 2$, $\sigma \in \mathcal{P}([n])$ *and* $A_1, ..., A_n \in \mathcal{Z}_\nu$,

$$\mathrm{St}_{\geq \sigma}^{\widehat{N},[n]}(A_1 \times \cdots \times A_n) \tag{5.12.90}$$

$$= \prod_{b=\{i_1,...,i_\ell\} \in \mathbf{B}_2(\sigma)} N(A_{i_1} \cap \cdots \cap A_{i_\ell}) \prod_{a=1}^{k} \widehat{N}(A_{j_a}),$$

where $\{\{j_1\}, ..., \{j_k\}\} = \mathbf{B}_1(\sigma)$, *and also*

$$\mathrm{St}_{\sigma}^{\widehat{N},[n]}(A_1 \times \cdots \times A_n) \tag{5.12.91}$$

$$= \sum_{(R_1;R_2) \in \mathbf{PB}_2(\sigma)} \prod_{b=\{i_1,...,i_\ell\} \in R_1} \nu(A_{i_1} \cap \cdots \cap A_{i_\ell}) \times \tag{5.12.92}$$

$$\times \mathrm{St}_{\widehat{0}}^{\widehat{N},[|R_2|+k]}\left(\underset{b=\{e_1,...,e_m\} \in R_2}{\mathrm{X}} (A_{e_1} \cap \cdots \cap A_{e_m}) \times A_{j_1} \times \cdots \times A_{j_k} \right) \tag{5.12.93}$$

$$+ \mathrm{St}_{\widehat{0}}^{\widehat{N},[|\mathbf{B}_2(\sigma)|+k]}\left(\underset{b=\{i_1,...,i_\ell\} \in \mathbf{B}_2(\sigma)}{\mathrm{X}} (A_{i_1} \cap \cdots \cap A_{i_\ell}) \times A_{j_1} \times \cdots \times A_{j_k} \right) \tag{5.12.94}$$

$$+ \prod_{b=\{i_1,...,i_\ell\} \in \mathbf{B}_2(\sigma)} \nu(A_{i_1} \cap \cdots \cap A_{i_\ell}) \, \mathrm{St}_{\widehat{0}}^{\widehat{N},[k]}(A_{j_1} \times \cdots \times A_{j_k}), \tag{5.12.95}$$

where $\{\{j_1\}, ..., \{j_k\}\} = \mathbf{B}_1(\sigma)$ *and where (by convention)* $\Sigma_\varnothing \equiv 0$, $\Pi_\varnothing \equiv 0$ *and* $\mathrm{St}_{\widehat{0}}^{\widehat{N},[0]} \equiv 1$. *Also,* $|R_2|$ *and* $|\mathbf{B}_2(\sigma)|$ *stand, respectively, for the cardinality of* R_2 *and* $\mathbf{B}_2(\sigma)$, *and in formula (5.12.93) we used the notation*

$$\underset{b=\{e_1,...,e_m\} \in R_2}{\mathrm{X}} (A_{e_1} \cap \cdots \cap A_{e_m}) = \underset{b \in R_2}{\mathrm{X}} (\cap_{e \in b} A_e) = (\cap_{e \in b_1} A_e) \times \cdots \times (\cap_{e \in b_{|R_2|}} A_e),$$

where $b_1, ..., b_{|R_2|}$ is some enumeration of R_2 (note that, due to the symmetry of $\mathrm{St}_{\hat{0}}^{\widehat{N},[|R_2|+k]}$, the choice of the enumeration is immaterial). The summand appearing in formula (5.12.94) is defined via the same conventions.

Remarks. (a) When writing formula (5.12.94), we implicitly use the following convention: if $\mathbf{B}_2(\sigma) = \varnothing$, then the symbol $\underset{b=\{i_1,...,i_\ell\}\in\mathbf{B}_2(\sigma)}{\mathrm{X}}(A_{i_1}\cap\cdots\cap A_{i_\ell})$ is immaterial, and one should read

$$\underset{b=\{i_1,...,i_\ell\}\in\mathbf{B}_2(\sigma)}{\mathrm{X}}(A_{i_1}\cap\cdots\cap A_{i_\ell})\times A_{j_1}\times\cdots\times A_{j_k} = A_{j_1}\times\cdots\times A_{j_k} = A_1\times\cdots\times A_n,$$

(5.12.96)

where the last equality follows from the fact that, in this case, $k = n$ and

$$\{j_1,...,j_k\} = \{1,...,n\} = [n].$$

To see how the convention (5.12.96) works, suppose that $\mathbf{B}_2(\sigma) = \varnothing$. Then, $\mathbf{PB}_2(\sigma) = \varnothing$ and consequently, according to the conventions stated in Theorem 5.12.2, the lines (5.12.92)–(5.12.93) and (5.12.95) are equal to zero (they correspond, respectively, to a sum and a product over the empty set). The equality (5.12.91) reads therefore

$$\mathrm{St}_\sigma^{\widehat{N},[n]}(A_1\times\cdots\times A_n) =$$

$$\mathrm{St}_{\hat{0}}^{\widehat{N},[n]}\left(\underset{b=\{i_1,...,i_\ell\}\in\mathbf{B}_2(\sigma)}{\mathrm{X}}(A_{i_1}\cap\cdots\cap A_{i_\ell})\times A_{j_1}\times\cdots\times A_{j_k}\right). \quad (5.12.97)$$

By using (5.12.96) one obtains

$$\mathrm{St}_{\hat{0}}^{\widehat{N},[n]}\left(\underset{b=\{i_1,...,i_\ell\}\in\mathbf{B}_2(\sigma)}{\mathrm{X}}(A_{i_1}\cap\cdots\cap A_{i_\ell})\times A_{j_1}\times\cdots\times A_{j_k}\right)$$

$$= \mathrm{St}_{\hat{0}}^{\widehat{N},[n]}(A_{j_1}\times\cdots\times A_{j_k}) = \mathrm{St}_\sigma^{\widehat{N},[n]}(A_1\times\cdots\times A_n),$$

entailing that, in this case, relation (5.12.91) is equivalent to the identity $\mathrm{St}_{\hat{0}}^{\widehat{N},[n]} = \mathrm{St}_{\hat{0}}^{\widehat{N},[n]}$.

(b) If $k = 0$ (that is, if $\mathbf{B}_1(\sigma)$ equals the empty set), then, according to the conventions stated in Theorem 5.12.2, one has

$$\mathrm{St}_{\hat{0}}^{\widehat{N},[k]} = \mathrm{St}_{\hat{0}}^{\widehat{N},[0]} = 1.$$

This yields that, in this case, one should read line (5.12.95) as follows:

$$\prod_{b=\{i_1,...,i_\ell\}\in\mathbf{B}_2(\sigma)}\nu(A_{i_1}\cap\cdots\cap A_{i_\ell})\,\mathrm{St}_{\hat{0}}^{\widehat{N},[k]}(A_{j_1}\times\cdots\times A_{j_k})$$

$$= \prod_{b=\{i_1,...,i_\ell\}\in\mathbf{B}_2(\sigma)}\nu(A_{i_1}\cap\cdots\cap A_{i_\ell}).$$

(c) If $k = 0$, one should also read line (5.12.93) as

$$\mathrm{St}_{\hat{0}}^{\widehat{N},[|R_2|+k]}\left(\sum_{b=\{e_1,\ldots,e_m\}\in R_2}(A_{e_1}\cap\cdots\cap A_{e_m})\times A_{j_1}\times\cdots\times A_{j_k}\right)$$

$$= \mathrm{St}_{\hat{0}}^{\widehat{N},[|R_2|]}\left(\sum_{b=\{e_1,\ldots,e_m\}\in R_2}(A_{e_1}\cap\cdots\cap A_{e_m})\right),$$

and line (5.12.94) as

$$\mathrm{St}_{\hat{0}}^{\widehat{N},[|\mathbf{B}_2(\sigma)|+k]}\left(\sum_{b=\{i_1,\ldots,i_\ell\}\in\mathbf{B}_2(\sigma)}(A_{i_1}\cap\cdots\cap A_{i_\ell})\times A_{j_1}\times\cdots\times A_{j_k}\right)$$

$$= \mathrm{St}_{\hat{0}}^{\widehat{N},[|\mathbf{B}_2(\sigma)|]}\left(\sum_{b=\{i_1,\ldots,i_\ell\}\in\mathbf{B}_2(\sigma)}(A_{i_1}\cap\cdots\cap A_{i_\ell})\right).$$

Proof of Theorem 5.12.2. To see that (5.12.89) must necessarily hold, use the fact that ν is non-atomic by assumption. Therefore,

$$\mathrm{St}_{\hat{1}}^{\widehat{N},[n]}(A_1\times\cdots\times A_n) = \mathrm{St}_{\hat{0}}^{\widehat{N},[1]}(A_1\cap\cdots\cap A_n)+\nu(A_1\cap\cdots\cap A_n)$$

$$= \widehat{N}(A_1\cap\cdots\cap A_n)+\nu(A_1\cap\cdots\cap A_n)=N(A_1\cap\cdots\cap A_n).$$

Observe also that (5.12.89) implies (5.12.88). Equation (5.12.90) is an immediate consequence of (5.7.67), (5.12.88) and (5.12.89). To prove (5.12.91), use (5.12.90) and the relation

$$N = \widehat{N}+\nu,$$

to write

$$\mathrm{St}_{\geq\sigma}^{\widehat{N},[n]}(A_1\times\cdots\times A_n)$$

$$= \prod_{\ell=2}^{n}\prod_{b=\{i_1,\ldots,i_\ell\}\in\sigma}\left[\widehat{N}(A_{i_1}\cap\cdots\cap A_{i_\ell})+\nu(A_{i_1}\cap\cdots\cap A_{i_\ell})\right]\prod_{a=1}^{k}\widehat{N}(A_{j_a}),$$

and the last expression equals

$$\sum_{(R_1;R_2)\in\mathbf{PB}_2(\sigma)}\prod_{b=\{i_1,\ldots,i_\ell\}\in R_1}\nu(A_{i_1}\cap\cdots\cap A_{i_\ell})\times$$

$$\times\mathrm{St}_{\geq\hat{0}}^{\widehat{N},[|R_2|+k]}\left(\sum_{b=\{e_1,\ldots,e_m\}\in R_2}(A_{e_1}\cap\cdots\cap A_{e_m})\times A_{j_1}\times\cdots\times A_{j_k}\right)$$

$$+\mathrm{St}_{\geq\hat{0}}^{\widehat{N},[|\mathbf{B}_2(\sigma)|+k]}\left(\sum_{b=\{i_1,\ldots,i_\ell\}\in\mathbf{B}_2(\sigma)}(A_{i_1}\cap\cdots\cap A_{i_\ell})\times A_{j_1}\times\cdots\times A_{j_k}\right)$$

$$+\prod_{b=\{i_1,\ldots,i_\ell\}\in\mathbf{B}_2(\sigma)}\nu(A_{i_1}\cap\cdots\cap A_{i_\ell})\mathrm{St}_{\geq\hat{0}}^{\widehat{N},[k]}(A_{j_1}\times\cdots\times A_{j_k}),$$

since

$$\mathrm{St}_{\geq\hat{0}}^{\widehat{N},[n]}(A_1\times\cdots\times A_n) = \widehat{N}(A_1)\cdots\widehat{N}(A_n).$$

The term before last in the displayed equation corresponds to $R_1 = \varnothing$, $R_2 = \mathbf{B}_2(\sigma)$, and the last term to $R_1 = \mathbf{B}_2(\sigma)$ and $R_2 = \varnothing$. By definition, these two cases are not involved in $\mathbf{PB}_2(\sigma)$. The last displayed equation yields

$$
\mathrm{St}_\sigma^{\widehat{N},[n]}(A_1 \times \cdots \times A_n) = \mathrm{St}_{\geq\sigma}^{\widehat{N},[n]}\left((A_1 \times \cdots \times A_n)\mathbf{1}_{Z_\sigma^n}\right)
$$

$$
= \sum_{(R_1;R_2)\in\mathbf{PB}_2(\sigma)} \prod_{b=\{i_1,\dots,i_\ell\}\in R_1} \nu(A_{i_1} \cap \cdots \cap A_{i_\ell}) \times
$$

$$
\times\, \mathrm{St}_{\geq\hat{0}}^{\widehat{N},[|R_2|+k]}
$$

$$
\left(\left[\underset{b=\{e_1,\dots,e_m\}\in R_2}{\mathrm{X}}(A_{e_1} \cap \cdots \cap A_{e_m}) \times A_{j_1} \times \cdots \times A_{j_k}\right]\mathbf{1}_{Z_{\hat{0}}^{|R_2|+k}}\right)
$$

$$
+\, \mathrm{St}_{\geq\hat{0}}^{\widehat{N},[|\mathbf{B}_2(\sigma)|+k]}
$$

$$
\left(\left[\underset{b=\{i_1,\dots,i_\ell\}\in\mathbf{B}_2(\sigma)}{\mathrm{X}}(A_{i_1} \cap \cdots \cap A_{i_\ell}) \times A_{j_1} \times \cdots \times A_{j_k}\right]\mathbf{1}_{Z_{\hat{0}}^{|\mathbf{B}_2(\sigma)|+k}}\right)
$$

$$
+ \prod_{b=\{i_1,\dots,i_\ell\}\in\mathbf{B}_2(\sigma)} \nu(A_{i_1} \cap \cdots \cap A_{i_\ell})\, \mathrm{St}_{\geq\hat{0}}^{\widehat{N},[k]}\left([A_{j_1} \times \cdots \times A_{j_k}]\mathbf{1}_{Z_{\hat{0}}^k}\right).
$$

Since, by definition,

$$
\mathrm{St}_{\geq\hat{0}}^{\widehat{N},[|R_2|+k]}\left(\left[\underset{b=\{e_1,\dots,e_m\}\in R_2}{\mathrm{X}}(A_{e_1} \cap \cdots \cap A_{e_m}) \times A_{j_1} \times \cdots \times A_{j_k}\right]\mathbf{1}_{Z_{\hat{0}}^{|R_2|+k}}\right)
$$

$$
= \mathrm{St}_{\hat{0}}^{\widehat{N},[|R_2|+k]}\left(\underset{b=\{e_1,\dots,e_m\}\in R_2}{\mathrm{X}}(A_{e_1} \cap \cdots \cap A_{e_m}) \times A_{j_1} \times \cdots \times A_{j_k}\right),
$$

and

$$
\mathrm{St}_{\geq\hat{0}}^{\widehat{N},[|\mathbf{B}_2(\sigma)|+k]}
$$

$$
\left(\left[\underset{b=\{i_1,\dots,i_\ell\}\in\mathbf{B}_2(\sigma)}{\mathrm{X}}(A_{i_1} \cap \cdots \cap A_{i_\ell}) \times A_{j_1} \times \cdots \times A_{j_k}\right]\mathbf{1}_{Z_{\hat{0}}^{|\mathbf{B}_2(\sigma)|+k}}\right)
$$

$$
= \mathrm{St}_{\hat{0}}^{\widehat{N},[|\mathbf{B}_2(\sigma)|+k]}\left(\underset{b=\{i_1,\dots,i_\ell\}\in\mathbf{B}_2(\sigma)}{\mathrm{X}}(A_{i_1} \cap \cdots \cap A_{i_\ell}) \times A_{j_1} \times \cdots \times A_{j_k}\right),
$$

and

$$
\mathrm{St}_{\geq\hat{0}}^{\widehat{N},[k]}\left([A_{j_1} \times \cdots \times A_{j_k}]\mathbf{1}_{Z_{\hat{0}}^k}\right) = \mathrm{St}_{\hat{0}}^{\widehat{N},[k]}(A_{j_1} \times \cdots \times A_{j_k}),
$$

one obtains immediately the desired conclusion.   ∎

Remark on the integral notation. It is instructive to express the results of Theorem 5.12.2 in integral notation. With $f(z_1, ..., z_n) = g(z_1) \cdots g(z_n)$, formula (5.12.90) becomes

$$\int_{\cup_{\pi \geq \sigma} Z_\pi^n} g(z_1) \cdots g(z_n) \, \widehat{N}(dz_1) \cdots \widehat{N}(dz_n)$$

$$= \left(\prod_{b \in \sigma, |b| \geq 2} \int_Z g(z)^{|b|} \, N(dz) \right) \times \left(\int_Z g(z) \, \widehat{N}(dz) \right)^k,$$

where $k = |\mathbf{B}_1(\sigma)|$. Again with $f(z_1, .., z_n) = g(z_1) \cdots g(z_n)$, (5.12.91)–(5.12.95) become

$$\int_{Z_\sigma^n} g(z_1) \cdots g(z_n) \, \widehat{N}(dz_1) \cdots \widehat{N}(dz_n)$$

$$= \sum_{(R_1;R_2) \in \mathbf{PB}_2(\sigma)} \prod_{b \in R_1} \int_Z g(z)^{|b|} \, \nu(dz) \times 1_{\left\{ R_2 = \left\{ b_1, ..., b_{|R_2|} \right\} \right\}} \times$$

$$\times \int_{z_1 \neq \cdots \neq z_{|R_2|+k}} g(z_1)^{|b_1|} \cdots g(z_{|R_2|})^{|b_{|R_2|}|} g(z_{|R_2|+1}) \cdots g(z_{|R_2|+k})$$

$$\widehat{N}(dz_1) \cdots \widehat{N}(dz_{|R_2|+k})$$

$$+ 1_{\left\{ \mathbf{B}_2(\sigma) = \left\{ b_1, ..., b_{|\mathbf{B}_2(\sigma)|} \right\} \right\}} \times$$

$$\times \int_{z_1 \neq \cdots \neq z_{|\mathbf{B}_2(\sigma)|+k}} g(z_1)^{|b_1|} \cdots g(z_{|\mathbf{B}_2(\sigma)|})^{|b_{|\mathbf{B}_2(\sigma)|}|} g(z_{|\mathbf{B}_2(\sigma)|+1}) \cdots$$

$$g(z_{|\mathbf{B}_2(\sigma)|+k}) \, \widehat{N}(dz_1) \cdots \widehat{N}(dz_{|R_2|+k})$$

$$+ \prod_{b \in \mathbf{B}_2(\sigma)} \int_Z g(z)^{|b|} \, \nu(dz) \times \int_{z_1 \neq \cdots \neq z_k} g(z_1) \cdots g(z_k) \, \widehat{N}(dz_1) \cdots \widehat{N}(dz_k),$$

where $k = |\mathbf{B}_1(\sigma)|$.

Example 5.12.3 The examples below apply to a compensated Poisson measure $\widehat{N}$, and should be compared with those discussed after Theorem 5.11.1. We suppose throughout that $n \geq 2$.

(i) When $\sigma = \hat{0} = \{\{1\}, ..., \{n\}\}$ one has, as in the Gaussian case,

$$\mathrm{St}_{\geq \hat{0}}^{\widehat{N},[n]} (A_1 \times \cdots \times A_n) = \widehat{N}(A_1) \cdots \widehat{N}(A_n)$$

because the symbol " $\geq \hat{0}$ " entails no restriction on the considered partitions. In integral notation, this becomes

$$\mathrm{St}_{\geq \hat{0}}^{\widehat{N},[n]} (f) = \int_{Z^n} f(z_1, ..., z_n) \, \widehat{N}(dz_1) \cdots \widehat{N}(dz_n).$$

The case of equality (5.12.91) has already been discussed: indeed, since $\sigma = \hat{0}$, and according to the conventions stated therein, one has that

$$\mathbf{B}_2 (\sigma) = \mathbf{PB}_2 (\sigma) = \varnothing,$$

and therefore (5.12.91) becomes an identity given by (5.12.95), namely $\mathrm{St}_{\hat{0}}^{\widehat{N},[n]} (\cdot) = \mathrm{St}_{\hat{0}}^{\widehat{N},[n]} (\cdot)$. Observe that, in integral notation, $\mathrm{St}_{\hat{0}}^{\widehat{N},[n]} (\cdot)$ is expressed as

$$\mathrm{St}_{\hat{0}}^{\widehat{N},[n]} (f) = \int_{z_1 \neq \cdots \neq z_n} f (z_1, ..., z_n) \, \widehat{N} (dz_1) \cdots \widehat{N} (dz_n).$$

(ii) Suppose now $\sigma = \hat{1} = \{\{1, ..., n\}\}$. Then, (5.12.91) reduces to (5.12.89). To see this, note that $k = 0$ and that $\mathbf{B}_2 \left(\hat{1} \right)$ contains only the block $\{1, ..., n\}$, so that $\mathbf{PB}_2 (\sigma) = \varnothing$. Hence the sum appearing in (5.12.92) vanishes and one has

$$\mathrm{St}_{\hat{1}}^{\widehat{N},[n]} (A_1 \times \cdots \times A_n) = \mathrm{St}_{\hat{0}}^{\widehat{N},[1]} (A_1 \cap \cdots \cap A_n) + \nu (A_1 \cap \cdots \cap A_n)$$
$$= \widehat{N} (A_1 \cap \cdots \cap A_n) + \nu (A_1 \cap \cdots \cap A_n) = N (A_1 \cap \cdots \cap A_n).$$

In integral notation,

$$\mathrm{St}_{\hat{1}}^{\widehat{N},[n]} (f) = \int_{z_1 = \cdots = z_n} f (z_1, ..., z_n) \, \widehat{N} (dz_1) \cdots \widehat{N} (dz_n)$$
$$= \int_Z f (z, ..., z) \, N (dz).$$

This last relation makes sense heuristically, in view of the computational rule

$$\left(\widehat{N} (dx) \right)^2 = (N (dx))^2 - 2N (dx) \nu (dx) + (\nu (dx))^2 = N (dx),$$

since $(N (dx))^2 = N (dx)$ and ν is non-atomic.

(iii) Let $n = 3$ and $\sigma = \{\{1\}, \{2, 3\}\}$, so that $\mathbf{B}_2 (\sigma) = \{\{2, 3\}\}$ and $\mathbf{PB}_2 (\sigma) = \varnothing$. According to (5.12.90),

$$\mathrm{St}_{\geq \sigma}^{\widehat{N},[3]} (A_1 \times A_2 \times A_3) = \widehat{N} (A_1) \, N (A_2 \cap A_3). \tag{5.12.98}$$

On the other hand, (5.12.91) yields

$$\mathrm{St}_{\sigma}^{\widehat{N},[3]} (A_1 \times A_2 \times A_3) = \mathrm{St}_{\hat{0}}^{\widehat{N},[2]} (A_1 \times (A_2 \cap A_3)) + \widehat{N} (A_1) \, \nu (A_2 \cap A_3). \tag{5.12.99}$$

In integral form, relation (5.12.98) becomes

$$\mathrm{St}_{\geq \sigma}^{\widehat{N},[3]} (f) = \int_Z \int_Z f (z_1, z_2, z_2) \, \widehat{N} (dz_1) \, N (dz_2),$$

and (5.12.99) becomes

$$\text{St}_\sigma^{\widehat{N},[3]}(f) = \int_{z_1 \neq z_2,\, z_2 = z_3} f(z_1, z_2, z_3)\, \widehat{N}(dz_1)\, \widehat{N}(dz_2)\, \widehat{N}(dz_3)$$

$$= \int_{z_1 \neq z_2} f(z_1, z_2, z_2)\, \widehat{N}(dz_1)\, \widehat{N}(dz_2)$$

$$+ \int_{Z^2} f(z_1, z_2, z_2)\, \widehat{N}(dz_1)\, \nu(dz_2).$$

This last relation makes sense heuristically, by noting that

$$\widehat{N}(dz_1)\, \widehat{N}(dz_2)\, \widehat{N}(dz_2) = \widehat{N}(dz_1)\, N(dz_2)$$
$$= \widehat{N}(dz_1)\, \widehat{N}(dz_2) + \widehat{N}(dz_1)\, \nu(dz_2).$$

We also stress that $\text{St}_\sigma^{\widehat{N},[3]}(f)$ can be also be expressed as

$$\text{St}_\sigma^{\widehat{N},[3]}(f) = I_2^{\widehat{N}}(g_1) + I_1^{\widehat{N}}(g_2), \qquad (5.12.100)$$

where

$$g_1(x, y) = f(x, y, y)$$

and

$$g_2(x) = \int_Z f(x, y, y)\, \nu(dy).$$

The form (5.12.100) will be needed later. Since, by (5.12.89), $\text{St}_{\hat{1}}^{\widehat{N},[3]}(A_1 \times A_2 \times A_3) = N(A_1 \cap A_2 \cap A_3)$ and since, for our σ, one has $\text{St}_{\geq\sigma}^{\widehat{N},[3]} = \text{St}_\sigma^{\widehat{N},[3]} + \text{St}_{\hat{1}}^{\widehat{N},[3]}$, one also deduces the relation

$$\widehat{N}(A_1)\, N(A_2 \cap A_3) =$$
$$\text{St}_{\hat{0}}^{\widehat{N},[2]}(A_1 \times (A_2 \cap A_3)) + N(A_1 \cap A_2 \cap A_3) + \widehat{N}(A_1)\, \nu(A_2 \cap A_3),$$

or, equivalently, since $\widehat{N} = N + \nu$,

$$\widehat{N}(A_1)\, \widehat{N}(A_2 \cap A_3) =$$
$$\text{St}_{\hat{0}}^{\widehat{N},[2]}(A_1 \times (A_2 \cap A_3)) + \nu(A_1 \cap A_2 \cap A_3) + \widehat{N}(A_1 \cap A_2 \cap A_3).$$
$$(5.12.101)$$

We will see that (5.12.101) is consistent with the multiplication formulae of next chapter.

(iv) Let $n = 6$, and $\sigma = \{\{1, 2\}, \{3\}, \{4\}, \{5, 6\}\}$, so that

$$\mathbf{B}_2(\sigma) = \{\{1, 2\}, \{5, 6\}\}$$

and the class $\mathbf{PB}_2\,(\sigma)$ contains the two pairs

$$(\{1,2\};\{5,6\}) \quad \text{and} \quad (\{5,6\};\{1,2\}).$$

First, (5.12.90) gives

$$\mathrm{St}_{\geq\sigma}^{\widehat{N},[6]}\,(A_1\times ...\times A_6) = N\,(A_1\cap A_2)\,N\,(A_5\cap A_6)\,\widehat{N}\,(A_3)\,\widehat{N}\,(A_4).$$

Moreover, we deduce from (5.12.91) that

$$\begin{aligned}
\mathrm{St}_{\sigma}^{\widehat{N},[6]}\,(A_1\times ...\times A_6) = &\;\nu\,(A_1\cap A_2)\,\mathrm{St}_{\hat{0}}^{\widehat{N},[3]}\,((A_5\cap A_6)\times A_3\times A_4)\\
&+\nu\,(A_5\cap A_6)\,\mathrm{St}_{\hat{0}}^{\widehat{N},[3]}\,((A_1\cap A_2)\times A_3\times A_4)\\
&+\mathrm{St}_{\hat{0}}^{\widehat{N},[4]}\,((A_1\cap A_2)\times(A_5\cap A_6)\times A_3\times A_4)\\
&+\nu\,(A_1\cap A_2)\,\nu\,(A_5\cap A_6)\,\mathrm{St}_{\hat{0}}^{\widehat{N},[2]}\,(A_3\times A_4).
\end{aligned}$$

The last displayed equation becomes in integral form

$$\begin{aligned}
&\mathrm{St}_{\sigma}^{\widehat{N},[6]}\,(f)\\
&= \int_{\substack{z_1=z_2,\ z_5=z_6,\ z_3\neq z_4\\ z_3\neq z_1,\ z_4\neq z_1\\ z_3\neq z_5,\ z_4\neq z_5}} f\,(z_1,...,z_6)\,\widehat{N}\,(dz_1)\cdots\widehat{N}\,(dz_6)\\
&= \int_{w,x\neq y\neq z} f\,(w,w,x,y,z,z)\,\nu\,(dw)\,\widehat{N}\,(dx)\,\widehat{N}\,(dy)\,\widehat{N}\,(dz)\\
&\quad + \int_{w\neq x\neq y,z} f\,(w,w,x,y,z,z)\,\widehat{N}\,(dw)\,\widehat{N}\,(dx)\,\widehat{N}\,(dy)\,\nu\,(dz)\\
&\quad + \int_{w\neq x\neq y\neq z} f\,(w,w,x,y,z,z)\,\widehat{N}\,(dw)\,\widehat{N}\,(dx)\,\widehat{N}\,(dy)\,\widehat{N}\,(dz)\\
&\quad + \int_{w,x\neq y,z} f\,(w,w,x,y,z,z)\,\nu\,(dw)\,\widehat{N}\,(dx)\,\widehat{N}\,(dy)\,\nu\,(dz).
\end{aligned}$$

Indeed, let us denote the RHS of the last expression as $(I)+(II)+(III)+(IV)$. For (I) and (II), we use (5.12.92)–(5.12.93) with $R_1 = \{\{1,2\}\}$ and $R_2 = \{\{5,6\}\}$ and $k = 2$, which corresponds to the number of singletons $\{3\}$, $\{4\}$. For (III), we use (5.12.94), with $k + |\mathbf{B}_2\,(\sigma)| = 2+2 = 4$, and $\mathbf{B}_2\,(\sigma) = \{\{1,2\},\{5,6\}\}$. For (5.12.95) we use $k = 2$ and our $\mathbf{B}_2\,(\sigma)$.

(v) Let $n = 6$, and $\sigma = \{\{1,2\},\{3\},\{4\},\{5\},\{6\}\}$. Here, $\mathbf{B}_2\,(\sigma) = \{\{1,2\}\}$ and the class $\mathbf{PB}_2\,(\sigma)$ is empty. Then,

$$\mathrm{St}_{\geq\sigma}^{\widehat{N},[6]}\,(A_1\times ...\times A_6) = N\,(A_1\cap A_2)\,\widehat{N}\,(A_3)\,\widehat{N}\,(A_4)\,\widehat{N}\,(A_5)\,\widehat{N}\,(A_6),$$

and also

$$\begin{aligned}
\mathrm{St}_{\sigma}^{\widehat{N},[6]}\,(A_1\times ...\times A_6) = &\;\nu\,(A_1\cap A_2)\,\mathrm{St}_{\hat{0}}^{\widehat{N},[4]}\,(A_3\times A_4\times A_5\times A_6)\\
&+\mathrm{St}_{\hat{0}}^{\widehat{N},[4]}\,((A_1\cap A_2)\times A_3\times A_4\times A_5\times A_6).
\end{aligned}$$

In integral form,

$$\mathrm{St}_{\sigma}^{\widehat{N},[6]}(f)$$

$$= \int_{\substack{z_1=z_2, \\ z_1 \neq z_j, \, j=3,\ldots,6 \\ z_i \neq z_j, \, 3 \leq i \neq j \leq 6}} f(z_1, z_2, z_3, z_4, z_5, z_6) \prod_{j=1}^{6} \widehat{N}(dz_j)$$

$$= \int_{\substack{z_1 \neq z_j, \, j=3,\ldots,6 \\ z_i \neq z_j, \, 3 \leq i \neq j \leq 6}} f(z_1, z_1, z_3, z_4, z_5, z_6)\, \nu(dz_1) \prod_{j=3}^{6} \widehat{N}(dz_j)$$

$$+ \int_{\substack{z_1 \neq z_j, \, j=3,\ldots,6 \\ z_i \neq z_j, \, 3 \leq i \neq j \leq 6}} f(z_1, z_1, z_3, z_4, z_5, z_6)\, \widehat{N}(dz_1) \prod_{j=3}^{6} \widehat{N}(dz_j).$$

Corollary 5.12.4 *Suppose that the Assumptions of Theorem 5.12.2 hold. Fix $\sigma \in \mathcal{P}([n])$ and assume that $|b| \geq 2$, for every $b \in \sigma$. Then,*

$$\mathbb{E}\left[\mathrm{St}_{\sigma}^{\widehat{N},[n]}(A_1 \times \cdots \times A_n)\right] = \prod_{m=2}^{n} \prod_{b \in \{j_1,\ldots,j_m\} \in \sigma} \nu(A_{j_1} \cap \cdots \cap A_{j_m}).$$

$$(5.12.102)$$

Proof. Use (5.12.91)–(5.12.95) and note that, by assumption, $k = 0$ (the partition σ does not contain any singleton $\{j\}$). It follows that the sum in (5.12.92) vanishes, and one is left with

$$\mathbb{E}\left[\mathrm{St}_{\sigma}^{\widehat{N},[n]}(A_1 \times \cdots \times A_n)\right]$$

$$= \mathbb{E}\left[\mathrm{St}_{\hat{0}}^{\widehat{N},[k+|\mathbf{B}_2(\sigma)|]}\left(\underset{b=\{i_1,\ldots,i_\ell\} \in \mathbf{B}_2(\sigma)}{\mathrm{X}}(A_{i_1} \cap \cdots \cap A_{i_\ell}) \times A_{j_1} \times \cdots \times A_{j_k}\right)\right]$$

$$+ \prod_{b=\{i_1,\ldots,i_\ell\} \in \mathbf{B}_2(\sigma)} \nu(A_{i_1} \cap \cdots \cap A_{i_\ell})$$

$$= \prod_{b=\{i_1,\ldots,i_\ell\} \in \mathbf{B}_2(\sigma)} \nu(A_{i_1} \cap \cdots \cap A_{i_\ell}),$$

which is equal to the RHS of (5.12.102).   ∎

5.13 A stochastic Fubini theorem

A stochastic Fubini theorem allows one to interchange deterministic and stochastic integrals. The following result is taken from [117].

Theorem 5.13.1 *Let φ be a completely random measure on $(Z, \mathcal{Z})$ with non-atomic control ν, let (S, μ) be a measure space and $f(s, x_1, \ldots, x_k)$ be a deterministic func-*

tion such that

$$\int_S \|f(s,\cdot)\|_{L^2(\mathbb{R}^k)}\,\mu(ds) =$$

$$\int_S \left(\int_{\mathbb{R}^k} f(s,x_1,\ldots,x_k)^2 \nu(dx_1)\ldots\nu(dx_k) \right)^{\frac{1}{2}} \mu(ds) < \infty. \qquad (5.13.103)$$

Then,

$$I_k^\varphi \left(\int_S f(s,\cdot)\mu(ds) \right) = \int_{\mathbb{R}^k}' \left\{ \int_S f(s,x_1,\ldots,x_k)\mu(ds) \right\} \varphi(dx_1)\ldots\varphi(dx_k) \qquad (5.13.104)$$

$$= \int_S \left\{ \int_{\mathbb{R}^k}' f(s,x_1,\ldots,x_k)\varphi(dx_1)\ldots\varphi(dx_k) \right\} \mu(ds)$$

$$= \int_S I_k^\varphi(f(s,\cdot))\mu(ds), \quad a.s.\text{-}\mathbb{P}. \qquad (5.13.105)$$

Remark. The sense in which the right-hand side of (5.13.105) is defined, is explained in the proof below. If $\mu(S) < \infty$, note also that the condition (5.13.103) is implied by

$$\int_S \int_{\mathbb{R}^k} f(s,x_1,\ldots,x_k)^2 \nu(dx_1)\ldots\nu(dx_k)\mu(ds) < \infty. \qquad (5.13.106)$$

For convenience, we include a "prime" in (5.13.104) and (5.13.105) to indicate that one does not integrate on hyperdiagonals $x_i = x_\ell$ for $i \neq \ell$.

Proof. Under (5.13.103), the right-hand side $\int_S I_k^\varphi(f(s,\cdot))\mu(ds)$ of (5.13.105) can be defined in a natural sense through approximating elementary functions. One can choose a sequence of elementary functions

$$f_n(s,x_1,\ldots,x_k) = \sum_{m,n_1,\ldots,n_k=1}^{k_n} a_{m,n_1,\ldots,n_k}^n 1_{B^n \times A_{n_1}^n \times \ldots \times A_{n_k}^n}(s,x_1,\ldots,x_k),$$

where $a_{m,n_1,\ldots,n_k}^n = 0$ if any two indices n_i and n_j are equal, such that

$$\int_S \|f_n(s,\cdot)\|_{L^2(\mathbb{R}^k)}\mu(ds) < \infty$$

and

$$\int_S \|f(s,\cdot) - f_n(s,\cdot)\|_{L^2(\mathbb{R}^k)}\mu(ds) \to 0.$$

Indeed, to show that such a sequence exists, it is enough to construct it for a function

$$1_{B \times A}(s,x_1,\ldots,x_n) = 1_B(s)1_A(x_1,\ldots,x_n).$$

This is the same as constructing elementary functions

$$g_n(x_1, \ldots, x_k) = \sum_{n_1, \ldots, n_k = 1}^{k_n} a_{n_1, \ldots, n_k}^n 1_{A_{n_1}^n \times \ldots \times A_{n_k}^n}(x_1, \ldots, x_k),$$

where $a_{n_1, \ldots, n_k}^n = 0$ if any two indices n_i and n_j are equal, such that $\|g_n\|_{L^2(\mathbb{R}^k)} < \infty$ and $\|1_A - g_n\|_{L^2(\mathbb{R}^k)} \to \infty$. How this can be done is explained, for example, in Nualart [94, p. 10]. The integral $\int_S I_k^\varphi(f(s, \cdot))\mu(ds)$ is now defined as the limit of $\int_S I_k^\varphi(f_n(s, \cdot))\mu(ds)$ in the $L^1(\mathbb{P})$-sense. The limit exists since

$$E \left| \int_S I_k^\varphi(f_m(s, \cdot))\mu(ds) - \int_S I_k^\varphi(f_n(s, \cdot))\mu(ds) \right|$$

$$\leq \int_S E|I_k^\varphi(f_m(s, \cdot) - f_n(s, \cdot))|\mu(ds)$$

$$\leq \int_S \left(E|I_k^\varphi(f_m(s, \cdot) - f_n(s, \cdot))|^2 \right)^{1/2} \mu(ds)$$

$$\leq (k!)^{1/2} \int_S \|f_m(s, \cdot) - f_n(s, \cdot)\|_{L^2(\mathbb{R}^k)}\mu(ds) \to 0,$$

as $m, n \to 0$, and it does not depend on the approximating sequence.

With the above definition of $\int_S I_k^\varphi(f(s, \cdot))\mu(ds)$, the proof of (5.13.105) is now elementary. The relation (5.13.105) clearly holds for the elementary functions f_n above. As $n \to \infty$, its right-hand side converges to $\int_S I_k^\varphi(f(s, \cdot))\mu(ds)$ by the definition above. For the left-hand side, observe that

$$\left\| \int_S f(s, \cdot)\mu(ds) - \int_S f_n(s, \cdot)\mu(ds) \right\|_{L^2(\mathbb{R}^k)} = \left\| \int_S (f(s, \cdot) - f_n(s, \cdot))\mu(ds) \right\|_{L^2(\mathbb{R}^k)}$$

$$\leq \int_S \|(f(s, \cdot) - f_n(s, \cdot))\|_{L^2(\mathbb{R}^k)}\mu(ds) \to 0,$$

where we have used the generalized Minkowski's inequality. Therefore, as $n \to \infty$, the left-hand side converges to $I_k^\varphi(\int_S f(s, \cdot)\mu(ds))$ in the $L^2(\mathbb{P})$-sense. ∎

See Section 11.5 for an example involving an explicit use of stochastic Fubini theorems.

6

Multiplication formulae

6.1 The general case

The forthcoming Theorem 6.1.1 applies to every good completely random measure φ. It gives a universal combinatorial rule, according to which every product of multiple stochastic integrals can be represented as a sum over diagonal measures that are indexed by non-flat diagrams (as defined in Section 4.1). We will see that product formulae are crucial in order to deduce explicit expressions for the cumulants and the moments of multiple integrals. As discussed later in this chapter, Theorem 6.1.1 contains (as special cases) two celebrated *product formulae* for integrals with respect to Gaussian and Poisson random measures. We provide two proofs of Theorem 6.1.1: the first one is new and it is based on a decomposition of partially diagonal sets; the second consists in a slight variation of the combinatorial arguments displayed in the proofs of [132, Th. 3 and Th. 4], and is included for the sake of completeness. The theorem is formulated for simple kernels to ensure that the integrals are always defined, in particular the quantity $\mathrm{St}_\sigma^{\varphi,[n]}$, which appears on the RHS of (6.1.2).

Theorem 6.1.1 (Rota and Wallstrom) *Let φ be a good completely random measure with non-atomic control ν. For $n_1, n_2, ..., n_k \geq 1$, we write*

$$n = n_1 + \cdots + n_k,$$

and we denote by π^* the partition of $[n]$ given by

$$\pi^* = \{\{1, ..., n_1\}, \{n_1 + 1, ..., n_1 + n_2\}, ..., \{n_1 + ... + n_{k-1} + 1, ..., n\}\}.$$
$$(6.1.1)$$

Then, if the kernels $f_1, ..., f_k$ are such that $f_j \in \mathcal{E}_{s,0}(\nu^{n_j})$ $(j = 1, ..., k)$, one has that

$$\prod_{j=1}^k I_{n_j}^\varphi(f_j) = \prod_{j=1}^k \mathrm{St}_{\hat{0}}^{\varphi,[n_j]}(f_j) = \sum_{\sigma \in \mathcal{P}([n]):\sigma \wedge \pi^* = \hat{0}} \mathrm{St}_\sigma^{\varphi,[n]}(f_1 \otimes_0 f_2 \otimes_0 \cdots \otimes_0 f_k),$$
$$(6.1.2)$$

G. Peccati, M.S. Taqqu: Wiener Chaos: Moments, Cumulants and Diagrams –
A survey with computer implementation.
© Springer-Verlag Italia 2011

*where, by definition, the function in n variables $f_1 \otimes_0 f_2 \otimes_0 \cdots \otimes_0 f_k \in \mathcal{E}(\nu^n)$ (sometimes called the **tensor product** of $f_1, ..., f_k$ is defined as*

$$f_1 \otimes_0 f_2 \otimes_0 \cdots \otimes_0 f_k (x_1, x_2, ..., x_n) = \prod_{j=1}^{k} f_j (x_{n_1 + ... + n_{j-1} + 1}, ..., x_{n_1 + ... + n_j}),$$

$(n_0 = 0).$ \hfill (6.1.3)

First proof. From the discussion of the previous chapter, one deduces that

$$\prod_{j=1}^{k} \mathrm{St}_{\hat{0}}^{\varphi, [n_j]} (f_j) = \mathrm{St}_{\geq \hat{0}}^{\varphi, [n]} [(f_1 \otimes_0 \cdots \otimes_0 f_k) \mathbf{1}_{A^*}],$$

where

$$A^* = \{(z_1, ..., z_n) \in \mathcal{Z}^n : z_i \neq z_j, \forall i \neq j \text{ such that } i \sim_{\pi^*} j\},$$

that is, A^* is obtained by excluding all diagonals within each block of π^*. We shall prove that

$$A^* = \bigcup_{\sigma \in \mathcal{P}([n]) : \sigma \wedge \pi^* = \hat{0}} Z_\sigma^n. \hfill (6.1.4)$$

Suppose first σ is such that $\sigma \wedge \pi^* = \hat{0}$, that is, the meet of σ and π^* is given by the singletons. For every $(z_1, ..., z_n) \in Z_\sigma^n$ the following implication holds: if $i \neq j$ and $i \sim_{\pi^*} j$, then i and j are in two different blocks of σ, and therefore $z_i \neq z_j$. This implies that $Z_\sigma^n \subset A^*$. For the converse, take $(z_1, ..., z_n) \in A^*$, and construct a partition $\sigma \in \mathcal{P}([n])$ by the following rule: $i \sim_\sigma j$ if and only if $z_i = z_j$. For every pair $i \neq j$ such that $i \sim_{\pi^*} j$, one has (by definition of A^*) $z_i \neq z_j$, so that $\sigma \wedge \pi^* = \hat{0}$, and hence (6.1.4). To conclude the proof of the theorem, just use the additivity of $\mathrm{St}_{\geq \hat{0}}^{\varphi, [n]}$ to write

$$\mathrm{St}_{\geq \hat{0}}^{\varphi, [n]} [(f_1 \otimes_0 \cdots \otimes_0 f_k) \mathbf{1}_{A^*}] = \sum_{\sigma \wedge \pi^* = \hat{0}} \mathrm{St}_{\geq \hat{0}}^{\varphi, [n]} [(f_1 \otimes_0 \cdots \otimes_0 f_k) \mathbf{1}_{Z_\sigma^n}]$$

$$= \sum_{\sigma \wedge \pi^* = \hat{0}} \mathrm{St}_\sigma^{\varphi, [n]} (f_1 \otimes_0 \cdots \otimes_0 f_k),$$

by using the relation

$$\mathrm{St}_{\geq \hat{0}}^{\varphi, [n]} [(\cdot) \mathbf{1}_{Z_\sigma^n}] = \mathrm{St}_\sigma^{\varphi, [n]} [\cdot].$$

Second proof (see [132]). This proof uses Proposition 2.6.5 and Proposition 2.6.6. To simplify the discussion (and without loss of generality) we can assume that $n_1 \geq n_2 \geq \cdots \geq n_k$. For $j = 1, ..., k$ we have that

$$\mathrm{St}_{\hat{0}}^{\varphi, [n_j]} (f_j) = \sum_{\sigma \in \mathcal{P}([n_j])} \mu(\hat{0}, \sigma) \mathrm{St}_{\geq \sigma}^{\varphi, [n_j]} (f_j),$$

where we have used (5.4.50) with $\pi = \hat{0}$. From this relation one obtains

$$\prod_{j=1}^{k} \mathrm{St}_{\hat{0}}^{\varphi,[n_j]}(f_j) = \sum_{\sigma_1 \in \mathcal{P}([n_1])} \cdots \sum_{\sigma_k \in \mathcal{P}([n_k])} \prod_{j=1}^{k} \mu\left(\hat{0}, \sigma_j\right) \mathrm{St}_{\geq \sigma_j}^{\varphi,[n_j]}(f_j)$$

$$= \sum_{\rho \in \mathcal{P}([n]):\rho \leq \pi^*} \mu\left(\hat{0}, \rho\right) \mathrm{St}_{\geq \rho}^{\varphi,[n]}(f_1 \otimes_0 \cdots \otimes_0 f_k), \tag{6.1.5}$$

where π^* is defined in (6.1.1). To prove equality (6.1.5), recall the definition of "class" in Section 2.3, as well as Example 2.3.3-(v). Observe that the segment $\left[\hat{0}, \pi^*\right]$ has class $(n_1, ..., n_k)$, thus yielding (thanks to Proposition 2.6.6) that $\left[\hat{0}, \pi^*\right]$ is isomorphic to the lattice product of the $\mathcal{P}([n_j])$'s. This implies that each vector

$$(\sigma_1, ..., \sigma_k) \in \mathcal{P}([n_1]) \times \cdots \times \mathcal{P}([n_k])$$

has indeed the form

$$(\sigma_1, ..., \sigma_k) = \psi^{-1}(\rho)$$

for a unique $\rho \in \left[\hat{0}, \pi^*\right]$, where ψ is a bijection defined as in (2.6.37). Now use, in order, Part 2 and Part 1 of Proposition 2.6.5 to deduce that

$$\prod_{j=1}^{k} \mu\left(\hat{0}, \sigma_j\right) = \mu\left(\hat{0}, (\sigma_1, ..., \sigma_k)\right) = \mu\left(\hat{0}, \psi^{-1}(\rho)\right) = \mu\left(\hat{0}, \rho\right). \tag{6.1.6}$$

Observe that

$$\hat{0} = \{\{1\}, ..., \{n_j\}\} \quad \text{in } \mu\left(\hat{0}, \sigma_j\right),$$

whereas

$$\hat{0} = \{\{1\}, ..., \{n\}\} \quad \text{in } \mu(\hat{0}, \rho).$$

Also, one has the relation

$$\prod_{j=1}^{k} \mathrm{St}_{\geq \sigma_j}^{\varphi,[n_j]}(f_j) = \mathrm{St}_{\geq \rho}^{\varphi,[n]}(f_1 \otimes_0 \cdots \otimes_0 f_k), \tag{6.1.7}$$

by the definition of $\mathrm{St}_{\geq \rho}^{\varphi,[n]}$ as the measure charging all the diagonals associated, through the map ψ, with the blocks of all the partitions σ_j $(j = 1, ..., k)$. Then, (6.1.6) and (6.1.7) yield immediately (6.1.5). To conclude the proof, write

$$\sum_{\rho \in \mathcal{P}([n]):\rho \leq \pi^*} \mu\left(\hat{0}, \rho\right) \mathrm{St}_{\geq \rho}^{\varphi,[n]}(f_1 \otimes_0 \cdots \otimes_0 f_k)$$

$$= \sum_{\rho \in \mathcal{P}([n]):\rho \leq \pi^*} \mu\left(\hat{0}, \rho\right) \sum_{\gamma \geq \rho} \mathrm{St}_{\gamma}^{\varphi,[n]}(f_1 \otimes_0 \cdots \otimes_0 f_k)$$

$$= \sum_{\gamma \in \mathcal{P}([n])} \mathrm{St}_{\gamma}^{\varphi,[n]}(f_1 \otimes_0 \cdots \otimes_0 f_k) \sum_{\hat{0} \leq \rho \leq \pi^* \wedge \gamma} \mu\left(\hat{0}, \rho\right).$$

Since, by (2.6.33),

$$\sum_{\hat{0} \leq \rho \leq \pi^* \wedge \gamma} \mu\left(\hat{0}, \rho\right) = \begin{cases} 0 \text{ if } \pi^* \wedge \gamma \neq \hat{0} , \\ 1 \text{ if } \pi^* \wedge \gamma = \hat{0}. \end{cases},$$

relation (6.1.2) is obtained.   ∎

Remark. The RHS of (6.1.2) can also be reformulated in terms of diagrams and in terms of graphs, as follows:

$$\sum_{\sigma \in \mathcal{P}([n]):\Gamma(\pi^*,\sigma) \text{ is non-flat}} \text{St}_\sigma^{\varphi,[n]}\left(f_1 \otimes_0 f_2 \otimes_0 \cdots \otimes_0 f_k\right),$$

where $\Gamma(\pi^*, \sigma)$ is the diagram of (π^*, σ), as defined in Section 4.1, or, whenever every $\Gamma(\pi^*, \sigma)$ involved in the previous sum is Gaussian,

$$\sum_{\sigma \in \mathcal{P}([n]):\hat{\Gamma}(\pi^*,\sigma) \text{ has no loops}} \text{St}_\sigma^{\varphi,[n]}\left(f_1 \otimes_0 f_2 \otimes_0 \cdots \otimes_0 f_k\right),$$

where $\hat{\Gamma}(\pi^*, \sigma)$ is the graph of (π^*, σ) defined in Section 4.3. This is because, thanks to Proposition 4.3.2, the relation $\pi^* \wedge \sigma = \hat{0}$ indicates that $\Gamma(\pi^*, \sigma)$ is non-flat or, equivalently in the case of Gaussian diagrams, that $\hat{\Gamma}(\pi^*, \sigma)$ has no loops.

Example 6.1.2   (i) Set $k = 2$ and $n_1 = n_2 = 1$ in Theorem 6.1.1. Then, $n = 2$, $\mathcal{P}([2]) = \{\hat{0}, \hat{1}\}$ and $\pi^* = \{\{1\}, \{2\}\} = \hat{0}$. Since $\hat{0} \wedge \hat{0} = \hat{0}$ and $\hat{1} \wedge \hat{0} = \hat{0}$, (6.1.2) gives immediately that, for every pair of elementary functions f_1, f_2,

$$I_1^\varphi(f_1) \times I_1^\varphi(f_2) = \text{St}_{\hat{0}}^{\varphi,[2]}\left(f_1 \otimes_0 f_2\right) + \text{St}_{\hat{1}}^{\varphi,[2]}\left(f_1 \otimes_0 f_2\right)$$
$$= I_2^\varphi\left(f_1 \otimes_0 f_2\right) + \text{St}_{\hat{1}}^{\varphi,[2]}\left(f_1 \otimes_0 f_2\right). \qquad (6.1.8)$$

Note that, if $\varphi = G$ is Gaussian, then relation (5.11.82) yields that

$$\text{St}_{\hat{1}}^{G,[2]}\left(f_1 \otimes_0 f_2\right) = \int_Z f_1(z) f_2(z) \nu(dz),$$

so that, in integral notation,

$$I_1^G(f_1) \times I_1^G(f_2) =$$
$$\int \int_{z_1 \neq z_2} f_1(z_1) f_2(z_2) G(dz_1) G(dz_2) + \int_Z f_1(z) f_2(z) \nu(dz).$$

On the other hand, if φ is compensated Poisson, then

$$\text{St}_{\hat{1}}^{\varphi,[2]}\left(f_1 \otimes_0 f_2\right) = \int_Z f_1(z) f_2(z) N(dz),$$

so that, by using the relation $N = \widehat{N} + \nu$, (6.1.8) reads

$$I_1^{\widehat{N}}(f_1) \times I_1^{\widehat{N}}(f_2)$$

$$= I_2^{\widehat{N}}(f_1 \otimes_0 f_2) + \int_z f_1(z) f_2(z) \widehat{N}(dz) + \int_z f_1(z) f_2(z) \nu(dz)$$

$$= \int\int_{z_1 \neq z_2} f_1(z_1) f_2(z_2) \widehat{N}(dz_1) \widehat{N}(dz_2)$$

$$+ \int_z f_1(z) f_2(z) \widehat{N}(dz) + \int_z f_1(z) f_2(z) \nu(dz)$$

$$= I_2^{\widehat{N}}(f_1 \otimes_0 f_2) + I_1^{\widehat{N}}(f_1 f_2) + \int_z f_1(z) f_2(z) \nu(dz).$$

(ii) Consider the case $k = 2$, $n_1 = 2$ and $n_2 = 1$. Then, $n = 3$, and $\pi^* = \{\{1,2\},\{3\}\}$. There are three elements $\sigma_1, \sigma_2, \sigma_3 \in \mathcal{P}([3])$ such that $\sigma_i \wedge \pi^* = \hat{0}$, namely $\sigma_1 = \hat{0}$, $\sigma_2 = \{\{1,3\},\{2\}\}$ and $\sigma_3 = \{\{1\},\{2,3\}\}$. Then, (6.1.2) gives that, for every pair $f_1 \in \mathcal{E}_{s,0}(\nu^2)$, $f_2 \in \mathcal{E}(\nu)$,

$$I_2^{\varphi}(f_1) \times I_1^{\varphi}(f_2) = St_{\hat{0}}^{\varphi,[3]}(f_1 \otimes_0 f_2) + St_{\sigma_2}^{\varphi,[3]}(f_1 \otimes_0 f_2) + St_{\sigma_3}^{\varphi,[3]}(f_1 \otimes_0 f_2).$$

$$= St_{\hat{0}}^{\varphi,[3]}(f_1 \otimes_0 f_2) + 2St_{\sigma_2}^{\varphi,[3]}(f_1 \otimes_0 f_2),$$

where we have used the fact that, by the symmetry of f_1, $St_{\sigma_2}^{\varphi,[3]}(f_1 \otimes_0 f_2) = St_{\sigma_3}^{\varphi,[3]}(f_1 \otimes_0 f_2)$.

When $\varphi = G$ is a Gaussian measure, one can use (5.11.84) applied to σ_2 to deduce that

$$St_{\sigma_2}^{G,[3]}(f_1 \otimes_0 f_2) = I_1^G \left[\int_Z f_1(\cdot, z) f_2(z) \nu(dz) \right],$$

or, more informally,

$$St_{\sigma_2}^{G,[3]}(f_1 \otimes_0 f_2) = \int_Z \int_Z f_1(z', z) f_2(z') \nu(dz) G(dz'),$$

so that one gets

$$I_2^G(f_1) \times I_1^G(f_2)$$

$$= St_{\hat{0}}^{G,[3]}(f_1 \otimes_0 f_2) + 2I_1^G \left[\int_Z f_1(\cdot, z) f_2(z) \nu(dz) \right]$$

$$= \int\int\int_{z_1 \neq z_2 \neq z_3} f_1(z_1, z_2) f_2(z_3) G(dz_1) G(dz_2) G(dz_3)$$

$$+ 2 \int_Z \int_Z f_1(z', z) f_2(z) \nu(dz) G(dz').$$

When $\varphi = \widehat{N}$ is compensated Poisson, as shown in (5.12.100), formula (5.12.91), applied to σ_2, yields

$$\text{St}_{\sigma_2}^{\widehat{N},[3]} (f_1 \otimes_0 f_2) = I_1^{\widehat{N}} \left[\int_Z f_1(\cdot,z) f_2(z) \nu(dz) \right] + I_2^{\widehat{N}} \left[f_1 \otimes_1^0 f_2 \right],$$

where $f_1 \otimes_1^0 f_2(z',z) = f_1(z',z) f_2(z)$.

(iii) Consider the case $k = 3$, $n_1 = n_2 = n_3 = 1$. Then, $n = 3$, and $\pi^* = \{\{1\},\{2\},\{3\}\} = \hat{0}$. For every $\sigma \in \mathcal{P}([3])$ one has that $\sigma \wedge \pi^* = \hat{0}$. Note also that

$$\mathcal{P}([3]) = \left\{ \hat{0}, \rho_1, \rho_2, \rho_3, \hat{1} \right\},$$

where

$$\rho_1 = \{\{1,2\},\{3\}\}, \quad \rho_2 = \{\{1,3\},\{2\}\}, \quad \rho_3 = \{\{1\},\{2,3\}\},$$

so that (6.1.2) gives that, for every $f_1, f_2, f_3 \in \mathcal{E}(\nu)$,

$$I_1^\varphi(f_1) I_1^\varphi(f_2) I_1^\varphi(f_3) = \text{St}_{\hat{0}}^{\varphi,[3]}(f_1 \otimes_0 f_2 \otimes_0 f_3) + \text{St}_{\rho_1}^{\varphi,[3]}(f_1 \otimes_0 f_2 \otimes_0 f_3)$$
$$+ \text{St}_{\rho_2}^{\varphi,[3]}(f_1 \otimes_0 f_2 \otimes_0 f_3) + \text{St}_{\rho_3}^{\varphi,[3]}(f_1 \otimes_0 f_2 \otimes_0 f_3)$$
$$+ \text{St}_{\hat{1}}^{\varphi,[3]}(f_1 \otimes_0 f_2 \otimes_0 f_3).$$

In particular, by taking $f_1 = f_2 = f_3 = f$ and by symmetry,

$$I_1^\varphi(f)^3 =$$
$$\text{St}_{\hat{0}}^{\varphi,[3]}(f \otimes_0 f \otimes_0 f) + \text{St}_{\hat{1}}^{\varphi,[3]}(f \otimes_0 f \otimes_0 f) + 3\text{St}_{\rho_1}^{\varphi,[3]}(f \otimes_0 f \otimes_0 f).$$

$$(6.1.9)$$

When $\varphi = G$ is Gaussian, then $\text{St}_{\hat{1}}^{G,[3]} = 0$ by (5.11.82) and $\text{St}_{\rho_1}^{\varphi,[3]}(f \otimes_0 f \otimes_0 f) = \|f\|^2 I_1^G(f)$, so that (6.1.9) becomes

$$I_1^G(f)^3 = I_3^G(f \otimes_0 f \otimes_0 f) + 3\|f\|^2 I_1^G(f)$$
$$= \int \int \int_{z_1 \neq z_2 \neq z_3} f(z_1) f(z_2) f(z_3) G(dz_1) G(dz_2) G(dz_3)$$
$$+ 3 \int_Z f(z)^2 \nu(dz) \times \int_Z f(z) G(dz).$$

When $\varphi = \widehat{N}$ is compensated Poisson, from (5.12.88), then

$$\text{St}_{\hat{1}}^{\widehat{N},[3]}(f \otimes_0 f \otimes_0 f) = \int_Z f(z)^3 N(dz) = I_1^{\widehat{N}}(f^3) + \int_Z f^3(z) \nu(dz)$$

by (5.12.100), and also

$$\text{St}_{\rho_1}^{\widehat{N},[3]}(f \otimes_0 f \otimes_0 f) = \|f\|^2 I_1^{\widehat{N}}(f) + I_2^{\widehat{N}}(f^2 \otimes_0 f),$$

where

$$f^2 \otimes_0 f (z, z') = f^2 (z) f (z'),$$

so that (6.1.9) becomes

$$
\begin{aligned}
I_1^{\widehat{N}} (f)^3 &= I_3^{\widehat{N}} (f \otimes_0 f \otimes_0 f) + I_1^{\widehat{N}} (f^3) + \int_Z f^3 (z) \nu (dz) \\
&\quad + 3 \|f\|^2 I_1^{\widehat{N}} (f) + I_2^{\widehat{N}} (f^2 \otimes_0 f) \\
&= \int \int \int_{z_1 \neq z_2 \neq z_3} f (z_1) f (z_2) f (z_3) \widehat{N} (dz_1) \widehat{N} (dz_2) \widehat{N} (dz_3) \\
&\quad + \int_Z f (z)^3 \left(\widehat{N} + \nu \right) (dz) + 3 \int_Z f (z)^2 \nu (dz) \times \int_Z f (z) \widehat{N} (dz) \\
&\quad + \int \int_{z_1 \neq z_2} f^2 (z_1) f (z_2) \widehat{N} (dz_1) \widehat{N} (dz_2).
\end{aligned}
$$

General applications to the Gaussian and Poisson cases are discussed, respectively, in Section 6.4 and Section 6.5.

6.2 Contractions

As anticipated, the statement of Theorem 6.1.1 contains two well-known *multiplication formulae*, associated with the Gaussian and Poisson cases. In order to state these results, we shall start with a standard definition of the *contraction kernels* associated with two symmetric functions f and g.

Roughly speaking, given $f \in L_s^2 (\nu^p)$ and $g \in L_s^2 (\nu^q)$, the contraction of f and g on $Z^{p+q-r-l}$ ($r = 0, ..., q \wedge p$ and $l = 1, ..., r$), denoted

$$\boxed{f \star_r^l g}$$

is obtained by reducing the number of variables in the tensor product $f (x_1, ..., x_p) g (x_{p+1}, ..., x_{p+q})$ as follows:

$$r \text{ variables are identified, and of these, } l \text{ are integrated out with respect to } \nu.$$

The formal definition of $f \star_r^l g$ is given below.

Definition 6.2.1 *Let ν be a σ-finite measure on $(Z, \mathcal{Z})$. For every $q, p \geq 1$, $f \in L^2 (\nu^p)$, $g \in L^2 (\nu^q)$ (not necessarily symmetric), $r = 0, ..., q \wedge p$ and $l = 1, ..., r$, the contraction (of index (r, l)) of f and g on $Z^{p+q-r-l}$, is the function $f \star_r^l g$ of $p+q-r-l$*

variables defined as follows:

$$f \star_r^l g(\gamma_1, \ldots, \gamma_{r-l}, t_1, \ldots, t_{p-r}, s_1, \ldots, s_{q-r})$$

$$= \int_{Z^l} f(z_1, \ldots, z_l, \gamma_1, \ldots, \gamma_{r-l}, t_1, \ldots, t_{p-r}) \times \qquad (6.2.10)$$

$$\times \; g(z_1, \ldots, z_l, \gamma_1, \ldots, \gamma_{r-l}, s_1, \ldots, s_{q-r}) \nu^l \, (dz_1 \ldots dz_l)$$

and, for $l = 0$,

$$f \star_r^0 g(\gamma_1, \ldots, \gamma_r, t_1, \ldots, t_{p-r}, s_1, \ldots, s_{q-r}) \qquad (6.2.11)$$

$$= f(\gamma_1, \ldots, \gamma_r, t_1, \ldots, t_{p-r}) g(\gamma_1, \ldots, \gamma_r, s_1, \ldots, s_{q-r}),$$

so that

$$f \star_0^0 g(t_1, \ldots, t_p, s_1, \ldots, s_q) = f(t_1, \ldots, t_p) g(s_1, \ldots, s_q).$$

For instance, if $p = q = 2$, one gets

$$f \star_1^0 g\,(\gamma, t, s) = f\,(\gamma, t)\, g\,(\gamma, s), \quad f \star_1^1 g\,(t, s)$$

$$= \int_Z f\,(z, t)\, g\,(z, s)\, \nu\,(dz) \qquad (6.2.12)$$

$$f \star_2^1 g\,(\gamma) = \int_Z f\,(z, \gamma)\, g\,(z, \gamma)\, \nu\,(dz), \qquad (6.2.13)$$

$$f \star_2^2 g = \int_Z \int_Z f\,(z_1, z_2)\, g\,(z_1, z_2)\, \nu\,(dz_1)\, \nu\,(dz_2). \qquad (6.2.14)$$

One also has

$$f \star_r^r g\,(x_1, \ldots, x_{p+q-2r}) \qquad (6.2.15)$$

$$= \int_{Z^r} f\,(z_1, \ldots, z_r, x_1, \ldots, x_{p-r})\, g\,(z_1, \ldots, z_r, x_{p-r+1}, \ldots, x_{p+q-2r})\, \nu\,(dz_1) \cdots \nu\,(dz_r),$$

but, in analogy with (6.1.3), we set

$$\star_r^r = \otimes_r,$$

and consequently write

$$f \star_r^r g\,(x_1, \ldots, x_{p+q-2r}) = f \otimes_r g\,(x_1, \ldots, x_{p+q-2r}), \qquad (6.2.16)$$

so that, in particular,

$$f \star_0^0 g = f \otimes_0 g.$$

The following elementary result is proved by using the Cauchy-Schwarz inequality. It ensures that the contractions of the type (6.2.16) are still square-integrable kernels.

Lemma 6.2.2 *Let $f \in L^2(\nu^p)$ and $g \in L^2(\nu^q)$. Then, for every $r = 0, ..., p \wedge q$, one has that $f \otimes_r g \in L^2(\nu^{p+q-2r})$.*

Proof. Just write

$$\int_{Z^{p+q-2r}} (f \otimes_r g)^2 \, d\nu^{p+q-2r}$$

$$= \int_{Z^{p+q-2r}} \left(\int_{Z^r} f(a_1, ..., a_r, z_1, ..., z_{p-r}) \right.$$

$$\left. g(a_1, ..., a_r, z_{p-r+1}, ..., z_{p+q-r}) \, \nu^r (da_1, ...da_r) \right)^2 \nu^{p+q-2r}(dz_1, ..., dz_{p+q-r})$$

$$\leq \|f\|^2_{L^2(\nu^p)} \times \|g\|^2_{L^2(\nu^q)} .$$

∎

6.3 Symmetrization of contractions

Suppose that $f \in L^2(\nu^p)$ and $g \in L^2_s(\nu^q)$, and let "~" denote symmetrization. Then $f = \tilde{f}$ and $g = \tilde{g}$. However, in general, the fact that f and g are symmetric *does not* imply that the contraction $f \otimes_r g$ is symmetric. For instance, if $p = q = 1$,

$$\widetilde{f \otimes_0 g}(s, t) = \frac{1}{2} [f(s)g(t) + g(s)f(t)] ;$$

if $p = q = 2$

$$\widetilde{f \otimes_1 g}(s, t) = \frac{1}{2} \int_Z [f(x, s)g(x, t) + g(x, s)f(x, t)] \nu(dx) .$$

In general, due to the symmetry of f and g, for every $p, q \geq 1$ and every $r = 0, ..., p \wedge q$ one has the relation

$$\widetilde{f \otimes_r g}(t_1, ..., t_{p+q-2r}) = \frac{1}{\binom{p+q-2r}{p-r}} \times$$

$$\times \sum_{1 \leq i_1 < \cdots < i_{p-r} \leq p+q-2r} \int_{Z^r} f\left(\mathbf{t}_{(i_1,...,i_{p-r})}, \mathbf{a}_r\right) g\left(\mathbf{t}_{(i_1,...,i_{p-r})^c}, \mathbf{a}_r\right) \nu^r(d\mathbf{a}_r),$$

where we used the shorthand notation

$$\mathbf{t}_{(i_1,...,i_{p-r})} = \left(t_{i_1}, ..., t_{i_{p-r}}\right)$$

$$\mathbf{t}_{(i_1,...,i_{p-r})^c} = (t_1, ..., t_{p+q-2r}) \setminus \left(t_{i_1}, ..., t_{i_{p-r}}\right)$$

$$\mathbf{a}_r = (a_1, ..., a_r)$$

$$\nu^r(d\mathbf{a}_r) = \nu^r(da_1, ..., da_r) .$$

Using the definition (6.2.10), one has also that $\widetilde{f \star_r^l g}$ indicates the symmetrization of $f \star_r^l g$, where $l < r$. For instance, if $p = 3$, $q = 2$, $r = 2$ and $l = 1$, one has that

$$f \star_r^l g (s,t) = f \star_2^1 g (s,t) = \int_Z f (z,s,t) g (z,s) \nu (dz),$$

and consequently, since f is symmetric,

$$\widetilde{f \star_r^l g} (s,t) = \widetilde{f \star_2^1 g} (s,t) = \frac{1}{2} \int_Z [f (z,s,t) g (z,s) + f (z,s,t) g (z,t)] \nu (dz).$$

6.4 The product of two integrals in the Gaussian case

The main result of this section is the following general formula for products of Gaussian multiple integrals.

Proposition 6.4.1 *Let $\varphi = G$ be a centered Gaussian measure with σ-finite and non-atomic control measure ν. Then, for every $q, p \geq 1$, $f \in L_s^2 (\nu^p)$ and $g \in L_s^2 (\nu^q)$,*

$$I_p^G (f) I_q^G (g) = \sum_{r=0}^{p \wedge q} r! \binom{p}{r} \binom{q}{r} I_{p+q-2r}^G (f \otimes_r g), \qquad (6.4.17)$$

where the contraction $f \otimes_r g$ is defined in (6.2.16), and for $p = q = r$, we write

$$I_0^G (f \otimes_p g)$$
$$= f \otimes_p g = \int_{Z^p} f (z_1, ..., z_p) g (z_1, ..., z_p) \nu (dz_1) \cdots \nu (dz_p)$$
$$= (f, g)_{L^2 (\nu^p)}.$$

Remark. In formula (6.4.17), f and g are assumed symmetric, while the contraction $f \otimes_r g$ is in general not symmetric. Since, however, that $I_n^G (h) = I_n^G (\tilde{h})$ (see formula (5.5.63)) this does not matter. One could, in fact, replace $f \otimes_r g$ with $\widetilde{f \otimes_r g}$. As usual, the symbol " $\sim$ " indicates a symmetrization.

Proof of Proposition 6.4.1. We start by assuming that $f \in \mathcal{E}_{s,0} (\nu^p)$ and $g \in \mathcal{E}_{s,0} (\nu^q)$, and we denote by π^* the partition of the set

$$[p + q] = \{1, ..., p + q\}$$

given by

$$\pi^* = \{\{1, ..., p\}, \{p + 1, ..., p + q\}\}.$$

According to formula (6.1.2)

$$I_p^G (f) I_q^G (g) = \sum_{\sigma \in \mathcal{P}([p+q]): \sigma \wedge \pi^* = \hat{0}} \mathrm{St}_\sigma^{G,[n]} (f \otimes_0 g).$$

Every partition $\sigma \in \mathcal{P}([p+q])$ such that $\sigma \wedge \pi^* = \hat{0}$ is necessarily composed of r ($0 \leq r \leq p \wedge q$) two-elements blocks of the type $\{i, j\}$ where $i \in \{1, ..., p\}$ and $j \in \{p+1, ..., p+q\}$, and $p + q - 2r$ singletons. Moreover, for every fixed $r \in \{0, ..., p \wedge q\}$, there are exactly $r! \binom{p}{r} \binom{q}{r}$ partitions of this type. To see this, observe that, to build such a partition, one should first select one of the $\binom{p}{r}$ subsets of size r of $\{1, ..., p\}$, say A_r, and a one of the $\binom{q}{r}$ subset of size r of $\{p+1, ..., p+q\}$, say B_r, and then choose one of the $r!$ bijections between A_r and B_r. When $r = 0$, and therefore $\sigma = \hat{0}$, one obtains immediately

$$\mathrm{St}_\sigma^{G,[p+q]} (f \otimes_0 g) = I_{p+q}^G (f \otimes_0 g).$$

On the other hand, we claim that every partition $\sigma \in \mathcal{P}([p+q])$ such that $\sigma \wedge \pi^* = \hat{0}$ and σ contains $r \geq 1$ two-elements blocks of the type $b = \{i, j\}$ (with $i \in \{1, ..., p\}$ and $j \in \{p+1, ..., p+q\}$) and $p + q - 2r$ singletons, satisfies also

$$\mathrm{St}_\sigma^{G,[p+q]} (f \otimes_0 g) = \mathrm{St}_{\hat{0}}^{G,[p+q-2r]} (f \otimes_r g) = I_{p+q-2r}^G (f \otimes_r g). \tag{6.4.18}$$

We give a proof of (6.4.18). Consider first the (not necessarily symmetric) functions

$$f^\circ = \mathbf{1}_{A_1 \times \cdots \times A_p} \quad \text{and} \quad g^\circ = \mathbf{1}_{A_{p+1} \times \cdots \times A_{p+q}},$$

where $A_l \in \mathcal{Z}_\nu$, $l = 1, ..., p + q$. Then, one may use (5.11.84), in order to obtain

$$\mathrm{St}_\sigma^{G,[p+q]} (f^\circ \otimes_0 g^\circ) = \prod_{b=\{i,j\} \in \sigma} \nu(A_i \cap A_j) \, \mathrm{St}_{\hat{0}}^{G,[p+q-2r]} \left(A_{j_1} \times \cdots \times A_{j_{p+q-2r}} \right)$$

$$= \mathrm{St}_{\hat{0}}^{G,[p+q-2r]} \left(\prod_{b=\{i,j\} \in \sigma} \nu(A_i \cap A_j) \mathbf{1}_{A_{j_1} \times \cdots \times A_{j_{p+q-2r}}} \right),$$

where $\{\{j_1\}, ..., \{j_{p+q-2r}\}\}$ are the singletons of σ (by the symmetry of $\mathrm{St}_{\hat{0}}^{G,[p+q-2r]}$ we can always suppose, here and in the following, that the singletons $\{j_1\}, ..., \{j_{p-r}\}$ are contained in $\{1, ..., p\}$ and that the singletons $\{j_{p-r+1}\}, ..., \{j_{p+q-2r}\}$ are in $\{p+1, ..., p+q\}$).

We now need to consider elementary symmetric functions, that is, functions in $\mathcal{E}_{s,0}$ (as defined in (5.5.59)). For every pair of permutations $w \in \mathfrak{S}_{[p]} = \mathfrak{S}_p$ (the group of permutations of $[p] = \{1, ..., p\}$) and $w' \in \mathfrak{S}_{[p+1,p+q]}$ (the group of permutations of the set $[p+1, p+q] = \{p+1, ..., p+q\}$), we define

$$f^{\circ,w} = \mathbf{1}_{A_1^w \times \cdots \times A_p^w}$$

and

$$g^{\circ,w'} = \mathbf{1}_{A_{p+1}^{w'} \times \cdots \times A_{p+q}^{w'}},$$

where

$$A_j^w = A_{w(j)}, \quad j = 1, ..., p,$$

(and analogously for w'). Observe that $f^{\circ,w}$ and $g^{\circ,w'}$ are not in $\mathcal{E}_{s,0}$, but

$$f = \sum_{w \in \mathfrak{S}_{[p]}} f^{\circ,w} \quad \text{and} \quad g = \sum_{w \in \mathfrak{S}_{[p+1,p+q]}} g^{\circ,w},$$

are. To express $\mathrm{St}_\sigma^{G,[p+q]} (f \otimes_0 g)$, note that

$$\mathrm{St}_\sigma^{G,[p+q]} \left(f^{\circ,w} \otimes_0 g^{\circ,w'} \right) \tag{6.4.19}$$

$$= \mathrm{St}_{\hat{0}}^{G,[p+q-2r]} \left(\prod_{b=\{i,j\}\in\sigma} \nu \left(A_i^w \cap A_j^{w'} \right) 1_{A_{j_1}^w \times \cdots \times A_{j_{p-r}}^w \times A_{j_{p-r+1}}^{w'} \times \cdots \times A_{j_{p+q-2r}}^{w'}} \right)$$

$$= \mathrm{St}_{\hat{0}}^{G,[p+q-2r]} \left(\prod_{b=\{i,j\}\in\sigma} \nu \left(A_i^w \cap A_j^{w'} \right) \widetilde{1}_{A_{j_1}^w \times \cdots \times A_{j_{p-r}}^w \times A_{j_{p-r+1}}^{w'} \times \cdots \times A_{j_{p+q-2r}}^{w'}} \right),$$

by (5.5.56), and where $\widetilde{1}_U$ stands for the symmetrization of the indicator function of a set U. Observe that (by using (6.2.16))

$$\sum_{w \in \mathfrak{S}_{[p]}} \sum_{w' \in \mathfrak{S}_{[p+1,p+q]}} \prod_{b=\{i,j\}\in\sigma} \nu \left(A_i^w \cap A_j^{w'} \right) \widetilde{1}_{A_{j_1}^w \times \cdots \times A_{j_{p-r}}^w \times A_{j_{p-r+1}}^{w'} \times \cdots \times A_{j_{p+q-2r}}^{w'}}$$

$$= \widetilde{f \otimes_r g}.$$

Since (6.4.19) gives

$$\mathrm{St}_\sigma^{G,[p+q]} (f \otimes_0 g)$$

$$= \mathrm{St}_{\hat{0}}^{G,[p+q-2r]} \left(\sum_{w \in \mathfrak{S}_{[p]}} \sum_{w' \in \mathfrak{S}_{[p+1,p+q]}} \prod_{b=\{i,j\}\in\sigma} \nu \left(A_i^w \cap A_j^{w'} \right) \right.$$

$$\left. \widetilde{1}_{A_{j_1}^w \times \cdots \times A_{j_{p-r}}^w \times A_{j_{p-r+1}}^{w'} \times \cdots \times A_{j_{p+q-2r}}^{w'}} \right)$$

$$= \mathrm{St}_{\hat{0}}^{G,[p+q-2r]} \left(\widetilde{f \otimes_r g} \right) = I_{p+q-2r}^G \left(\widetilde{f \otimes_r g} \right) = I_{p+q-2r}^G (f \otimes_r g),$$

we obtain (6.4.18), so that, in particular, (6.4.17) is proved for symmetric simple functions vanishing on diagonals.

The general result is obtained by using the fact that the linear spaces $\mathcal{E}_{s,0}(\nu^p)$ and $\mathcal{E}_{s,0}(\nu^q)$ are dense, respectively, in $L_s^2(\nu^p)$ and $L_s^2(\nu^q)$. Indeed, to conclude the proof it is sufficient to observe that, if $\{f_k\} \subset \mathcal{E}_{s,0}(\nu^p)$ and $\{g_k\} \subset \mathcal{E}_{s,0}(\nu^q)$ are such that $f_k \to f$ in $L_s^2(\nu^p)$ and $g_k \to g$ in $L_s^2(\nu^q)$, then, for instance by Cauchy-Schwarz, $I_p^G(f_k) I_q^G(g_k) \to I_p^G(f) I_q^G(g)$ in any norm $L^s(\mathbb{P})$, $s \geq 1$ (use, for example, (5.9.79)), and also

$$\widetilde{f_k \otimes_r g_k} \to \widetilde{f \otimes_r g}$$

in $L_s^2\left(\nu^{p+q-2r}\right)$, so that

$$I_{p+q-2r}^G\left(\widetilde{f_k \otimes_r g_k}\right) \to I_{p+q-2r}^G\left(\widetilde{f \otimes_r g}\right)$$

in $L^2\left(\mathbb{P}\right)$. Observe finally that

$$I_{p+q-2r}^G\left(\widetilde{f \otimes_r g}\right) = I_{p+q-2r}^G\left(f \otimes_r g\right),$$

by virtue of formula (5.5.63).   ∎

Other proofs of Proposition 6.4.1 can be found, for example, in [66], [20] or [94, Proposition 1.1.3].

Example 6.4.2   (i) When $p = q = 1$, one obtains

$$I_1^G\left(f\right) I_1^G\left(g\right) = I_2^G\left(f \otimes_0 g\right) + I_0^G\left(f \otimes_1 g\right) = I_2^G\left(f \otimes_0 g\right) + \langle f, g \rangle_{L^2(\nu)},$$

which is consistent with (6.1.8).

(ii) When $p = q = 2$, one obtains

$$I_2^G\left(f\right) I_2^G\left(g\right) = I_4^G\left(f \otimes_0 g\right) + 4 I_2^G\left(f \otimes_1 g\right) + \langle f, g \rangle_{L^2(\nu)}.$$

(iii) When $p = 3$ and $q = 2$, one obtains

$$I_3^G\left(f\right) I_2^G\left(g\right) = I_5^G\left(f \otimes_0 g\right) + 6 I_3^G\left(f \otimes_1 g\right) + 6 I_1^G\left(f \otimes_2 g\right),$$

where $f \otimes_2 g\left(z\right) = \int_{Z^2} f\left(z, x, y\right) g\left(x, y\right) \nu\left(dx\right) \nu\left(dy\right)$.

Example 6.4.3 Let's compute $\mathbb{E}\left[I_p^G(f)\right]^4$ for $f \in L_s^2(\nu^p)$. By (6.4.17) and (5.5.65), we have

$$\left(I_p^G(f)\right)^2 = \sum_{r=0}^p r!\binom{p}{r}^2 I_{2p-2r}^G\left(\widetilde{f \otimes_r f}\right),$$

and therefore

$$\mathbb{E}\left[I_p^G(f)\right]^4 = \sum_{r=0}^p \sum_{r'=0}^p r!\binom{p}{r}^2 r'!\binom{p}{r'}^2 \mathbb{E}I_{2p-2r}^G\left(\widetilde{f \otimes_r f}\right) I_{2(p-r')}^G\left(\widetilde{f \otimes_{r'} f}\right)$$

$$= \sum_{r=0}^p \left[r!\binom{p}{r}^2\right]^2 (2p-2r)! \left\|\widetilde{f \otimes_r f}\right\|_{L^2(\nu^{2p-2r})}^2$$

by the orthogonality relation (5.5.62). The term $r = 0$ equals $(2p)! \left\|\widetilde{f \otimes_0 f}\right\|_{L^2(\nu^{2p})}^2$ and the term with $r = p$ equals

$$(p!)^2 \left\|\widetilde{f \otimes_p f}\right\|_{L^2(\nu^0)}^2$$

$$= (p!)^2 \left[\int_{Z^p} f^2(z_1, \cdots, z_p)\nu(dz_1)\cdots\nu(dz_p)\right]^2 = (p!)^2 \|f\|_{L^2(\nu^p)}^4,$$

since $\widetilde{f \otimes_p f} = f \otimes_p f$ is a scalar. Hence,

$$
\mathbb{E}\left[I_p^G(f)\right]^4 = (p!)^2 \|f\|_{L^2(\nu^p)}^4 + (2p!) \left\|\widetilde{f \otimes_0 f}\right\|_{L^2(\nu^{2p})}^2
$$

$$
+ \sum_{r=1}^{p-1} \left[r!\binom{p}{r}^2\right]^2 (2p-2r)! \left\|\widetilde{f \otimes_r f}\right\|_{L^2(\nu^{2p-2r})}^2 . \tag{6.4.20}
$$

Note that while f is symmetric, the function

$$
f \otimes_0 f(x_1, \cdots, x_p, y_1, \cdots, y_p) = f(x_1, \cdots, x_p) f(y_1, \cdots, y_p)
$$

is not necessarily symmetric. For example, if $f(x_1, x_2) = x_1 + x_2$, then $f \otimes_0 f(x_1, x_2, y_1, y_2) = (x_1 + x_2)(y_1 + y_2)$ is not symmetric in the four variables.

6.5 The product of two integrals in the Poisson case

We now focus on the product of two Poisson integrals.

Proposition 6.5.1 Let $\varphi = \widehat{N}$ be a compensated Poisson measure, with σ-finite and non-atomic control measure ν. Then, for every $q, p \geq 1$, $f \in \mathcal{E}_{s,0}(\nu^p)$ and $g \in \mathcal{E}_{s,0}(\nu^q)$,

$$
I_p^{\widehat{N}}(f) I_q^{\widehat{N}}(g) = \sum_{r=0}^{p \wedge q} r! \binom{p}{r}\binom{q}{r} \sum_{l=0}^{r} \binom{r}{l} I_{p+q-r-l}^{\widehat{N}}(f \star_r^l g). \tag{6.5.21}
$$

Formula (6.5.21) continues to hold for functions $f \in L_s^2(\nu^p)$ and $g \in L_s^2(\nu^q)$ such that

$$
f \star_r^l g \in L^2\left(\nu^{q+p-r-l}\right), \; \forall r = 0, ..., p \wedge q, \; \forall l = 0, ..., r.
$$

Sketch of the proof. We shall only prove formula (6.5.21) in the simple case where $p = q = 2$. The generalization to general indices $p, q \geq 1$ (left to the reader) does not present any particular additional difficulty, except for the need of a rather heavy notation. We shall therefore prove that

$$
I_2^{\widehat{N}}(f) I_2^{\widehat{N}}(g) = \sum_{r=0}^{2} r! \binom{2}{r}\binom{2}{r} \sum_{l=0}^{r} \binom{r}{l} I_{4-r-l}^{\widehat{N}}(f \star_r^l g) \tag{6.5.22}
$$

$$
= I_4^{\widehat{N}}(f \star_0^0 g) \tag{6.5.23}
$$

$$
+ 4 \left[I_3^{\widehat{N}}(f \star_1^0 g) + I_2^{\widehat{N}}(f \star_1^1 g)\right] \tag{6.5.24}
$$

$$
+ 2 \left[I_2^{\widehat{N}}(f \star_2^0 g) + 2I_1^{\widehat{N}}(f \star_2^1 g) + \langle f, g \rangle_{L^2(\nu^2)}\right] . \tag{6.5.25}
$$

Moreover, by linearity, we can also assume that

$$f = 1_{A_1 \times A_2} + 1_{A_2 \times A_1} \text{ and } g = 1_{B_1 \times B_2} + 1_{B_2 \times B_1},$$

where $A_1 \cap A_2 = B_1 \cap B_2 = \varnothing$. Denote by π^* the partition of $[4] = \{1, ..., 4\}$ given by

$$\pi^* = \{\{1, 2\}, \{3, 4\}\},$$

and apply the general result (6.1.2) to deduce that

$$I_2^{\widehat{N}}(f) I_2^{\widehat{N}}(g) = \sum_{\sigma \in \mathcal{P}([4]): \sigma \wedge \pi^* = \hat{0}} \mathrm{St}_{\sigma}^{\widehat{N}, [n]} (f \otimes_0 g).$$

We shall prove that

$$\sum_{\sigma \in \mathcal{P}([4]): \sigma \wedge \pi^* = \hat{0}} \mathrm{St}_{\sigma}^{\widehat{N}, [n]} (f \otimes_0 g) = (6.5.23) + (6.5.24) + (6.5.25).$$

To see this, observe that the class

$$\left\{ \sigma \in \mathcal{P}([4]) : \sigma \wedge \pi^* = \hat{0} \right\}$$

contains exactly 7 elements, that is:

(i)   the trivial partition $\hat{0}$, containing only singletons;

(ii) four partitions $\sigma_1, ..., \sigma_4$ containing one block of two elements and two singletons, namely

$$\sigma_1 = \{\{1, 3\}, \{2\}, \{4\}\}, \quad \sigma_2 = \{\{1, 4\}, \{2\}, \{3\}\}$$
$$\sigma_3 = \{\{1\}, \{2, 3\}, \{4\}\} \text{ and } \sigma_4 = \{\{1\}, \{2, 4\}, \{3\}\};$$

(iii) two partitions σ_5, σ_6 composed of two blocks of two elements, namely

$$\sigma_5 = \{\{1, 3\}, \{2, 4\}\} \text{ and } \sigma_6 = \{\{1, 4\}, \{2, 3\}\}.$$

By definition, one has that

$$\mathrm{St}_{\hat{0}}^{\widehat{N}, [n]} (f \otimes_0 g) = I_4^{\widehat{N}}(\widetilde{f \star_0^0 g}) = I_4^{\widehat{N}}(f \star_0^0 g),$$

giving (6.5.23). Now consider the partition σ_1, as defined in Point **(II)** above. By using the notation (5.12.86), one has that $\mathbf{B}_2(\sigma_1) = \{\{1, 3\}\}$, $|\mathbf{B}_2(\sigma_1)| = 1$ and $\mathbf{PB}_2(\sigma_1) = \varnothing$. It follows from formula (5.12.91) that

$$\mathrm{St}_{\sigma_1}^{\widehat{N}, [4]} (f \otimes_0 g) = \mathrm{St}_{\sigma_1}^{\widehat{N}, [4]} \left((1_{A_1 \times A_2} + 1_{A_2 \times A_1}) \otimes_0 (1_{B_1 \times B_2} + 1_{B_2 \times B_1}) \right)$$

$$= \mathrm{St}_{\hat{0}}^{\widehat{N}, [3]} \left(1_{(A_1 \cap B_1) \times A_2 \times B_2} + 1_{(A_1 \cap B_2) \times A_2 \times B_1} + 1_{(A_2 \cap B_1) \times A_1 \times B_2} \right.$$

$$+ 1_{(A_2 \cap B_2) \times A_1 \times B_1} \big) + \nu(A_1 \cap B_1) \mathrm{St}_{\hat{0}}^{\widehat{N}, [2]} (1_{A_2 \times B_2})$$

$$+ \nu(A_1 \cap B_2) \mathrm{St}_{\hat{0}}^{\widehat{N}, [2]} (1_{A_2 \times B_1}) + \nu(A_2 \cap B_1) \mathrm{St}_{\hat{0}}^{\widehat{N}, [2]} (1_{A_1 \times B_2})$$

$$+ \nu(A_2 \cap B_2) \mathrm{St}_{\hat{0}}^{\widehat{N}, [2]} (1_{A_1 \times B_1}).$$

Observe that

$$\text{St}_{\hat{0}}^{\hat{N},[3]}\left(\mathbf{1}_{(A_1\cap B_1)\times A_2\times B_2}+\mathbf{1}_{(A_1\cap B_2)\times A_2\times B_1}+\mathbf{1}_{(A_2\cap B_1)\times A_1\times B_2}\right.$$
$$\left.+\mathbf{1}_{(A_2\cap B_2)\times A_1\times B_1}\right)=I_3^{\hat{N}}(f\star_1^0 g),$$

and moreover,

$$\nu(A_1\cap B_1)\,\text{St}_{\hat{0}}^{\hat{N},[2]}\left(\mathbf{1}_{A_2\times B_2}\right)+\nu(A_1\cap B_2)\,\text{St}_{\hat{0}}^{\hat{N},[2]}\left(\mathbf{1}_{A_2\times B_1}\right)+$$
$$\nu(A_2\cap B_1)\,\text{St}_{\hat{0}}^{\hat{N},[2]}\left(\mathbf{1}_{A_1\times B_2}\right)+\nu(A_2\cap B_2)\,\text{St}_{\hat{0}}^{\hat{N},[2]}\left(\mathbf{1}_{A_1\times B_1}\right)$$
$$=I_2^{\hat{N}}(f\star_1^1 g)=I_2^{\hat{N}}(f\star_1^1 g).$$

By repeating exactly same argument, one sees immediately that

$$\text{St}_{\sigma_1}^{\hat{N},[4]}(f\otimes_0 g)=\text{St}_{\sigma_i}^{\hat{N},[4]}(f\otimes_0 g),$$

for every for $i=2,3,4$ (the partitions σ_i being defined as in Point (II) above) so that the quantity

$$\sum_{i=1,\dots,4}\text{St}_{\sigma_i}^{\hat{N},[n]}(f\otimes_0 g)$$

equals necessarily the expression appearing in (6.5.24). Now we focus on the partition σ_5 appearing in Point (III). Plainly (by using once again the notation introduced in (5.12.86)),

$$\mathbf{B}_2(\sigma_5)=\{\{1,3\},\{2,4\}\},$$

$|\mathbf{B}_2(\sigma_5)|=2$, and the set $\mathbf{PB}_2(\sigma_5)$ contains two elements, namely

$$(\{\{1,3\}\};\{\{2,4\}\})\quad\text{and}\quad(\{\{2,4\}\};\{\{1,3\}\})$$

(note that we write $\{\{1,3\}\}$ (with two accolades), since the elements of $\mathbf{PB}_2(\sigma_5)$ are pairs of collections of blocks of σ_2, so that $\{\{1,3\}\}$ is indeed the singleton whose only element is $\{1,3\}$). We can now apply formula (5.12.91) to deduce that

$$\text{St}_{\sigma_5}^{\hat{N},[4]}(f\otimes_0 g)\qquad\qquad\qquad\qquad\qquad\qquad\qquad\qquad\qquad(6.5.26)$$
$$=\text{St}_{\sigma_5}^{\hat{N},[4]}\left((\mathbf{1}_{A_1\times A_2}+\mathbf{1}_{A_2\times A_1})\otimes_0(\mathbf{1}_{B_1\times B_2}+\mathbf{1}_{B_2\times B_1})\right)$$
$$=2\left[\nu(A_1\cap B_1)\,\text{St}_{\hat{0}}^{\hat{N},[1]}\left(\mathbf{1}_{A_2\cap B_2}\right)+\nu(A_1\cap B_2)\,\text{St}_{\hat{0}}^{\hat{N},[1]}\left(\mathbf{1}_{A_2\cap B_1}\right)\right.$$
$$\left.+\nu(A_2\cap B_1)\,\text{St}_{\hat{0}}^{\hat{N},[1]}\left(\mathbf{1}_{A_1\cap B_2}\right)+\nu(A_2\cap B_2)\,\text{St}_{\hat{0}}^{\hat{N},[1]}\left(\mathbf{1}_{A_1\cap B_1}\right)\right]$$
$$+\,\text{St}_{\hat{0}}^{\hat{N},[2]}\left(\mathbf{1}_{(A_1\cap B_1)\times(A_2\cap B_2)}+\mathbf{1}_{(A_1\cap B_2)\times(A_2\cap B_1)}+\mathbf{1}_{(A_2\cap B_1)\times(A_1\cap B_2)}\right.$$
$$\left.+\mathbf{1}_{(A_2\cap B_2)\times(A_1\cap B_1)}\right)$$
$$+\,\nu(A_1\cap B_1)\,\nu(A_2\cap B_2)+\nu(A_1\cap B_2)\,\nu(A_2\cap B_1)$$
$$+\,\nu(A_2\cap B_1)\,\nu(A_1\cap B_2)+\nu(A_2\cap B_2)\,\nu(A_1\cap B_1).$$

One easily verifies that

$$2\left[\nu\left(A_1\cap B_1\right)\mathrm{St}_0^{\widehat{N},[1]}\left(\mathbf{1}_{A_2\cap B_2}\right)+\nu\left(A_1\cap B_2\right)\mathrm{St}_0^{\widehat{N},[1]}\left(\mathbf{1}_{A_2\cap B_1}\right)\right.$$
$$\left.+\nu\left(A_2\cap B_1\right)\mathrm{St}_0^{\widehat{N},[1]}\left(\mathbf{1}_{A_1\cap B_2}\right)+\nu\left(A_2\cap B_2\right)\mathrm{St}_0^{\widehat{N},[1]}\left(\mathbf{1}_{A_1\cap B_1}\right)\right]$$
$$=2I_1^{\widehat{N}}(f\star_2^1 g)=2I_1^{\widehat{N}}(\widetilde{f\star_2^1 g}),\tag{6.5.27}$$

and moreover

$$\mathbf{1}_{(A_1\cap B_1)\times(A_2\cap B_2)}+\mathbf{1}_{(A_1\cap B_2)\times(A_2\cap B_1)}+\mathbf{1}_{(A_2\cap B_1)\times(A_1\cap B_2)}+\mathbf{1}_{(A_2\cap B_2)\times(A_1\cap B_1)}$$
$$=f\star_2^0 g=\widetilde{f\star_2^0 g},\tag{6.5.28}$$

since f and g are symmetric, and

$$\langle f,g\rangle_{L^2(\nu^2)}=\nu\left(A_1\cap B_1\right)\nu\left(A_2\cap B_2\right)+\nu\left(A_1\cap B_2\right)\nu\left(A_2\cap B_1\right)\tag{6.5.29}$$
$$+\nu\left(A_2\cap B_1\right)\nu\left(A_1\cap B_2\right)+\nu\left(A_2\cap B_2\right)\nu\left(A_1\cap B_1\right).$$

Since, trivially, $\mathrm{St}_{\sigma_5}^{\widehat{N},[4]}\left(f\otimes_0 g\right)=\mathrm{St}_{\sigma_6}^{\widehat{N},[4]}\left(f\otimes_0 g\right)$, we deduce immediately from (6.5.26)–(6.5.29) that the sum $\mathrm{St}_{\sigma_5}^{\widehat{N},[4]}\left(f\otimes_0 g\right)+\mathrm{St}_{\sigma_6}^{\widehat{N},[4]}\left(f\otimes_0 g\right)$ equals the expression appearing in (6.5.25). This proves the first part of the Proposition. The last assertion in the statement can be proved by a density argument, similar to the one used in order to conclude the proof of Proposition 6.4.1. ∎

Other proofs of Proposition 6.5.1 can be found for instance in [49, 150, 158].

Example 6.5.2   (i) When $p=q=1$, one obtains

$$I_1^{\widehat{N}}(f)\,I_1^{\widehat{N}}(g)=I_2^{\widehat{N}}(f\otimes_0 g)+I_1^{\widehat{N}}(f\star_1^0 g)+\langle f,g\rangle_{L^2(\nu)}.$$

(ii) When $p=2$, and $q=1$, one has

$$I_2^{\widehat{N}}(f)\,I_1^{\widehat{N}}(g)=I_3^{\widehat{N}}(f\otimes_0 g)+2I_2^{\widehat{N}}(f\star_1^0 g)+2I_1^{\widehat{N}}(f\star_1^1 g)$$
$$=\int\int\int_{z_1\neq z_2\neq z_3} f(z_1,z_2)\,g(z_3)\,\widehat{N}(dz_1)\,\widehat{N}(dz_2)\,\widehat{N}(dz_3)$$
$$+2\int\int_{z_1\neq z_2} f(z_1,z_2)\,g(z_1)\,\widehat{N}(dz_1)\,\widehat{N}(dz_2)$$
$$+2\int\left(\int f(z_1,x)\,g(z_1)\,\nu(dx)\right)\widehat{N}(dz_1).$$

7

Diagram formulae

We now want to derive a general formula for computing cumulants and expectations of products of multiple integrals, that is, formulae for objects of the type

$$\mathbb{E}\left[I_{n_1}^{\varphi}\left(f_1\right) \times \cdots \times I_{n_k}^{\varphi}\left(f_k\right)\right] \quad \text{and} \quad \chi\left(I_{n_1}^{\varphi}\left(f_1\right), ..., I_{n_k}^{\varphi}\left(f_k\right)\right).$$

7.1 Formulae for moments and cumulants

As in the previous sections, we shall focus on completely random measures φ that are also *good* (in the sense of Definition 5.1.6), so that moments and cumulants of stochastic measures over Cartesian products are well-defined. As usual, we shall assume that the control measure ν is non-atomic. This last assumption is not enough, however, because while the measure $\nu(A) = \mathbb{E}\varphi(A)^2$ may be non-atomic, for some $n \geq 2$ the mean measure (concentrated on the "full diagonal")

$$\langle\Delta_n^{\varphi}(A)\rangle \triangleq \mathbb{E}\left[\mathrm{St}_{\hat{1}}^{\varphi,[n]}(A)\right] = \mathbb{E}\left[\varphi^{\otimes n}\left\{(z_1, ..., z_n) \in A : z_1 = \cdots = z_n\right\}\right]$$

may be atomic. We shall therefore assume that φ is "multiplicative", that is, that this phenomenon does not take place for any $n \geq 2$.

Proceeding formally, let φ be a good random measure on Z, and fix $n \geq 2$. Recall that $\mathcal{Z}_{\nu}^n$ denotes the collection of all sets B in $\mathcal{Z}^{\otimes n}$ such that $\nu^{\otimes n}(B) = \nu^n(B) < \infty$ (see (5.1.3)). As before, for every partition $\pi \in \mathcal{P}([n])$, the class $\mathcal{Z}_{\pi}^n$ is the collection of all π-diagonal elements of $\mathcal{Z}^n$ (see (5.1.1)). Recall also that $\mathrm{St}_{\pi}^{\varphi,[n]}$ is the restriction of the measure $\varphi^{\otimes n} = \varphi^{[n]}$ on Z_{π}^n (see (5.4.43)). Now let

$$\left\langle\mathrm{St}_{\pi}^{\varphi,[n]}\right\rangle(C) = \mathbb{E}\left[\mathrm{St}_{\pi}^{\varphi,[n]}(C)\right], \quad C \in \mathcal{Z}_{\nu}^n, \tag{7.1.1}$$

$$\Delta_1^{\varphi}(A) = \varphi(A), \tag{7.1.2}$$

$$\Delta_n^{\varphi}(A) = \mathrm{St}_{\hat{1}}^{\varphi,[n]}(\underbrace{A \times \cdots \times A}_{n \text{ times}}), \quad A \in \mathcal{Z}_{\nu}, \tag{7.1.3}$$

$$\langle\Delta_n^{\varphi}\rangle(A) = \mathbb{E}\left[\Delta_n^{\varphi}(A)\right], \quad A \in \mathcal{Z}_{\nu}. \tag{7.1.4}$$

G. Peccati, M.S. Taqqu: Wiener Chaos: Moments, Cumulants and Diagrams –
A survey with computer implementation.
© Springer-Verlag Italia 2011

Thus, $\Delta_n^\varphi(A)$ denotes the random measure concentrated on the full diagonal $z_1 = \cdots = z_n$ of the $n$tuple product $A \times \cdots \times A$, and $\langle \cdot \rangle$ denotes expectation.

Definition 7.1.1 *We say that the good completely random measure φ is **multiplicative** if the deterministic measure $A \mapsto \langle \Delta_n^\varphi \rangle(A)$ is non-atomic for every $n \geq 2$. We show in the examples below that a Gaussian or compensated Poisson measure, with non-atomic control measure ν, is always multiplicative.*

The term "multiplicative" (which we take from [132]) originates from the fact that φ is multiplicative (in the sense of the previous definition) if and only if for every partition π the non-random measure $\left\langle \mathrm{St}_\pi^{\varphi,[n]} \right\rangle (\cdot)$ can be written as a product measure. In particular (see Proposition 8 in [132]), the completely random measure φ is multiplicative if and only if for every $\pi \in \mathcal{P}([n])$ and every $A_1, ..., A_n \in \mathcal{Z}_\nu$,

$$\left\langle \mathrm{St}_\pi^{\varphi,[n]} \right\rangle (A_1 \times \cdots \times A_n) = \prod_{b \in \pi} \left\langle \mathrm{St}_1^{\varphi,[|b|]} \right\rangle \left(\underset{j \in b}{\mathrm{X}} A_j \right), \qquad (7.1.5)$$

where, for every $b = \{j_1, ..., j_k\} \in \pi$, we used once again the notation $\underset{j \in b}{\mathrm{X}} A_j \triangleq A_{j_1} \times \cdots \times A_{j_k}$. Note that the RHS of (7.1.5) involves products over blocks of the partition π, in which there is concentration over the diagonals associated with the blocks. Thus, in view of (5.1.2) and (7.1.3), one has that

$$\left\langle \mathrm{St}_\pi^{\varphi,[n]} \right\rangle (A_1 \times \cdots \times A_n) = \prod_{b \in \pi} \left\langle \Delta_{|b|}^\varphi \right\rangle (\cap_{j \in b} A_j), \qquad (7.1.6)$$

that is, we can express the LHS of (7.1.5) as a product of measures involving sets in $\mathcal{Z}_\nu$. Observe that one can rewrite relation (7.1.5) in the following (compact) way:

$$\left\langle \mathrm{St}_\pi^{\varphi,[n]} \right\rangle = \bigotimes_{b \in \pi} \left\langle \mathrm{St}_1^{\varphi,[|b|]} \right\rangle. \qquad (7.1.7)$$

Example 7.1.2 (i) When φ is Gaussian with non-atomic control measure ν, relation (7.1.5) implies that $\left\langle \mathrm{St}_\pi^{\varphi,[n]} \right\rangle$ is 0 if π contains at least one block b such that $|b| \neq 2$. If, on the other hand, every block of π contains exactly two elements, we deduce from (5.11.82) and (5.11.84) that

$$\left\langle \mathrm{St}_\pi^{\varphi,[n]} \right\rangle (A_1 \times \cdots \times A_n) = \prod_{b=\{i,j\} \in \pi} \nu(A_i \cap A_j), \qquad (7.1.8)$$

which is not atomic. We conclude that φ is multiplicative.

(ii) If φ is a compensated Poisson measure with non-atomic control measure ν, then $\left\langle \mathrm{St}_\pi^{\varphi,[n]} \right\rangle$ is 0 whenever π contains at least one block b such that $|b| = 1$ (indeed, that block would have measure 0, since φ is centered). If, on the other hand, every block of π has more than two elements, then, by Corollary 5.12.4

$$\left\langle St_{\pi}^{\varphi,[n]} \right\rangle (A_1 \times \cdots \times A_n) = \prod_{k=2}^{n} \prod_{b=\{j_1,\ldots,j_k\}\in\pi} \nu \left(A_{j_1} \cap \cdots \cap A_{j_k} \right), \quad (7.1.9)$$

which is non-atomic. Hence, φ is multiplicative. See [132] for (quite pathological) examples of non-multiplicative measures.

Notation. In what follows, the notation

$$\int_{Z^n} f(z_1,\ldots,z_n) \bigotimes_{b\in\pi} \left\langle St_{\hat{1}}^{\varphi,[|b|]} \right\rangle (dz_1,\ldots,dz_n) \triangleq \bigotimes_{b\in\pi} \left\langle St_{\hat{1}}^{\varphi,[|b|]} \right\rangle (f) \quad (7.1.10)$$

will be used for every function $f \in \mathcal{E}(\nu^n)$. [1]

The next result gives a new universal combinatorial formula for the computation of the cumulants and the moments associated with the multiple Wiener-Itô integrals with respect to a completely random *multiplicative* measure.

Theorem 7.1.3 (Diagram formulae) *Let φ be a good completely random measure, with non-atomic control measure ν, and suppose that φ is also multiplicative in the sense of Definition 7.1.1. For every $n_1,\ldots,n_k \geq 1$, we write $n = n_1 + \cdots + n_k$, and we denote by π^* the partition of $[n] = \{1,\ldots,n\}$ given by (6.1.1). Then, for every collection of kernels $f_1,\ldots,f_k$ such that $f_j \in \mathcal{E}_{s,0}(\nu^{n_j})$, one has that*

$$\mathbb{E}\left[I_{n_1}^{\varphi}(f_1) \cdots I_{n_k}^{\varphi}(f_k) \right] = \sum_{\{\sigma\in\mathcal{P}([n]):\sigma\wedge\pi^*=\hat{0}\}} \bigotimes_{b\in\sigma} \left\langle St_{\hat{1}}^{\varphi,[|b|]} \right\rangle (f_1 \otimes_0 f_2 \otimes_0 \cdots \otimes_0 f_k),$$

$$(7.1.11)$$

and

$$\chi\left(I_{n_1}^{\varphi}(f_1),\cdots,I_{n_k}^{\varphi}(f_k) \right) = \sum_{\substack{\sigma\wedge\pi^*=\hat{0} \\ \sigma\vee\pi^*=\hat{1}}} \bigotimes_{b\in\sigma} \left\langle St_{\hat{1}}^{\varphi,[|b|]} \right\rangle (f_1 \otimes_0 f_2 \otimes_0 \cdots \otimes_0 f_k),$$

$$(7.1.12)$$

---

[1] The integral $\int_{Z^n} f\{d \bigotimes_{b\in\pi} \left\langle St_{\hat{1}}^{\varphi,[|b|]} \right\rangle\}$ in (7.1.10) is well defined, since the set function $\bigotimes_{b\in\pi} \left\langle St_{\hat{1}}^{\varphi,[|b|]} \right\rangle (\cdot)$ is a σ-additive signed measure (thanks to (5.1.10)) on the algebra generated by the products of the type $A_1 \times \cdots \times A_n$, where each A_j is in $\mathcal{Z}_\nu$.

Proof. Formula (7.1.11) is a consequence of Theorem 6.1.1 and (7.1.5). In order to prove (7.1.12), we shall first show that the following two equalities hold

$$\chi\left(I_{n_1}^{\varphi}(f_1),\cdots,I_{n_k}^{\varphi}(f_k)\right)$$

$$= \sum_{\pi^*\leq\rho=(r_1,\ldots,r_l)\in\mathcal{P}([n])} \mu\left(\rho,\hat{1}\right) \prod_{j=1}^{l} \mathbb{E}\left[\prod_{a:\{n_1+\cdots+n_{a-1}+1,\ldots,n_1+\cdots+n_a\}\subseteq r_j} I_{n_a}^{\varphi}(f_a)\right] \tag{7.1.13}$$

$$= \sum_{\pi^*\leq\rho=(r_1,\ldots,r_l)\in\mathcal{P}([n])} \mu\left(\rho,\hat{1}\right) \sum_{\substack{\gamma\leq\rho\\ \gamma\wedge\pi^*=\hat{0}}} \bigotimes_{b\in\gamma} \left\langle \mathrm{St}_{\hat{1}}^{\varphi,[|b|]} \right\rangle (f_1\otimes_0\cdots\otimes_0 f_k), \tag{7.1.14}$$

where $n_1 + n_0 = 0$ by convention.

The proof of (7.1.13) uses arguments analogous to those in the proof of Malyshev's formula (4.1.1). Indeed, one can use relation (3.2.7) to deduce that

$$\chi\left(I_{n_1}^{\varphi}(f_1),\cdots,I_{n_k}^{\varphi}(f_k)\right) = \sum_{\sigma=\{x_1,\ldots,x_l\}\in\mathcal{P}([k])} (-1)^{l-1}(l-1)! \prod_{j=1}^{l} \mathbb{E}\left[\prod_{a\in x_j} I_{n_a}^{\varphi}(f_a)\right]. \tag{7.1.15}$$

Now observe that there exists a *bijection*

$$\mathcal{P}([k]) \to \left[\pi^*,\hat{1}\right] : \sigma\mapsto\rho^{(\sigma)},$$

between $\mathcal{P}([k])$ and the segment $\left[\pi^*,\hat{1}\right]$, which is defined as the set of those $\rho\in\mathcal{P}([n])$ such that $\pi^*\leq\rho$, where π^* is given by (6.1.1). Such a bijection is realized as follows: for every $\sigma=\{x_1,\ldots,x_l\}\in\mathcal{P}([k])$, define $\rho^{(\sigma)}\in\left[\pi^*,\hat{1}\right]$ by merging two blocks

$$\{n_1+\cdots+n_{a-1}+1,\ldots,n_1+\cdots+n_a\} \qquad\text{and}$$
$$\{n_1+\cdots+n_{b-1}+1,\ldots,n_1+\cdots+n_b\}$$

of π^* ($1\leq a\neq b\leq k$) if and only if $a\sim_\sigma b$. Note that this construction implies that $|\sigma|=\left|\rho^{(\sigma)}\right|=l$, so that (2.5.23) yields

$$(-1)^{l-1}(l-1)! = \mu\left(\sigma,\hat{1}\right) = \mu\left(\rho^{(\sigma)},\hat{1}\right) \tag{7.1.16}$$

(observe that the two Möbius functions appearing in (7.1.16) refer to two different lattices of partitions). Now use the notation $\rho^{(\sigma)} = \left\{r_1^{(\sigma)},\ldots,r_l^{(\sigma)}\right\}$ to indicate the blocks of $\rho^{(\sigma)}$: since, by construction,

$$\prod_{j=1}^{l} \mathbb{E}\left[\prod_{a\in x_j} I_{n_a}^{\varphi}(f_a)\right] = \prod_{j=1}^{l} \mathbb{E}\left[\prod_{a:\{n_1+\cdots+n_{a-1}+1,\ldots,n_1+\cdots+n_a\}\subseteq r_j^{(\sigma)}} I_{n_a}^{\varphi}(f_a)\right], \tag{7.1.17}$$

we immediately obtain (7.1.13) by plugging (7.1.16) and (7.1.17) into (7.1.15).

To prove (7.1.14), fix $\rho = \{r_1, ..., r_l\}$ such that $\pi^* \le \rho$. For $j = 1, ..., l$, we write $\pi^*(j)$ to indicate the partition of the block r_j whose blocks are the sets $\{n_1 + \cdots + n_{a-1} + 1, ..., n_1 + \cdots + n_a\}$ such that

$$\{n_1 + \cdots + n_{a-1} + 1, ..., n_1 + \cdots + n_a\} \subseteq r_j. \tag{7.1.18}$$

According to (7.1.11),

$$
\mathbb{E}\left[\prod_{a:\{n_1+\cdots+n_{a-1}+1,...,n_1+\cdots+n_a\}\subseteq r_j} I_{n_a}^\varphi (f_a) \right]
$$

$$
= \sum_{\{\sigma \in \mathcal{P}(r_j): \sigma \wedge \pi^*(j) = \hat{0}\}} \bigotimes_{b \in \sigma} \left\langle \mathrm{St}\,_{\hat{1}}^{\varphi,[|b|]} \right\rangle (\{\otimes_{r_j,0} f\}),
$$

where the function $\{\otimes_{r_j,0} f\}$ is obtained by juxtaposing the $|\pi^*(j)|$ functions f_a such that the index a verifies (7.1.18). Now observe that $\gamma \in \mathcal{P}([n])$ satisfies

$$\gamma \le \rho \quad \text{and} \quad \gamma \wedge \pi^* = \hat{0},$$

if and only if γ admits a (unique) representation as a union of the type

$$\gamma = \bigcup_{j=1}^{l} \sigma(j),$$

where each $\sigma(j)$ is an element of $\mathcal{P}(r_j)$ such that $\sigma(j) \wedge \pi^*(j) = \hat{0}$. This yields

$$
\prod_{j=1}^{l} \sum_{\{\sigma \in \mathcal{P}(r_j): \sigma \wedge \pi^*(j) = \hat{0}\}} \bigotimes_{b \in \sigma} \left\langle \mathrm{St}\,_{\hat{1}}^{\varphi,[|b|]} \right\rangle (\{\otimes_{r_j,0} f\})
$$

$$
= \sum_{\substack{\gamma \le \rho \\ \gamma \wedge \pi^* = \hat{0}}} \bigotimes_{b \in \gamma} \left\langle \mathrm{St}_{\hat{1}}^{\varphi,[|b|]} \right\rangle (f_1 \otimes_0 \cdots \otimes_0 f_k).
$$

This relation, together with (7.1.16) and (7.1.17), shows that (7.1.13) implies (7.1.14). To conclude the proof, just observe that, by inverting the order of summation in (7.1.14), one obtains that

$$
\chi \left(I_{n_1}^\varphi (f_1), \cdots, I_{n_k}^\varphi (f_k) \right) = \sum_{\gamma \wedge \pi^* = \hat{0}} \bigotimes_{b \in \gamma} \left\langle \mathrm{St}_{\hat{1}}^{\varphi,[|b|]} \right\rangle (f_1 \otimes_0 \cdots \otimes_0 f_k) \sum_{\pi^* \vee \gamma \le \rho \le \hat{1}} \mu \left(\rho, \hat{1} \right)
$$

$$
= \sum_{\substack{\gamma \wedge \pi^* = \hat{0} \\ \pi^* \vee \gamma = \hat{1}}} \bigotimes_{b \in \gamma} \left\langle \mathrm{St}_{\hat{1}}^{\varphi,[|b|]} \right\rangle (f_1 \otimes_0 \cdots \otimes_0 f_k),
$$

where the last equality is a consequence of the relation

$$
\sum_{\pi^* \vee \gamma \le \rho \le \hat{1}} \mu \left(\rho, \hat{1} \right) = \begin{cases} 1 & \text{if } \pi^* \vee \gamma = \hat{1} \\ 0 & \text{otherwise}, \end{cases}
$$

which is in turn a special case of (2.6.33).   ∎

Remark. Observe that the only difference between the moment formula (7.1.11) and the cumulant formula (7.1.12) is that in the first case the sum is over all σ such that $\sigma \wedge \pi^* = \hat{0} = \{\{1\}, ..., \{n\}\}$, and that in the second case σ must satisfy in addition that $\sigma \vee \pi^* = \hat{1} = \{[n]\}$.

Moreover, the relations (7.1.11) and (7.1.12) can be restated in terms of diagrams by rewriting the sums as

$$\sum_{\{\sigma \in \mathcal{P}([n]): \sigma \wedge \pi^* = \hat{0}\}} = \sum_{\sigma \in \mathcal{P}([n]): \Gamma(\pi^*, \sigma) \text{ is non-flat}} \quad ; \quad \sum_{\substack{\sigma \wedge \pi^* = \hat{0} \\ \sigma \vee \pi^* = \hat{1}}} = \sum_{\substack{\sigma \in \mathcal{P}([n]): \Gamma(\pi^*, \sigma) \text{ is non-flat} \\ \text{and connected}}} ,$$

where $\Gamma(\pi^*, \sigma)$ is the diagram of (π^*, σ), as defined in Section 4.1.

7.2 MSets and MZeroSets

We shall now provide a version of Theorem 7.1.3 in the case where φ is, respectively, Gaussian and Poisson. For convenience, using the same notation as that in (6.1.1), let $n_1, ..., n_k \geq 1$ be such that $n_1 + \cdots + n_k = n$ and

$$\pi^* = \{\{1, ..., n_1\}, \{n_1 + 1, ..., n_1 + n_2\}, ..., \{n_1 + ... + n_{k-1} + 1, ..., n\}\} \in \mathcal{P}([n]).$$

Now define

$$\mathcal{M}([n], \pi^*) \triangleq \left\{\sigma \in \mathcal{P}([n]) : \sigma \vee \pi^* = \hat{1} \text{ and } \sigma \wedge \pi^* = \hat{0}\right\} \quad (7.2.19)$$

$$\mathcal{M}^0([n], \pi^*) \triangleq \left\{\sigma \in \mathcal{P}([n]) : \sigma \wedge \pi^* = \hat{0}\right\} \quad (7.2.20)$$

and

$$\mathcal{M}_2([n], \pi^*) \triangleq \{\sigma \in \mathcal{M}([n], \pi^*) : |b| = 2, \forall b \in \sigma\} \quad (7.2.21)$$

$$\mathcal{M}_2^0([n], \pi^*) \triangleq \{\sigma \in \mathcal{M}^0([n], \pi^*) : |b| = 2, \forall b \in \sigma\} \quad (7.2.22)$$

$$\mathcal{M}_{\geq 2}([n], \pi^*) \triangleq \{\sigma \in \mathcal{M}([n], \pi^*) : |b| \geq 2, \forall b \in \sigma\} \quad (7.2.23)$$

$$\mathcal{M}_{\geq 2}^0([n], \pi^*) \triangleq \{\sigma \in \mathcal{M}^0([n], \pi^*) : |b| \geq 2, \forall b \in \sigma\}. \quad (7.2.24)$$

Note that these definitions do not involve the specific form of $\pi^* \in \mathcal{P}([n])$. By using the formalism of diagrams Γ and multigraphs $\hat{\Gamma}$ introduced in Chapter 4, one has that

$$\mathcal{M}([n], \pi^*) = \{\sigma \in \mathcal{P}([n]) : \Gamma(\pi^*, \sigma) \text{ is connected and non-flat}\} \quad (7.2.25)$$

$$\mathcal{M}^0([n], \pi^*) = \{\sigma \in \mathcal{P}([n]) : \Gamma(\pi^*, \sigma) \text{ is connected}\} \quad (7.2.26)$$

$$\mathcal{M}_2([n], \pi^*) = \{\sigma \in \mathcal{P}([n]) : \Gamma(\pi^*, \sigma) \text{ is connected, non-flat and Gaussian}\} \quad (7.2.27)$$

$$= \left\{\sigma \in \mathcal{P}([n]) : \hat{\Gamma}(\pi^*, \sigma) \text{ is connected and has no loops}\right\}$$

$$\mathcal{M}_2^0([n], \pi^*) = \{\sigma \in \mathcal{P}([n]) : \Gamma(\pi^*, \sigma) \text{ is non-flat and Gaussian}\} \quad (7.2.28)$$

$$= \left\{\sigma \in \mathcal{P}([n]) : \hat{\Gamma}(\pi^*, \sigma) \text{ has no loops}\right\}.$$

Clearly,

$$\mathcal{M}_2\left([n],\pi^*\right) \subset \mathcal{M}_2^0\left([n],\pi^*\right),$$
$$\mathcal{M}_2\left([n],\pi^*\right) \subset \mathcal{M}_{\geq 2}\left([n],\pi^*\right),$$
$$\mathcal{M}_2^0\left([n],\pi^*\right) \subset \mathcal{M}_{\geq 2}^0\left([n],\pi^*\right).$$

The sets $\mathcal{M}_2\left([n],\pi^*\right)$ and $\mathcal{M}_2^0\left([n],\pi^*\right)$ appear in diagram formulae when φ is Gaussian (Section 7.3). The sets $\mathcal{M}_{\geq 2}\left([n],\pi^*\right)$ and $\mathcal{M}_{\geq 2}^0\left([n],\pi^*\right)$ appear when φ is the compensated Poisson measure $\widehat{N}$ (Section 7.4).

7.3 The Gaussian case

The following corollary to Theorem 7.1.3 provides the diagram formulae in the case where φ is a Gaussian measure.

Corollary 7.3.1 (Gaussian measures) *Suppose $\varphi = G$ is a centered Gaussian measure with non-atomic control measure ν, fix integers $n_1, ..., n_k \geq 1$ and let $n = n_1 + \cdots + n_k$. Write π^* for the partition of $[n]$ appearing in (6.1.1). Then, for any vector of functions $(f_1, ..., f_k)$ such that $f_j \in L_s^2(\nu^{n_j})$, $j = 1, ..., k$, the following relations hold:*

1. *if $\mathcal{M}_2\left([n],\pi^*\right) = \varnothing$ (in particular, if n is odd), then $\chi\left(I_{n_1}^G(f_1), \cdots, I_{n_k}^G(f_k)\right) = 0$;*

2. *if $\mathcal{M}_2\left([n],\pi^*\right) \neq \varnothing$, then*

$$\chi\left(I_{n_1}^G(f_1), \cdots, I_{n_k}^G(f_k)\right) = \sum_{\sigma \in \mathcal{M}_2([n],\pi^*)} \int_{Z^{n/2}} f_{\sigma,k} d\nu^{n/2}, \qquad (7.3.29)$$

where, for every $\sigma \in \mathcal{M}_2\left([n],\pi^*\right)$, the function $f_{\sigma,k}$, of $n/2$ variables, is obtained by identifying the variables x_i and x_j in the argument of $f_1 \otimes_0 \cdots \otimes_0 f_k$ (as given in (6.1.3)) if and only if $i \sim_\sigma j$;

3. *if $\mathcal{M}_2^0\left([n],\pi^*\right) = \varnothing$ (in particular, if n is odd), then $\mathbb{E}\left(I_{n_1}^G(f_1) \cdots I_{n_k}^G(f_k)\right) = 0$;*

4. *if $\mathcal{M}_2^0\left([n],\pi^*\right) \neq \varnothing$,*

$$\mathbb{E}\left(I_{n_1}^G(f_1) \cdots I_{n_k}^G(f_k)\right) = \sum_{\sigma \in \mathcal{M}_2^0([n],\pi^*)} \int_{Z^{n/2}} f_{\sigma,k} d\nu^{n/2}. \qquad (7.3.30)$$

Remark. As noted in (6.1.3), one has that

$$f_1 \otimes_0 \cdots \otimes_0 f_k(x_1, ..., x_n)$$
$$= f_1(x_1, ..., x_{n_1}) \times f_2(x_{n_1+1}, ..., x_{n_1+n_2}) \times \cdots \times f_k(x_{n_1+\cdots+n_{k-1}+1}, ..., x_n).$$
$$(7.3.31)$$

One obtains the function $f_{\sigma,k}$, of $n/2$ variables, by identifying in (7.3.31) the variables x_i and x_j if and only if $i \sim_\sigma j$.

Example 7.3.2 Consider the function of $n = 6$ variables given by

$$f_1 \otimes_0 f_2 \otimes_0 f_3(x_1, x_2, x_3, x_4, x_5, x_6) = f_1(x_1, x_2, x_3) f_2(x_4) f_3(x_5, x_6)$$

for which $\pi^* = \{\{1,2,3\}, \{4\}, \{5,6\}\}$ is the π in Fig. 4.7. If $\sigma = \{\{1,4\}, \{2,5\}, \{3,6\}\}$ as in Fig. 4.7, then

$$f_{\sigma,3} = f_1(x_1, x_2, x_3) f_2(x_1) f_3(x_2, x_3).$$

Proof of Corollary 7.3.1. First observe that, since $\varphi = G$ is Gaussian, then $\left\langle \text{St}_{\hat{1}}^{G,[|b|]} \right\rangle \equiv 0$ whenever $|b| \neq 2$. Assume for the moment that $f_j \in \mathcal{E}_{s,0}(\nu^{n_j})$, $j = 1, ..., k$. In this case, we can apply formula (7.1.12) and obtain that

$$\chi \left(I_{n_1}^G(f_1), \cdots, I_{n_k}^G(f_k) \right)$$

$$= \sum_{\substack{\{\sigma:\sigma \wedge \pi^* = \hat{0} \,; \\ \sigma \vee \pi^* = \hat{1} \,; \, |b|=2 \,\, \forall b \in \sigma\}}} \bigotimes_{b \in \sigma} \left\langle \text{St}_{\hat{1}}^{G,[|b|]} \right\rangle (f_1 \otimes_0 f_2 \otimes_0 \cdots \otimes_0 f_k)$$

$$= \sum_{\sigma \in \mathcal{M}_2([n],\pi^*)} \bigotimes_{b \in \sigma} \left\langle \text{St}_{\hat{1}}^{G,[|b|]} \right\rangle (f_1 \otimes_0 f_2 \otimes_0 \cdots \otimes_0 f_k),$$

where we have used (7.2.21). The last relation trivially implies Point 1 in the statement. Moreover, since, for every $B, C \in \mathcal{Z}_\nu$,

$$\left\langle \text{St}_{\hat{1}}^{G,[2]} \right\rangle (B \times C) = \left\langle \text{St}_{\hat{1}}^{G,[2]} \right\rangle ((B \cap C) \times (B \cap C)) = \left\langle \Delta_2^G \right\rangle (B \cap C),$$

$$(7.3.32)$$

one deduces immediately that the support of the deterministic measure $\bigotimes_{b \in \sigma} \left\langle \text{St}_{\hat{1}}^{G,[|b|]} \right\rangle$ is contained in the set

$$Z_{\geq \sigma}^n = \{(z_1, ..., z_n) : z_i = z_j \text{ for every } i, j \text{ such that } i \sim_\sigma j\}.$$

Since, by (5.11.82) and (7.3.32),

$$\left\langle \text{St}_{\hat{1}}^{G,[|b|]} \right\rangle (B \times C) = \nu(B \cap C), \qquad (7.3.33)$$

for every $B, C \in \mathcal{Z}_\nu$, we infer that

$$\bigotimes_{b \in \sigma} \left\langle \text{St}_{\hat{1}}^{G,[|b|]} \right\rangle (f_1 \otimes_0 f_2 \otimes_0 \cdots \otimes_0 f_k) = \bigotimes_{b \in \sigma} \left\langle \text{St}_{\hat{1}}^{G,[|b|]} \right\rangle (f_\sigma) = \int_{Z^{n/2}} f_{\sigma,k} d\nu^{n/2}.$$

$$(7.3.34)$$

where the function $f_{\sigma,k}$ is defined in the statement. To obtain the last equality in (7.3.34), one should start with functions f_j of type

$$f_j(z_1, ..., z_{n_j}) = \mathbf{1}_{C_1^{(j)} \times \cdots \times C_{n_j}^{(j)}}(z_1, ..., z_{n_j}),$$

where the $C_\ell^{(j)} \in \mathcal{Z}_\nu$ are disjoint, and then apply formula (7.1.8), so that the extension to general functions $f_j \in \mathcal{E}_{s,0}(\nu^{n_j})$ is obtained by the multilinearity of the application

$$(f_1, ..., f_k) \mapsto \int_{Z^{n/2}} f_{\sigma,k} d\nu^{n/2}.$$

To obtain (7.3.29) for general functions $f_1, ..., f_k$ such that $f_j \in L_s^2(\nu^{n_j})$, start by observing that $\mathcal{E}_{s,0}(\nu^{n_j})$ is dense in $L_s^2(\nu^{n_j})$, and then use the fact that, if a sequence $f_1^{(r)}, ..., f_k^{(r)}$, $r \geq 1$, is such that $f_j^{(r)} \in \mathcal{E}_{s,0}(\nu^{n_j})$ and $f_j^{(r)} \to f_j$ in $L_s^2(\nu^{n_j})$ ($j = 1, ..., k$), then

$$\chi\left(I_{n_1}^G\left(f_1^{(r)}\right), \cdots, I_{n_k}^G\left(f_k^{(r)}\right)\right) \to \chi\left(I_{n_1}^G(f_1), \cdots, I_{n_k}^G(f_k)\right),$$

by (5.9.79), and moreover

$$\int_{Z^{n/2}} f_{\sigma,k}^{(r)} d\nu^{n/2} \to \int_{Z^{n/2}} f_{\sigma,k} d\nu^{n/2},$$

where $f_{\sigma,k}^{(r)}$ is constructed from $f_1^{(r)}, ..., f_k^{(r)}$, as specified in the statement (a similar argument was needed in the proof of Proposition 6.4.1). Points 3 and 4 in the statement are obtained analogously, by using (7.1.11) and thus

$$\mathbb{E}\left(I_{n_1}^G(f_1) \cdots I_{n_k}^G(f_k)\right)$$
$$= \sum_{\substack{\{\sigma:\sigma \wedge \pi^* = \hat{0}\,;\\ |b|=2\ \forall b \in \sigma\}}} \bigotimes_{b \in \sigma} \left\langle \mathrm{St}_{\hat{1}}^{G,[|b|]} \right\rangle (f_1 \otimes_0 f_2 \otimes_0 \cdots \otimes_0 f_k)$$
$$= \sum_{\sigma \in \mathcal{M}_2^0([n],\pi^*)} \bigotimes_{b \in \sigma} \left\langle \mathrm{St}_{\hat{1}}^{G,[|b|]} \right\rangle (f_1 \otimes_0 f_2 \otimes_0 \cdots \otimes_0 f_k),$$

and then by applying the same line of reasoning as above.   ∎

Example 7.3.3   (i) We want to use Corollary 7.3.1 to compute the cumulant of the two integrals

$$I_{n_1}^G(f_1) = \int_{Z_{\hat{0}}^{n_1}} f_1(z_1, ..., z_{n_1}) G(dz_1) \cdots G(dz_{n_1})$$

$$I_{n_2}^G(f_2) = \int_{Z_{\hat{0}}^{n_2}} f_2(z_1, ..., z_{n_2}) G(dz_1) \cdots G(dz_{n_2}),$$

that is, the quantity

$$\chi\left(I_{n_1}^G(f_1), I_{n_2}^G(f_2)\right) = \mathbb{E}\left(I_{n_1}^G(f_1) I_{n_2}^G(f_2)\right).$$

Here, $\pi^* \in \mathcal{P}([n_1 + n_2])$ is given by

$$\pi^* = \{\{1, ..., n_1\}, \{n_1 + 1, ..., n_1 + n_2\}\}.$$

It is easily seen that

$$M_2\left(\left[n_1 + n_2\right], \pi^*\right) \neq \emptyset \quad \text{if and only if} \quad n_1 = n_2.$$

Indeed, each partition $M_2\left(\left[n_1 + n_2\right], \pi^*\right)$ is of the form

$$\sigma = \{\{i_1, i_2\} : i_1 \in \{1, ..., n_1\}, i_2 \in \{n_1 + 1, ..., n_1 + n_2\}\} \quad (7.3.35)$$

(this is the case because σ must have blocks of size $|b| = 2$ only, and no blocks can be constructed using only the indices $\{1, ..., n_1\}$ or $\{n_1 + 1, ..., n_1 + n_2\}$, since the corresponding diagram must be non-flat). In the case where $n_1 = n_2$, there are exactly $n_1!$ partitions as in (7.3.35), since to each element in $\{1, ..., n_1\}$ one attaches one element of $\{n_1 + 1, ..., n_1 + n_2\}$. Moreover, for any such σ one has that

$$\int_{Z^{n/2}} f_{\sigma, 2}\, d\nu^{n/2} = \int_{Z^{n_1}} f_1 f_2\, d\nu^{n_1}, \quad (7.3.36)$$

where $n = n_1 + n_2$ and we have used the symmetry of f_1 and f_2 to obtain that

$$f_{\sigma, 2}\left(z_1, ..., z_{\frac{n}{2}}\right) = f_{\sigma, 2}\left(z_1, ..., z_{n_1}\right) = f_1\left(z_1, ..., z_{n_1}\right) f_2\left(z_1, ..., z_{n_1}\right).$$

From (7.3.30) and (7.3.36), we deduce that

$$\mathbb{E}\left(I_{n_1}^G\left(f_1\right) I_{n_2}^G\left(f_2\right)\right) = 1_{n_1 = n_2} \times n_1! \int_{Z^{n_1}} f_1 f_2\, d\nu^{n_1},$$

as expected (see (5.5.62)). Note also that, since every diagram associated with π^* has two rows, one also has

$$M_2\left(\left[n_1 + n_2\right], \pi^*\right) = M_2^0\left(\left[n_1 + n_2\right], \pi^*\right),$$

that is, every non-flat diagram is also connected, thus yielding (thanks to (7.3.29) and (7.3.30))

$$\chi\left(I_{n_1}^G\left(f_1\right), I_{n_2}^G\left(f_2\right)\right) = \mathbb{E}\left(I_{n_1}^G\left(f_1\right) I_{n_2}^G\left(f_2\right)\right).$$

(ii) We fix an integer $k \geq 3$ and set $n_1 = ... = n_k = 1$, that is, we focus on functions $f_j, j = 1, ..., k$, of one variable, so that the integral $I_1^G\left(f_j\right)$ is Gaussian for every j, and we consider $\chi\left(I_1^G\left(f_1\right), ..., I_1^G\left(f_k\right)\right)$ and $\mathbb{E}\left[I_1^G\left(f_1\right) ... I_1^G\left(f_k\right)\right]$. In this case, $n_1 + \cdots + n_k = k$, and $\pi^* = \{\{1\}, ..., \{k\}\} = \hat{0}$. For instance, for $k = 6$, π^* is represented in Fig. 7.1. In that case $M_2\left(\left[k\right], \pi^*\right) = \emptyset$, because all diagrams will be disconnected. One of such diagrams is represented in Fig. 7.2. (**Exercise**: give an algebraic proof of the fact that $M_2\left(\left[k\right], \pi^*\right) = \emptyset$). It follows from Point 1 of Corollary 7.3.1 that

$$\chi\left(I_1^G\left(f_1\right), ..., I_1^G\left(f_k\right)\right) = 0$$

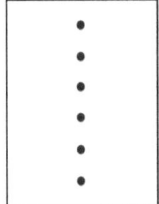

Fig. 7.1. A representation of the partition $\hat{0}$

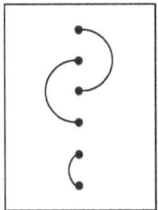

Fig. 7.2. A disconnected Gaussian diagram

(this is consistent with the properties of cumulants of Gaussian vectors noted in Chapter 3).

Now focus on $\mathcal{M}_2^0([k], \pi^*)$. Recall that, according to (7.2.28), a partition $\sigma \in \mathcal{P}([k])$ is an element of $\mathcal{M}_2^0([k], \pi^*)$ if and only if the diagram $\Gamma(\pi^*, \sigma)$ (see Section 4.1) is Gaussian and non-flat. If k is odd the class $\mathcal{M}_2^0([k], \pi^*)$ is empty, and, for k even, $\mathcal{M}_2^0([k], \pi^*)$ coincides with the collection of all partitions

$$\sigma = \left\{ \{i_1, j_1\}, \ldots, \left\{ i_{\frac{k}{2}}, j_{\frac{k}{2}} \right\} \right\} \in \mathcal{P}([k]) \tag{7.3.37}$$

whose blocks have size two (that is, $\mathcal{M}_2^0([k], \pi^*)$ is the class of all perfect matchings of the first k integers). For σ as in (7.3.37), we have

$$f_{\sigma,k}\left(z_1, \ldots, z_{\frac{k}{2}}\right) = \prod_{\substack{\{i_l, j_l\} \in \sigma \\ l = 1, \ldots, k/2}} f_{i_l}(z_l)\, f_{j_l}(z_l).$$

Points 3 and 4 of Corollary 7.3.1 yield therefore

$$\mathbb{E}\left(I_1^G(f_1) \cdots I_1^G(f_k) \right)$$
$$= \begin{cases} \sum_{\sigma = \{\{i_1, j_1\}, \ldots, \{i_{k/2}, j_{k/2}\}\} \in \mathcal{P}([k])} \int_Z f_{i_1} f_{j_1}\, d\nu \cdots \int_Z f_{i_{k/2}} f_{j_{k/2}}\, d\nu, & k \text{ even} \\ 0, & k \text{ odd}. \end{cases}$$

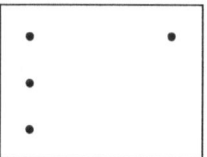

Fig. 7.3. A three-block partition

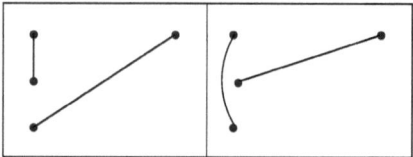

Fig. 7.4. The two elements of $\mathcal{M}_2([4], \pi^*)$

which is just a special case of (3.2.21), since $\mathbb{E}\left(I_1^G\left(f_i\right) I_1^G\left(f_j\right)\right) = \int_Z f_{i_1} f_{j_1} d\nu$. For instance, if $k = 4$, one has that

$$\mathbb{E}\left(I_1^G\left(f_1\right) \cdots I_1^G\left(f_4\right)\right) = \int_Z f_1 f_2 d\nu \times \int_Z f_3 f_4 d\nu + \int_Z f_1 f_3 d\nu \times \int_Z f_2 f_4 d\nu$$
$$\int_Z f_1 f_4 d\nu \times \int_Z f_2 f_3 d\nu.$$

(iii) Consider the case $k = 3$, $n_1 = 2$, $n_2 = n_3 = 1$. Here, $n = n_1 + n_2 + n_3 = 4$, and

$$\pi^* = \{\{1, 2\}, \{3\}, \{4\}\}.$$

The partition π^* is represented in Fig. 7.3.
The class $\mathcal{M}_2([4], \pi^*)$ contains only two elements, namely

$$\sigma_1 = \{\{1, 3\}, \{2, 4\}\} \quad \text{and} \quad \sigma_2 = \{\{1, 4\}, \{2, 3\}\},$$

whose diagrams are given in Fig. 7.4.
Since the rows of these diagrams cannot be divided into two subsets (see Section 4.1), they are connected, and one has $\mathcal{M}_2([4], \pi^*) = \mathcal{M}_2^0([4], \pi^*)$, that is, cumulants equal moments by (7.3.29) and (7.3.30). Moreover,

$$f_{\sigma_1, 3}(z_1, z_2) = f_1(z_1, z_2) f_2(z_1) f_3(z_2)$$
$$f_{\sigma_2, 3}(z_1, z_2) = f_1(z_1, z_2) f_2(z_2) f_3(z_1).$$

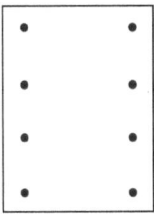

Fig. 7.5. A four-block partition

It follows that

$$\chi\left(I_2^G\left(f_1\right), I_1^G\left(f_2\right), I_1^G\left(f_3\right)\right) = \mathbb{E}\left(I_2^G\left(f_1\right) I_1^G\left(f_2\right) I_1^G\left(f_3\right)\right)$$

$$= \int_{Z^2}\{f_{\sigma_1,3}\left(z_1, z_2\right) + f_{\sigma_2,3}\left(z_1, z_2\right)\}\nu^2\left(dz_1, dz_2\right)$$

$$= 2\int_{Z^2} f_1\left(z_1, z_2\right) f_2\left(z_1\right) f_3\left(z_2\right) \nu^2\left(dz_1, dz_2\right),$$

where in the last equality we have used the symmetry of f_1.

(iv) We want to use Point 1 and 2 of Corollary 7.3.1 to compute the kth cumulant

$$\chi_k\left(I_2^G\left(f\right)\right) = \chi(\underbrace{I_2^G\left(f\right), ..., I_2^G\left(f\right)}_{k \text{ times.}}),$$

for every $k \geq 3$, that is, cumulants of multiple integrals of order 2. This can be done by specializing formula (7.3.29) to the case: $k \geq 3$ and $n_1 = n_2 = ... = n_k = 2$. Here, $n = 2k$ and

$$\pi^* = \{\{1, 2\}, \{3, 4\}, ..., \{2k - 1, 2k\}\};$$

for instance, for $k = 4$ the partition π^* can be represented as in Fig. 7.5.
Now consider the set $\mathcal{M}_2\left([2k], \pi^*\right)$. Recall that the blocks of its partitions have only two elements. It contains for example the partition

$$\sigma^* = \{\{1, 2k\}, \{2, 3\}, \{4, 5\}, ..., \{2k - 2, 2k - 1\}\} \in \mathcal{P}\left([2k]\right).$$

For $k = 4$, one has

$$\sigma^* = \{\{1, 8\}, \{2, 3\}, \{4, 5\}, \{6, 7\}\},$$

whose diagram appears in Fig. 7.6.
Note that such a diagram is circular, and that the corresponding multigraph looks like the one in Fig. 4.17. Therefore,

$$f_{\sigma^*,k}\left(z_1, ..., z_k\right) = f\left(z_1, z_2\right) f\left(z_2, z_3\right) \cdots f\left(z_{k-1}, z_k\right) f\left(z_k, z_1\right). \quad (7.3.38)$$

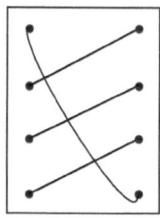

Fig. 7.6. A circular diagram with four rows

It is not difficult to see that $\mathcal{M}_2\left([2k],\pi^*\right)$ contains exactly $2^{k-1}\left(k-1\right)!$ elements and that the diagram $\Gamma\left(\pi^*,\sigma\right)$ associated to each $\sigma\in\mathcal{M}_2\left([2k],\pi^*\right)$ is equivalent (up to a permutation, or equivalently to a renumbering of the rows) to a circular diagram (see Section 4.1). It follows that, for every $\sigma\in\mathcal{M}_2\left([2k],\pi^*\right)$, one has that

$$f_{\sigma,k}\left(z_1,...,z_k\right)=f_{\sigma^*,k}\left(z_1,...,z_k\right),$$

where $f_{\sigma^*,2k}$ is given in (7.3.38). This yields the classic formula (see, for example, [31]):

$$\chi_k\left(I_2^G\left(f\right)\right) \tag{7.3.39}$$
$$=2^{k-1}\left(k-1\right)!\int_{Z^k}f\left(z_1,z_2\right)f\left(z_2,z_3\right)\cdots$$
$$\cdots f\left(z_{k-1},z_k\right)f\left(z_k,z_1\right)\nu\left(dz_1\right)\cdots\nu\left(dz_k\right).$$

7.4 The Poisson case

The following result provides diagram formulae for Wiener-Itô integrals with respect to compensated Poisson measures. It is stated only for elementary functions, so as not to have to deal with convergence issues. The proof is similar to the one of Corollary 7.3.1, and it is only sketched. We let $\mathcal{M}_{\geq2}\left([n],\pi^*\right)$ and $\mathcal{M}_{\geq2}^0\left([n],\pi^*\right)$ be defined as in Section 7.3.

Corollary 7.4.1 (Poisson measures) *Suppose* $\varphi=\widehat{N}$ *is a centered Poisson measure with non-atomic control measure* ν, *fix integers* $n_1,...,n_k\geq1$ *and let* $n=n_1+\cdots+n_k$. *Write* π^* *for the partition of* $[n]$ *appearing in (6.1.1). Then, for any vector of functions* $\left(f_1,...,f_k\right)$ *such that* $f_j\in\mathcal{E}_{s,0}\left(\nu^{n_j}\right)$, $j=1,...,k$, *the following relations hold:*

1. if $\mathcal{M}_{\geq2}\left([n],\pi^*\right)=\varnothing$, *then* $\chi\left(I_{n_1}^{\widehat{N}}\left(f_1\right),\cdots,I_{n_k}^{\widehat{N}}\left(f_k\right)\right)=0$;

2. *if $\mathcal{M}_{\geq 2}([n], \pi^*) \neq \varnothing$, then*

$$\chi\left(I_{n_1}^{\widehat{N}}(f_1), \cdots, I_{n_k}^{\widehat{N}}(f_k)\right) = \sum_{\sigma \in \mathcal{M}_{\geq 2}([n], \pi^*)} \int_{Z^{|\sigma|}} f_{\sigma, k} d\nu^{|\sigma|}, \qquad (7.4.40)$$

where, for every $\sigma \in \mathcal{M}_{\geq 2}([n], \pi^*)$, the function $f_{\sigma, k}$, in $|\sigma|$ variables, is obtained by identifying the variables x_i and x_j in the argument of $f_1 \otimes_0 \cdots \otimes_0 f_k$ (as defined in (6.1.3)) if and only if $i \sim_\sigma j$;

3. *if $\mathcal{M}_{\geq 2}^0([n], \pi^*) = \varnothing$, then $\mathbb{E}\left(I_{n_1}^{\widehat{N}}(f_1) \cdots I_{n_k}^{\widehat{N}}(f_k)\right) = 0$;*

4. *if $\mathcal{M}_{\geq 2}^0([n], \pi^*) \neq \varnothing$,*

$$\mathbb{E}\left(I_{n_1}^{\widehat{N}}(f_1) \cdots I_{n_k}^{\widehat{N}}(f_k)\right) = \sum_{\sigma \in \mathcal{M}_{\geq 2}^0([n], \pi^*)} \int_{Z^{|\sigma|}} f_{\sigma, k} d\nu^{|\sigma|}. \qquad (7.4.41)$$

Sketch of the Proof. The proof follows closely that of Corollary 7.3.1. The only difference is in evaluating (7.1.6). Instead of having (7.1.8) which requires considering $\mathcal{M}_2$ and $\mathcal{M}_2^0$, one has (7.1.9), which implies that one must use $\mathcal{M}_{\geq 2}$ and $\mathcal{M}_{\geq 2}^0$. ∎

Remark. Corollaries 7.3.1 and 7.4.1 are quite similar. In the Poisson case, however, $f_{\sigma, k}$ depends on $|\sigma|$ variables, whereas in the Gaussian case it depends on $n/2$ variables.

All kernels appearing in the following examples are symmetric, elementary and vanishing on diagonals. This ensures that multiple integrals have moments of all orders, because they are sums of products of independent Poisson random variables (for infinitely divisible random measures, however, one needs additional moment conditions on the Lévy-Khintchine measure, as noted below).

Example 7.4.2 We apply Corollary 7.4.1 in order to compute the cumulant

$$\chi\left(I_{n_1}^{\widehat{N}}(f_1), I_{n_2}^{\widehat{N}}(f_2)\right) = \mathbb{E}\left(I_{n_1}^{\widehat{N}}(f_1) I_{n_2}^{\widehat{N}}(f_2)\right),$$

where $n_1, n_2 \geq 1$ are arbitrary. In this case, $\pi^* \in \mathcal{P}([n_1 + n_2])$ is given by

$$\pi^* = \{\{1, ..., n_1\}, \{n_1 + 1, ..., n_1 + n_2\}\}.$$

Moreover,

$$\mathcal{M}_2^0([n_1 + n_2], \pi^*) = \mathcal{M}_2([n_1 + n_2], \pi^*)$$
$$= \mathcal{M}_{\geq 2}([n_1 + n_2], \pi^*) = \mathcal{M}_{\geq 2}^0([n_1 + n_2], \pi^*)$$

(indeed, since any diagram of π^* is composed of two rows, every non-flat diagram must be necessarily connected and Gaussian). This gives, in particular,

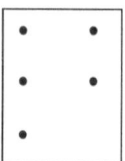

Fig. 7.7. A three-row partition

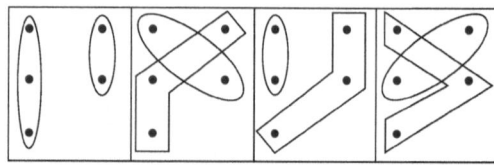

Fig. 7.8. The four elements of $\mathcal{M}_{\geq 2}([5], \pi^*)$

$\mathcal{M}_{\geq 2}([n_1 + n_2], \pi^*) \neq \varnothing$ if and only if $n_1 = n_2$. The computations performed in the Gaussian case thus apply and therefore yield

$$\chi\left(I_{n_1}^{\widehat{N}}(f_1), I_{n_2}^{\widehat{N}}(f_2)\right) = \mathbb{E}\left(I_{n_1}^{\widehat{N}}(f_1) I_{n_2}^{\widehat{N}}(f_2)\right) = \mathbf{1}_{n_1 = n_2} \times n_1! \int_{Z^{n_1}} f_1 f_2 d\nu^{n_1},$$

which is once again consistent with (5.5.62).

Example 7.4.3 Consider the case $k = 3$, $n_1 = n_2 = 2$, $n_3 = 1$. Here, $n = n_1 + n_2 + n_3 = 5$, and $\pi^* = \{\{1,2\}, \{3,4\}, \{5\}\}$. The partition π^* can be represented as in Fig. 7.7.

The class $\mathcal{M}_{\geq 2}^0([5], \pi^*)$, of σ's such that $\sigma \wedge \pi^* = \hat{0}$, contains four elements, that is,

$$\sigma_1 = \{\{1,3,5\}, \{2,4\}\}, \quad \sigma_2 = \{\{1,4\}, \{2,3,5\}\}$$
$$\sigma_3 = \{\{1,3\}, \{2,4,5\}\} \quad \text{and} \quad \sigma_4 = \{\{1,4,5\}, \{2,3\}\},$$

whose diagrams are given in Fig. 7.8.

Since all these diagrams are connected, the class $\mathcal{M}_{\geq 2}([5], \pi^*)$ coincides with $\mathcal{M}_{\geq 2}^0([5], \pi^*)$. Note also that, since the above diagrams have an odd number of vertices, $\mathcal{M}_{\geq 2}([5], \pi^*)$ does not contain partitions σ whose diagram is Gaussian. Thus,

$$f_{\sigma_1,3}(z_1, z_2) = f_1(z_1, z_2) f_2(z_1, z_2) f_3(z_1)$$
$$f_{\sigma_2,3}(z_1, z_2) = f_1(z_1, z_2) f_2(z_2, z_1) f_3(z_2)$$
$$f_{\sigma_3,3}(z_1, z_2) = f_1(z_1, z_2) f_2(z_1, z_2) f_3(z_2)$$
$$f_{\sigma_4,3}(z_1, z_2) = f_1(z_1, z_2) f_2(z_2, z_1) f_3(z_1).$$

For instance, $f_{\sigma_1,3}(z_1, z_2)$ has been obtained by identifying the variables of $f_1(x_1, x_2) f_2(x_3, x_4) f_3(x_5)$ as $x_1 = x_3 = x_5 = z_1$ and $x_2 = x_4 = z_2$. By exploiting

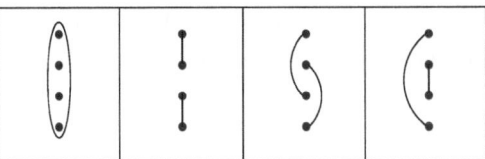

Fig. 7.9. The elements of $\mathcal{M}^0_{\geq 2}([5], \pi^*)$

the symmetry of f_1 and f_2, one deduces that the four quantities

$$\int_{Z^2} f_{\sigma_i,3}(z_1, z_2)\, \nu^2(dz_1, dz_2), \quad i = 1, ..., 4,$$

are equal. It follows from (7.4.40) and (7.4.41) that

$$\chi\left(I_2^{\widehat{N}}(f_1), I_2^{\widehat{N}}(f_2), I_1^{\widehat{N}}(f_3)\right) = \mathbb{E}\left(I_2^{\widehat{N}}(f_1)\, I_2^{\widehat{N}}(f_2)\, I_1^{\widehat{N}}(f_3)\right)$$

$$= 4\int_{Z^2} \{f_1(z_1, z_2)\, f_2(z_1, z_2)\, f_3(z_1)\}\, \nu^2(dz_1, dz_2).$$

Example 7.4.4 Consider the case $k = 4$ and $n_i = 1$, $i = 1, ..., 4$. Here, $\pi^* = \hat{0} = \{\{1\}, \{2\}, \{3\}, \{4\}\}$, and consequently π^* can be represented as a single column of four vertices. The class $\mathcal{M}_{\geq 2}([4], \pi^*)$ contains only the maximal partition $\hat{1} = \{\{1, 2, 3, 4\}\}$, whereas $\mathcal{M}^0_{\geq 2}([5], \pi^*)$ contains $\hat{1}$ and the three elements

$$\sigma_1 = \{\{1, 2\}, \{3, 4\}\}, \sigma_2 = \{\{1, 3\}, \{2, 4\}\}, \text{ and}$$
$$\sigma_3 = \{\{1, 4\}, \{2, 3\}\}.$$

The diagrams associated with the class $\mathcal{M}^0_{\geq 2}([5], \pi^*) = \left\{\hat{1}, \sigma_1, \sigma_2, \sigma_3\right\}$ are represented in Fig. 7.9.

Now take $f_i = f$ for $i = 1, ..., 4$, where f is an elementary kernel. One has that

$$f_{\hat{1},4}(z) = f(z)^4$$
$$f_{\sigma_j,4}(z_1, z_2) = f(z_1)^2\, f(z_2)^2, \quad j = 1, 2, 3.$$

It follows from (7.4.40) and (7.4.41) that

$$\chi\left(I_1^{\widehat{N}}(f), I_1^{\widehat{N}}(f), I_1^{\widehat{N}}(f), I_1^{\widehat{N}}(f)\right) = \chi_4\left(I_1^{\widehat{N}}(f)\right) = \int_Z f(z)^4\, \nu(dz)$$

$$\mathbb{E}\left(I_1^{\widehat{N}}(f)^4\right) = \int_Z f(z)^4\, \nu(dz)$$

$$+ 3\int_{Z^2} f(z_1)^2\, f(z_2)^2\, \nu^2(dz_1, dz_2).$$

Example 7.4.5 Let Y be a centered random variable with finite moments of all orders, and suppose that Y is infinitely divisible and such that

$$\mathbb{E}\left[\exp\left(i\theta Y\right)\right] = \exp\left[\int_{\mathbb{R}} \left(e^{i\theta u} - 1 - i\theta u\right) \rho\left(du\right)\right], \qquad (7.4.42)$$

where the measure ρ is such that $\rho\left(\{0\}\right) = 0$ and $\int_{\mathbb{R}} |u|^k \rho\left(du\right) < \infty$ for every $k \geq 1$ (see Lemma 5.3.4). Then, combining (7.4.42) and (3.1.2), one deduces that

$$\chi_k\left(Y\right) = \int_{\mathbb{R}} u^k \rho\left(du\right), \quad k \geq 2, \qquad (7.4.43)$$

(note that $\chi_1\left(Y\right) = \mathbb{E}\left(Y\right) = 0$). We shall prove that (7.4.43) is consistent with (7.4.40). Indeed, according to the discussion contained in Section 5.3, one has that

$$Y \stackrel{law}{=} \int_{\mathbb{R}} \int_0^1 u\widehat{N}\left(du, dx\right) = I_1^{\widehat{N}}\left(f\right),$$

where $f\left(u, x\right) = u1_{[0,1]}\left(x\right)$, and $\widehat{N}$ is a centered Poisson measure on $[0, 1] \times \mathbb{R}$, with control $\rho\left(du\right) dx$. It follows that

$$\chi_k\left(Y\right) = \chi_k(I_1^{\widehat{N}}\left(f\right)) = \chi(\underbrace{I_1^{\widehat{N}}\left(f\right), ..., I_1^{\widehat{N}}\left(f\right)}_{k \text{ times}}). \qquad (7.4.44)$$

The RHS of (7.4.44) can be computed by means of Corollary 7.4.1 in the special case where $n_j = 1$ $(\forall j = 1, ..., k)$, $n = \Sigma_j n_j = k$, and $\pi^* = \hat{0} = \{\{1\}, ..., \{k\}\}$. One has clearly that $\hat{1}$ is the only partition such that the diagram $\Gamma\left(\hat{0}, \hat{1}\right)$ is connected, so that

$$\mathcal{M}_{\geq 2}\left([k], \pi^*\right) = \left\{\hat{1}\right\} = \{\{1, ..., k\}\}.$$

Since

$$f_{\hat{1}, k}\left(u, x\right) = u^k 1_{[0,1]}\left(x\right),$$

we can now use (7.4.40) to deduce that

$$\chi_k(I_1^{\widehat{N}}\left(f\right)) = \int_{\mathbb{R}} \int_0^1 f_{\hat{1}, k}\left(u, x\right) \rho\left(du\right) dx = \int_0^1 dx \int_{\mathbb{R}} u^k \rho\left(du\right) = \int_{\mathbb{R}} u^k \rho\left(du\right).$$

Example 7.4.6 As an explicit example of (7.4.43), consider the case where Y is a centered Gamma random variable with shape parameter $a > 0$ and unitary scale parameter, that is,

$$\mathbb{E}\left[\exp\left(i\theta Y\right)\right] = \frac{e^{-i\theta a}}{\left(1 - i\theta\right)^a} = \exp\left[a \int_0^\infty \left(e^{i\theta u} - 1 - i\theta u\right) e^{-u} \frac{du}{u}\right].$$

Thus, $\rho(du) = a1_{\{u>0\}} u^{-1} e^{-u} du$. It follows that, $\chi_1\left(Y\right) = \mathbb{E}\left(Y\right) = 0$ and, for $k \geq 2$,

$$\chi_k\left(Y\right) = a \int_{\mathbb{R}} u^k e^{-u} \frac{du}{u} = a\Gamma\left(k\right) = a\left(k - 1\right)!.$$

8

From Gaussian measures to isonormal Gaussian processes

We now return to the Gaussian framework and start this chapter by relating multiple stochastic integrals to Hermite polynomials and prove a corresponding chaotic decomposition. We then generalize our setup, by replacing Gaussian measures by isonormal Gaussian processes.

8.1 Multiple stochastic integrals as Hermite polynomials

In this section, $\varphi = G$ is a completely random Gaussian measure over the Polish space $(Z, \mathcal{Z})$. We denote by ν the σ-finite control measure of G, and, for $n \geq 1$, I_n^G indicates the multiple stochastic integral of order n with respect to G (see (5.5.61)).

Definition 8.1.1 *The sequence of **Hermite polynomials** $\{H_q : q \geq 0\}$ on $\mathbb{R}$, is defined via the following relations: $H_0 \equiv 1$ and, for $q \geq 1$,*

$$H_q(x) = (-1)^q e^{\frac{x^2}{2}} \frac{d^q}{dx^q} e^{-\frac{x^2}{2}}, \quad x \in \mathbb{R}.$$

For instance, $H_1(x) = 1$, $H_2(x) = x^2 - 1$ and $H_3(x) = x^3 - 3x$.

Remark. Di Nunno, Øksendal and Proske [21], Karatzas and Shreve [52] and Kuo [57] define the Hermite polynomials as we do. Nualart [94] defines them differently, by dividing each of our H_q by $q!$.

Exercise. Prove that, for $q \geq 1$, the qth Hermite polynomial verifies the equation

$$H_q(x) = (\delta^q 1)(x), \quad x \in \mathbb{R}, \tag{8.1.1}$$

where 1 stands for the function which is constantly equal to one, and δ^q is the qth iteration of the operator δ, acting on smooth functions f as $\delta f(x) = x f(x) - f'(x)$. The operator δ, sometimes called the *divergence operator*, plays a crucial role in the

G. Peccati, M.S. Taqqu: Wiener Chaos: Moments, Cumulants and Diagrams – A survey with computer implementation.
© Springer-Verlag Italia 2011

so-called "Stein's method" for normal approximations. See, for example, the role of the operator δ in Stein's Lemma 3.2.4.

Recall that the sequence $\{(q!)^{-1/2} H_q : q \geq 0\}$ is an orthonormal basis of $L^2(\mathbb{R}, (2\pi)^{-1/2} e^{-x^2/2} dx)$. Several relevant properties of Hermite polynomials can be deduced from the following formula, valid for every $\theta, x \in \mathbb{R}$,

$$\exp\left(\theta x - \frac{\theta^2}{2}\right) = \sum_{n=0}^{\infty} \frac{\theta^n}{n!} H_n(x). \tag{8.1.2}$$

For instance, one deduces immediately from the previous expression that

$$\frac{d}{dx} H_n(x) = n H_{n-1}(x), \quad n \geq 1, \tag{8.1.3}$$

$$H_{n+1}(x) = x H_n(x) - n H_{n-1}(x), \quad n \geq 1. \tag{8.1.4}$$

One has also

$$H_n(x) = \sum_{k=0}^{\lfloor n/2 \rfloor} \binom{n}{2k} (2k-1)!!(-1)^k x^{n-2k}, \quad n \geq 0,$$

where we use the double factorial symbol as before, i.e. $(2k-1)!! = 1 \cdot 3 \cdot 5 \cdots (2k-1)$.

Remark. It is sometimes convenient to use a more general type of Hermite polynomials, denoted by $H_n(x, \rho)$, $n \geq 0$, involving a parameter $\rho > 0$. These polynomials are orthogonal with respect to the Gaussian density function with mean zero and variance ρ, namely

$$\int_{-\infty}^{\infty} H_n(x, \rho) H_m(x, \rho) \frac{e^{-\frac{x^2}{2\rho}}}{\sqrt{2\pi\rho}} dx = n! \rho^n \delta_{nm}, \quad n, m \geq 0, \tag{8.1.5}$$

where δ_{nm} is the Kronecker delta. Their generating function is

$$\exp\left(\theta x - \frac{1}{2}\rho\theta^2\right) = \sum_{n=0}^{\infty} \frac{H_n(x, \rho)}{n!} \theta^n. \tag{8.1.6}$$

The generalized Hermite polynomials also satisfy

$$H_n(x, \rho) = (-\rho)^n e^{\frac{x^2}{2\rho}} \frac{d^n}{dx^n} e^{-\frac{x^2}{2\rho}}, \quad n \geq 0,$$

with $\frac{d^0}{dx^0}$ equal to the identity operator, and are expressed as

$$H_n(x, \rho) = \sum_{k=0}^{\lfloor n/2 \rfloor} \binom{n}{2k} (2k-1)!!(-\rho)^k x^{n-2k}, \quad n \geq 0.$$

The first few are

$$H_0(x, \rho) = 1,$$
$$H_1(x, \rho) = x$$
$$H_2(x, \rho) = x^2 - \rho,$$
$$H_3(x, \rho) = x^3 - 3x\rho.$$

Observe that

$$a^{-n} H_n(ax, a^2) = H_n(x). \tag{8.1.7}$$

For more information, see Kuo [58]. Here, we shall work with the Hermite polynomials $H_n(x)$ that are obtained from $H_n(x, \rho)$ by setting $\rho = 1$.

The next result uses (6.4.17) and (8.1.4) in order to establish an explicit relation between multiple stochastic integrals and Hermite polynomials.

Proposition 8.1.2 *Let $h \in L^2(\nu)$ be such that $\|h\|_{L^2(\nu)} = 1$, and, for $n \geq 1$, define*

$$h^{\otimes n}(z_1, .., z_n) = h(z_1) \times \cdots \times h(z_n), \quad (z_1, ..., z_n) \in Z^n.$$

Then,

$$I_n^G(h^{\otimes n}) = H_n(I_1^G(h)) = H_n(G(h)). \tag{8.1.8}$$

Proof. Of course, $H_1(I_1^G(h)) = I_1^G(h) = G(h)$, so the statement holds for $n = 1$. By the multiplication formula (6.4.17), one has therefore that, for $n \geq 2$,

$$I_n^G(h^{\otimes n}) I_1^G(h) = I_{n+1}^G(h^{\otimes n+1}) + n I_{n-1}^G(h^{\otimes n-1}),$$

and the conclusion is obtained from (8.1.4), and by recursion on n. ∎

In the following corollary we do not suppose that the random variable $X = I_1^G(h)$ has unit variance.

Corollary 8.1.3 *For every $n \geq 1$ and every $h \in L^2(\nu)$, one has that*

$$H_n\big(I_1^G(h), \|h\|_{L^2(\nu)}^2\big) = I_n^G(h^{\otimes n}). \tag{8.1.9}$$

Proof. Let $X = I_1^G(h)$. Then, by (8.1.7) and Proposition 8.1.2,

$$H_n\big(X, \|h\|_{L^2(\nu)}^2\big) = \|h\|_{L^2(\nu)}^n H_n\left(\frac{X}{\|h\|_{L^2(\nu)}^n}\right)$$

$$= \|h\|_{L^2(\nu)}^n I_n^G\left(\frac{h^{\otimes n}}{\|h\|_{L^2(\nu)}^n}\right) = I_n^G(h^{\otimes n}). \qquad \blacksquare$$

Corollary 8.1.4 *For every jointly Gaussian random variables (U, V) with zero mean and unitary variance,*

$$\mathbb{E}\left[H_n\left(U\right)H_m\left(V\right)\right] = \begin{cases} 0 & \text{if } m \neq n \\ n!\mathbb{E}\left[UV\right]^n & \text{if } m = n. \end{cases}$$

Proof. By using the relation (5.5.62), we have $\mathbb{E}\left[I_n^G\left(h^{\otimes n}\right)I_n^G\left(g^{\otimes n}\right)\right] = n!\left\langle h^{\otimes n}, g^{\otimes n}\right\rangle_{L^2(\nu^n)} = n!\left\langle h, g\right\rangle_{L^2(\nu)}^n$, and hence the result follows from (8.1.8). ∎

Remark. Consider a discrete-time stationary Gaussian stochastic process $\{X_j : j \geq 0\}$ with mean zero, unit variance and covariance $r(k) = \mathbb{E}X_{j+k}X_k$, decreasing to 0 as $k \rightarrow \infty$. Then the covariance of the process $\{Y_j : j \geq 0\}$, defined for some fixed $n \geq 2$ as $Y_j = H_n(X_j)$, is

$$R_n(k) = \mathbb{E}H_n(X_{j+k})H_n(X_j) = n!r^n(k),$$

which decreases much faster as $k \rightarrow \infty$. Suppose, for example, that $\{X_j : j \geq 0\}$ has covariance $r(k) \sim k^{-D}$ as $k \rightarrow \infty$, with $D < 1$ so that $\sum_{j=-\infty}^{\infty} r(k) = \infty$. Such a sequence is said to have *long-range dependence* or *long memory* (see [154]). Then the covariance of $\{Y_j : j \geq 0\}$ is such that $R_n(k) \sim n!k^{-nD}$ as $k \rightarrow \infty$, and hence $\sum_{j=-\infty}^{\infty} R_n(k) = \infty$ only if $D < 1/n$, for example, $D < 1/2$ when $n = 2$. The original sequence $\{X_j : j \geq 0\}$ must have had particularly strong dependence to ensure that $\{Y_j : j \geq 0\}$ remains long-range dependent.

8.2 Chaotic decompositions

By combining (8.1.2) and (8.1.8), one obtains the following fundamental decomposition of the square-integrable functionals of G.

Theorem 8.2.1 (Chaotic decomposition) *For every $F \in L^2\left(\mathbb{P}, \sigma\left(G\right)\right)$ (that is, $\mathbb{E}(F^2) < \infty$), there exists a unique sequence $\{f_n : n \geq 1\}$, with $f_n \in L_s^2\left(\nu^n\right)$, such that*

$$F = \mathbb{E}\left[F\right] + \sum_{n=1}^{\infty} I_n^G\left(f_n\right), \tag{8.2.10}$$

where the series converges in $L^2\left(\mathbb{P}\right)$.

Proof. Fix $h \in L^2\left(\nu\right)$ such that $\|h\|_{L^2(\nu)} = 1$, as well as $\theta \in \mathbb{R}$. Then, $G(h)$ is Gaussian with mean zero and variance 1. By using (8.1.2) and (8.1.8), one obtains that

$$\exp\left(\theta G\left(h\right) - \frac{\theta^2}{2}\right) = \sum_{n=0}^{\infty} \frac{\theta^n}{n!}H_n\left(G\left(h\right)\right) = 1 + \sum_{n=1}^{\infty} \frac{\theta^n}{n!}I_n^G\left(h^{\otimes n}\right). \tag{8.2.11}$$

Since

$$\mathbb{E}\left[\exp\left(\theta G(h) - \frac{\theta^2}{2}\right)\right] = 1,$$

one deduces that (8.2.10) holds for every random variable of the form $F = \exp\left(\theta G(h) - \frac{\theta^2}{2}\right)$, with $f_n = \frac{\theta^n}{n!}h^{\otimes n}$. The conclusion is obtained by observing that the linear combinations of random variables of this type are dense in $L^2(\mathbb{P}, \sigma(G))$, as well as using the fact that random variables admitting the chaotic decomposition 8.2.10 are an Hilbert space (just use the orthogonality and isometric properties of multiple integrals). ∎

Remarks. (1) By inspection of the proof of Theorem 8.2.1, we deduce that the linear combinations of random variables of the type $I_n^G(h^{\otimes n})$, with $n \geq 1$ and $\|h\|_{L^2(\nu)} = 1$, are dense in $L^2(\mathbb{P}, \sigma(G))$. This implies in particular that the random variables $I_n^G(h^{\otimes n})$ generate the nth Wiener chaos $C_n(G)$.

(2) The first proof of (8.2.10) dates back to Wiener [162]. See also Nualart and Schoutens [99] and Stroock [149]. See, for example, [20, 46, 61] for further references and results on chaotic decompositions.

8.3 Isonormal Gaussian processes

We shall now generalize some of the previous results to the case of an *isonormal Gaussian process*. These objects have been introduced by R.M. Dudley in [24], and are a natural generalization of the Gaussian measures introduced in Section 5.1. In particular, the concept of isonormal Gaussian process can be very useful in the study of fractional fields. See, for example, Pipiras and Taqqu [115, 116, 118], or the second edition of Nualart's book [94]. For a general approach to Gaussian analysis by means of Hilbert space techniques, and for further details on the subjects discussed in this section, the reader is referred to Janson [46].

Let $\mathfrak{H}$ be a real separable Hilbert space with inner product $(\cdot, \cdot)_{\mathfrak{H}}$. In what follows, we will denote by

$$X = X(\mathfrak{H}) = \{X(h) : h \in \mathfrak{H}\}$$

an *isonormal Gaussian process* over $\mathfrak{H}$. This means that X is a centered real-valued Gaussian family, indexed by the elements of $\mathfrak{H}$ and such that

$$\mathbb{E}[X(h)X(h')] = (h, h')_{\mathfrak{H}}, \quad \forall h, h' \in \mathfrak{H}. \tag{8.3.12}$$

In other words, relation (8.3.12) means that X is a centered Gaussian Hilbert space (with respect to the inner product canonically induced by the covariance) isomorphic to $\mathfrak{H}$.

Example 8.3.1 (Euclidean spaces) Fix an integer $d \geq 1$, set $\mathfrak{H} = \mathbb{R}^d$ and let $(e_1, ..., e_d)$ be an orthonormal basis of $\mathbb{R}^d$ (with respect to the usual Euclidean inner product). Let $(Z_1, ..., Z_d)$ be a Gaussian vector whose components are i.i.d. $N(0, 1)$. For every $h = \sum_{j=1}^{d} c_j e_j$ (where the c_j are real and uniquely defined), set

$$X(h) = \sum_{j=1}^{d} c_j Z_j;$$

and define

$$X = \{X(h) : h \in \mathbb{R}^d\}.$$

Then, X is an isonormal Gaussian process over $\mathbb{R}^d$. Observe, in particular, that $\mathbb{E}X(h)^2 = \sum_{j=1}^{d} c_j^2 = \langle h, h \rangle_{\mathfrak{H}}$.

Example 8.3.2 (Gaussian measures) Let $(Z, \mathcal{Z}, \nu)$ be a measure space, where ν is positive, σ-finite and non atomic. Consider a completely random Gaussian measure $G = \{G(A) : A \in \mathcal{Z}_\nu\}$ (as defined in Section 5.1), where the class $\mathcal{Z}_\nu$ is given by (5.1.3). Set

$$\mathfrak{H} = L^2(Z, \mathcal{Z}, \nu);$$

and thus, for every $h, h' \in \mathfrak{H}$,

$$(h, h')_{\mathfrak{H}} = \int_Z h(z)h'(z)\nu(dz).$$

Also, for every $h \in \mathfrak{H}$, define

$$X(h) = I_1^G(h)$$

to be the Wiener-Itô integral of h with respect to G, as defined in (5.2.14). Recall that $X(h)$ is a centered Gaussian random variable with variance given by $\|h\|_{\mathfrak{H}}^2$. Then, relation (5.2.15) implies that the collection

$$X = \{X(h) : h \in L^2(Z, \mathcal{Z}, \nu)\}$$

is an isonormal Gaussian process over $L^2(Z, \mathcal{Z}, \nu)$.

Example 8.3.3 (Isonormal spaces built from covariances) Let $Y = \{Y_t : t \geq 0\}$ be a real-valued centered Gaussian process indexed by the positive axis, and set

$$R(s, t) = \mathbb{E}[Y_s Y_t]$$

to be the covariance function of Y. This process is not isonormal. However, one can embed Y into some isonormal Gaussian process as follows:

(i) define $\mathcal{E}$ as the collection of all finite linear combinations of indicator functions of the type $\mathbf{1}_{[0,t]}$, $t \geq 0$;

(ii) define $\mathfrak{H} = \mathfrak{H}_R$ to be the Hilbert space given by the closure of $\mathcal{E}$ with respect to the inner product
$$(f, h)_R := \sum_{i,j} a_i c_j R(s_i, t_j),$$
where $f = \sum_i a_i 1_{[0, s_i]}$ and $h = \sum_j c_j 1_{[0, t_j]}$ are two generic elements of $\mathcal{E}$;

(iii) for $h = \sum_j c_j 1_{[0, t_j]} \in \mathcal{E}$, set $X(h) = \sum_j c_j Y_{t_j}$;

(iv) for $h \in \mathfrak{H}_R$, set $X(h)$ to be the $L^2(\mathbb{P})$ limit of any sequence of the type $X(h_n)$, where $\{h_n\} \subset \mathcal{E}$ converges to h in $\mathfrak{H}_R$. Note that such a sequence $\{h_n\}$ necessarily exists and may not be unique (however, the definition of $X(h)$ does not depend on the choice of the sequence $\{h_n\}$).

Then, by construction, the Gaussian space
$$\{X(h) : h \in \mathfrak{H}\}$$
is an isonormal Gaussian process over $\mathfrak{H}_R$. See Janson [46, Ch. 1] or Nualart [94] for more details on this construction.

Example 8.3.4 (Fractional Brownian motion) As a particular case of the previous example, consider the fractional Brownian motion $\{B_H(t) : t \geq 0\}$, where $0 < H < 1$, and whose covariance is
$$R(s, t) = \mathbb{E}(B_H(s) B_H(t)) = \frac{\sigma^2}{2} \{s^{2H} + t^{2H} - |s - t|^{2H}\}, \qquad (8.3.13)$$
with $\sigma > 0$. It is a Brownian motion if $H = 1/2$. Define the Hilbert space $\mathfrak{H}$ and the process $\{X(h) : h \in \mathfrak{H}\}$ as in the previous example. Then, $\{X(h) : h \in \mathfrak{H}\}$ is isonormal. One customarily denotes it as $\{B_H(h) : h \in \mathfrak{H}\}$.

Example 8.3.5 (Even functions and symmetric control measures) Other classic examples of isonormal Gaussian processes (see, for example,, [12, 34, 66, 151]) are given by objects of the type
$$X_\beta = \{X_\beta(\psi) : \psi \in \mathfrak{H}_{E,\beta}\},$$
where β is a real non-atomic symmetric measure on $\mathbb{R}$ (that is, $\beta(dx) = \beta(-dx)$), and
$$\mathfrak{H}_{E,\beta} = L_E^2(\mathbb{R}, d\beta) \qquad (8.3.14)$$
stands for the collection of *real* linear combinations of complex-valued *even* functions that are square-integrable with respect to β (recall that a function ψ is "even" or "Hermitian" if $\overline{\psi(x)} = \psi(-x)$). The class $\mathfrak{H}_{E,\beta}$ (the "E" stands for even) is indeed a real Hilbert space, endowed with the inner product
$$(\psi_1, \psi_2)_\beta = \int_{\mathbb{R}} \psi_1(x) \psi_2(-x) \beta(dx) \in \mathbb{R}. \qquad (8.3.15)$$

This type of construction is used in the spectral theory of time series (see Chapter 9 for more details).

Example 8.3.6 (Gaussian Free Fields) Let $d \geq 2$ and let D be a domain in $\mathbb{R}^d$. Denote by $H_s(D)$ the space of real-valued continuous and continuously differentiable functions on $\mathbb{R}^d$ that are supported on a compact subset of D (note that this implies that the first derivatives of the elements of $H_s(D)$ are square-integrable with respect to the Lebesgue measure). Write $H(D)$ in order to indicate the real Hilbert space obtained as the closure of $H_s(D)$ with respect to the inner product

$$(f,g) = \int_{\mathbb{R}^d} \nabla f(x) \cdot \nabla g(x) dx,$$

where ∇ indicates the gradient. An isonormal Gaussian process of the type $X = \{X(h) : h \in H(D)\}$ is called a *Gaussian Free Field* (GFF). The reader is referred to the survey by Sheffield [139] for a discussion of the emergence of GFFs in several areas of modern probability. See Rider and Virág [129] for a connection with the "circular law" for Gaussian non-Hermitian random matrices.

Example 8.3.7 (Obtaining a Gaussian process on $\mathbb{R}_+$ from a an isonormal Gaussian process) Let $X = \{X(h) : h \in \mathfrak{H}\}$ be isonormal. For each $t \geq 0$ select a function $h_t \in \mathfrak{H}$. Then, the process $X_t = X(h_t)$ is Gaussian with covariance

$$\mathbb{E}(X_t X_s) = \langle h_t, h_s \rangle_{\mathfrak{H}}.$$

For instance, consider the case $\mathfrak{H} = L^2(\mathbb{R}, dx)$ and define, for $H \in (0,1)$,

$$h_t^H(x) = (t-x)_+^{H-1/2} - (-x)_+^{H-1/2}, \quad x \in \mathbb{R},$$

for $H \neq 1/2$, as well as $h_t^{1/2}(x) = \mathbf{1}_{[0,t]}(x)$ for $H = 1/2$. Then, for every $H \in (0,1)$, the centered Gaussian process $X(h_t^H)$, $t \geq 0$, has a covariance which is a multiple of the RHS of (8.3.13) (that is, $X(h_t^H)$ is a multiple of a fractional Brownian motion of Hurst index H). Check this as an exercise!

8.4 Wiener chaos

We shall now show how to extend the notion of *Wiener chaos* (as appearing in Section 5.9 Section 8.2) to the case of an isonormal Gaussian process. The reader is referred to [94, Sect. 1.1] for a complete discussion of this subject. We need some further (standard) definitions.

Definition 8.4.1 *From now on, the symbol $\mathcal{A}_\infty$ will denote the class of those sequences $\alpha = \{\alpha_i : i \geq 1\}$ such that: (i) each α_i is a nonnegative integer, (ii) α_i is* *different from zero only for a finite number of indices i. A sequence of this type is called a **multiindex**. For $\alpha \in \mathcal{A}_\infty$, we use the notation $|\alpha| = \sum_i \alpha_i$. For $q \geq 1$, we also write*

$$\mathcal{A}_{\infty,q} = \{\alpha \in \mathcal{A}_\infty : |\alpha| = q\}.$$

Notation. Fix $q \geq 2$. Given a real separable Hilbert space $\mathfrak{H}$, we denote by

$$\mathfrak{H}^{\otimes q} \quad \text{the } q\text{th } \textit{tensor power of } \mathfrak{H}$$

and by

$$\mathfrak{H}^{\odot q} \quad \text{the } q\text{th } \textit{symmetric tensor power of } \mathfrak{H}$$

(see, for example, [46]). We conventionally set $\mathfrak{H}^{\otimes 1} = \mathfrak{H}^{\odot 1} = \mathfrak{H}$.

We recall four classic facts concerning tensors powers of Hilbert spaces (see, for example, [46]).

(i) The spaces $\mathfrak{H}^{\otimes q}$ and $\mathfrak{H}^{\odot q}$ are real separable Hilbert spaces, such that $\mathfrak{H}^{\odot q} \subset \mathfrak{H}^{\otimes q}$.

(ii) Let $\{e_j : j \geq 1\}$ be an orthonormal basis of $\mathfrak{H}$; then, an orthonormal basis of $\mathfrak{H}^{\otimes q}$ is given by the collection of all tensors of the type

$$e_{j_1} \otimes \cdots \otimes e_{j_q}, \quad j_1, ..., j_d \geq 1.$$

(iii) Let $\{e_j : j \geq 1\}$ be an orthonormal basis of $\mathfrak{H}$ and endow $\mathfrak{H}^{\odot q}$ with the inner product $(\cdot, \cdot)_{\mathfrak{H}^{\otimes q}}$; then, an orthogonal (and, in general, *not* orthonormal) basis of $\mathfrak{H}^{\odot q}$ is given by all elements of the type

$$\mathbf{e}(j_1, ..., j_q) = \mathbf{sym}\{e_{j_1} \otimes \cdots \otimes e_{j_q}\}, \quad 1 \leq j_1 \leq ... \leq j_q < \infty, \quad (8.4.16)$$

where $\mathbf{sym}\{\cdot\}$ stands for a canonical symmetrization. **Exercise**: find an orthonormal basis of $\mathfrak{H}^{\odot q}$.

The following identification is often used (see, for example, [46, Appendix E] for a discussion of this point).

Lemma 8.4.2 *If $\mathfrak{H} = L^2(Z, \mathcal{Z}, \nu)$, where ν is σ-finite and non-atomic, then $\mathfrak{H}^{\otimes q}$ is isomorphic (as a Hilbert space) to $L^2(Z^q, \mathcal{Z}^q, \nu^q)$ and $\mathfrak{H}^{\odot q}$ is isomorphic to $L_s^2(Z^q, \mathcal{Z}^q, \nu^q)$, where $L_s^2(Z^q, \mathcal{Z}^q, \nu^q)$ is the subspace of $L^2(Z^q, \mathcal{Z}^q, \nu^q)$ composed of symmetric functions.*

Now observe that, once an orthonormal basis of $\mathfrak{H}$ is fixed and due to the symmetrization, each element $\mathbf{e}(j_1, ..., j_q)$ in (8.4.16) can be completely described in terms of a unique multiindex $\alpha \in \mathcal{A}_{\infty,q}$, as follows: (i) set $\alpha_i = 0$ if $i \neq j_r$ for every $r = 1, ..., q$, (ii) set $\alpha_j = k$ for every $j \in \{j_1, ..., j_q\}$ such that j is repeated exactly k times in the vector $(j_1, ..., j_q)$ $(k \geq 1)$.

Example 8.4.3   (i) The multiindex $(1, 0, 0, ...)$ is associated with the element of $\mathfrak{H}$ given by e_1.

(ii) Consider the element $e\,(1, 7, 7)$. In $(1, 7, 7)$ the number 1 is not repeated and 7 is repeated twice, hence $e\,(1, 7, 7)$ is associated with the multiindex $\alpha \in \mathcal{A}_{\infty, 3}$ such that $\alpha_1 = 1$, $\alpha_7 = 2$ and $\alpha_j = 0$ for every $j \neq 1, 7$, that is, $\alpha = (1, 0, 0, 0, 0, 0, 2, 0, 0, ...)$.

(iii) The multindex $\alpha = (1, 2, 2, 0, 5, 0, 0, 0, ...)$ is associated with the element of $\mathfrak{H}^{\odot 10}$ given by $e\,(1, 2, 2, 3, 3, 5, 5, 5, 5, 5)$.

In what follows, given $\alpha \in \mathcal{A}_{\infty, q}$ $(q \geq 1)$, we shall write $e\,(\alpha)$ in order to indicate the element of $\mathfrak{H}^{\odot q}$ uniquely associated with α.

Definition 8.4.4 *For every $h \in \mathfrak{H}$, we set $I_1^X\,(h) = X\,(h)$. Now fix an orthonormal basis $\{e_j : j \geq 1\}$ of $\mathfrak{H}$: for every $q \geq 2$ and every $h \in \mathfrak{H}^{\odot q}$ such that*

$$h = \sum_{\alpha \in \mathcal{A}_{\infty, q}} c_\alpha e\,(\alpha)$$

(with convergence in $\mathfrak{H}^{\odot q}$, endowed with the inner product $(\cdot, \cdot)_{\mathfrak{H}^{\otimes q}}$), we set

$$I_q^X\,(h) = \sum_{\alpha \in \mathcal{A}_{\infty, q}} c_\alpha \prod_j H_{\alpha_j}\,(X\,(e_j)), \qquad (8.4.17)$$

where the products only involve the non-zero terms of each multiindex α, and H_m indicates the mth Hermite polynomial. For $q \geq 1$, the collection of all random variables of the type $I_q^X\,(h)$, $h \in \mathfrak{H}^{\odot q}$, is called the qth Wiener chaos associated with X and is denoted by $C_q\,(X)$.. One sets conventionally $C_0\,(X) = \mathbb{R}$.

Example 8.4.5   (i) If $h = e\,(\alpha)$, where $\alpha = (1, 1, 0, 0, 0, ...) \in \mathcal{A}_{\infty, 2}$, then

$$I_2^X\,(h) = H_1\,(X\,(e_1))\,H_1\,(X\,(e_2)) = X\,(e_1)\,X\,(e_2).$$

(ii) If $\alpha = (1, 0, 1, 2, 0, ...) \in \mathcal{A}_{\infty, 4}$, then

$$\begin{aligned}
I_4^X\,(h) &= H_1\,(X\,(e_1))\,H_1\,(X\,(e_3))\,H_2\,(X\,(e_4)) \\
&= X\,(e_1)\,X\,(e_3)\,\left(X\,(e_4)^2 - 1\right) \\
&= X\,(e_1)\,X\,(e_3)\,X\,(e_4)^2 - X\,(e_1)\,X\,(e_3).
\end{aligned}$$

(iii) If $\alpha = (3, 1, 1, 0, 0, ...) \in \mathcal{A}_{\infty, 5}$, then

$$\begin{aligned}
I_5^X\,(h) &= H_3\,(X\,(e_1))\,H_1\,(X\,(e_2))\,H_1\,(X\,(e_3)) \\
&= \left(X\,(e_1)^3 - 3X\,(e_1)\right) X\,(e_2)\,X\,(e_3) \\
&= X\,(e_1)^3\,X\,(e_2)\,X\,(e_3) - 3X\,(e_1)\,X\,(e_2)\,X\,(e_3).
\end{aligned}$$

The following result collects some well-known facts concerning Wiener chaos and isonormal Gaussian processes. In particular: the first point characterizes the operators I_q^X as isomorphisms; the second point is an equivalent of the chaotic representation property for Gaussian measures, as stated in formula (5.9.78); the third point establishes a formal relation between random variables of the type $I_q^X(h)$ and the multiple Wiener-Itô integrals introduced in Section 5.5 (see [94, Ch. 1] for proofs and further discussions of all these facts). Compare with Section 5.9.

Proposition 8.4.6 *1. For every $q \geq 1$, the qth Wiener chaos $C_q(X)$ is a Hilbert subspace of $L^2(\mathbb{P})$, and the application*

$$h \mapsto I_q^X(h), \quad h \in \mathfrak{H}^{\odot q},$$

defines a Hilbert space isomorphism between $\mathfrak{H}^{\odot q}$, endowed with the inner product $q!(\cdot, \cdot)_{\mathfrak{H}^{\otimes q}}$, and $C_q(X)$.
2. *For every $q, q' \geq 0$ such that $q \neq q'$, the spaces $C_q(X)$ and $C_{q'}(X)$ are orthogonal in $L^2(\mathbb{P})$.*
3. *Let F be a functional of the isonormal Gaussian process X satisfying $\mathbb{E}[F(X)^2] < \infty$: then, there exists a unique sequence $\{f_q : q \geq 1\}$ such that $f_q \in \mathfrak{H}^{\odot q}$, and*

$$F = \mathbb{E}(F) + \sum_{q=1}^{\infty} I_q^X(f_q),$$

where the series converges in $L^2(\mathbb{P})$.
4. *Suppose that $\mathfrak{H} = L^2(Z, \mathcal{Z}, \nu)$, where ν is σ-finite and non-atomic. Then, for $q \geq 2$, the symmetric power $\mathfrak{H}^{\odot q}$ can be identified with $L_s^2(Z^q, \mathcal{Z}^q, \nu^q)$ and, for every $f \in \mathfrak{H}^{\odot q}$, the random variable $I_q^X(f)$ coincides with the Wiener-Itô integral (see Definition 5.5.4) of f with respect to the Gaussian measure given by $A \to X(\mathbf{1}_A)$, $A \in \mathcal{Z}_\nu$.*

Remark. The combination of Point 1. and Point 2. in the statement of Proposition 8.4.6 implies that, for every $q, q' \geq 1$,

$$\mathbb{E}\left[I_q^X(f) I_{q'}^X(f')\right] = \mathbf{1}_{q=q'} q! (f, f')_{\mathfrak{H}^{\otimes q}}$$

(compare with (5.5.62)). One can also prove (see, for example, Janson [46, Ch. VI]) that the hypercontractivity property (5.9.79) extends to the framework of isonormal Gaussina processes. This means that, for every $p > 2$ and every $n \geq 2$, there exists a universal constant $c_{p,n} > 0$, such that

$$\mathbb{E}\left[|I_n^X(f)|^p\right]^{1/p} \leq c_{n,p} \mathbb{E}\left[I_n^X(f)^2\right]^{1/2}, \tag{8.4.18}$$

for every $f \in \mathfrak{H}^{\odot n}$.

8.5 Contractions, products and some explicit formulae

We start by introducing the notion of *contraction* in the context of powers of Hilbert spaces.

Definition 8.5.1 *Consider a real separable Hilbert space $\mathfrak{H}$, and let $\{e_i : i \geq 1\}$ be an orthonormal basis of $\mathfrak{H}$. For every $n, m \geq 1$, every $r = 0, ..., n \wedge m$ and every $f \in \mathfrak{H}^{\odot n}$ and $g \in \mathfrak{H}^{\odot m}$, we define the* **contraction** *of order r, of f and g, as the element of $\mathfrak{H}^{\otimes n+m-2r}$ given by*

$$f \otimes_r g = \sum_{i_1,...,i_r=1}^{\infty} (f, e_{i_1} \otimes \cdots \otimes e_{i_r})_{\mathfrak{H}^{\otimes r}} \otimes (g, e_{i_1} \otimes \cdots \otimes e_{i_r})_{\mathfrak{H}^{\otimes r}}, \quad (8.5.19)$$

and we denote by $\widetilde{f \otimes_r g}$ its symmetrization. This definition does not depend on the chosen orthonormal basis $\{e_i : i \geq 1\}$. Observe that

$$(f, e_{i_1} \otimes \cdots \otimes e_{i_r})_{\mathfrak{H}^{\otimes r}} \in \mathfrak{H}^{\odot(n-r)} \text{ and } (g, e_{i_1} \otimes \cdots \otimes e_{i_r})_{\mathfrak{H}^{\otimes r}} \in \mathfrak{H}^{\odot(m-r)}.$$

Remark. One can prove (**Exercise!**) the following result: if $\mathfrak{H} = L^2(Z, \mathcal{Z}, \nu)$, $f \in \mathfrak{H}^{\odot n} = L_s^2(Z^n, \mathcal{Z}^n, \nu^n)$ and $g \in \mathfrak{H}^{\odot m} = L_s^2(Z^m, \mathcal{Z}^m, \nu^m)$, then the definition of the contraction $f \otimes_r g$ given in (8.5.19) and the one given in (6.2.16) coincide.

Example 8.5.2 Let $\mathfrak{H}$ be a real separable Hilbert space, and let $\{e_i : i \geq 1\}$ be an orthonormal basis of $\mathfrak{H}$. Suppose that f, g are two elements of $\mathfrak{H}^{\odot 2}$ given by

$$f = \sum_{i_1,i_2=1}^{\infty} a(i_1, i_2)(e_{i_1} \otimes e_{i_2}) \quad g = \sum_{i_1,i_2=1}^{\infty} b(i_1, i_2)(e_{i_1} \otimes e_{i_2})$$

where the applications $(i_1, i_2) \mapsto a(i_1, i_2), b(i_1, i_2)$ are symmetric. We want to compute $f \otimes_1 g$ according to Definition 8.5. First of all, one has that, for every $i \geq 1$

$$(f, e_i)_{\mathfrak{H}} = \sum_{j=1}^{\infty} a(i, j)e_j \in \mathfrak{H} \quad \text{and} \quad (g, e_i)_{\mathfrak{H}} = \sum_{j=1}^{\infty} b(i, j)e_j \in \mathfrak{H},$$

from which one infers that

$$f \otimes_1 g = \sum_{j_1=1}^{\infty} \sum_{j_2=1}^{\infty} (f, e_i)_{\mathfrak{H}} \otimes (g, e_j)_{\mathfrak{H}} = \sum_{j_1=1}^{\infty} \sum_{j_2=1}^{\infty} \left[\sum_{i=1}^{\infty} a(i, j_1)b(i, j_2)\right] (e_{j_1} \otimes e_{j_2}).$$

The following result extends the product formula (6.4.17) to the case of isonormal Gaussian processes. The proof (which is left to the reader) can be obtained from Proposition 6.4.1, by using the fact that every real separable Hilbert space is isomorphic to a space of the type $L^2(Z, \mathcal{Z}, \nu)$, where ν is σ-finite and non-atomic (see Lemma 8.4.2). An alternative proof (by induction) can be found in [94, Ch. 1].

Proposition 8.5.3 *Let X be an isonormal Gaussian process over some real separable Hilbert space $\mathfrak{H}$. Then, for every $n, m \geq 1$, $f \in \mathfrak{H}^{\odot n}$ and $g \in \mathfrak{H}^{\odot m}$,*

$$I_n^X(f) I_m^X(g) = \sum_{r=0}^{m \wedge n} r! \binom{m}{r} \binom{n}{r} I_{n+m-2r}^X(f \otimes_r g), \qquad (8.5.20)$$

where the contraction $f \otimes_r g$ is defined according to (8.5.19), and for $m = n = r$, we write

$$I_0^X(f \otimes_n g) = (f, g)_{\mathfrak{H}^{\otimes n}}.$$

We stress that one can obtain a generalization of the cumulant formulae (7.3.29) in the framework of isonormal Gaussian processes. To do this, one should represent each integral of the type $\int_{Z^{n/2}} f_\sigma d\nu^{n/2}$, appearing in (7.3.29), as the inner product between two iterated contractions of the kernels $\{f_{n_j}\}$, and then use the canonical isomorphism between $\mathfrak{H}$ and a space of the form $L^2(Z, \mathcal{Z}, \nu)$. However, the formalism associated with this extension is rather heavy (and not really useful for the discussion to follow), and is left to the reader.

Example 8.5.4 We focus once again on isonormal Gaussian processes of the type $X_\beta = \{X_\beta(\psi) : \psi \in \mathfrak{H}_{E,\beta}\}$, where the Hilbert space $\mathfrak{H}_{E,\beta}$ is given in (8.3.14). In this case, for $d \geq 2$, the symmetric power $\mathfrak{H}_{E,\beta}^{\odot d}$ can be identified with the real Hilbert space of those functions ψ_d that are symmetric on $\mathbb{R}^d$, square integrable with respect to β^d, and such that $\overline{\psi_d(x_1, ..., x_d)} = \psi_d(-x_1, ..., -x_d)$. The bar indicates complex conjugation. For every $n_1, ..., n_k \geq 1$, one can write explicitly a diagram formula as follows:

$$\chi\left(I_{n_1}^{X_\beta}(\psi_1), \cdots, I_{n_k}^{X_\beta}(\psi_k)\right) = \sum_{\sigma \in \mathcal{M}_2([n], \pi^*)} \int_{\mathbb{R}^{n/2}} \psi_\sigma d\beta^{n/2},$$

where $\mathcal{M}_2([n], \pi^*)$ is defined in (7.2.21) and ψ_σ is the function in $(n_1 + \cdots + n_k)/2$ variables obtained by setting $x_i = -x_j$ in $\psi_1 \otimes_0 \cdots \otimes_0 \psi_d$ if and only if $i \sim_\sigma j$. The field X_β is often defined in terms of a complex Gaussian measure (see [12, 34, 66]). This example is further expanded and discussed in the forthcoming Chapter 9.

9

Hermitian random measures and spectral representations

The aim of this chapter is to analyze more deeply the class of random variables encountered in Example 8.5.4. In particular, we shall consider complex-valued functions ψ, to which we associate real-valued (multiple) stochastic integrals. This type of construction is used in the spectral theory of time series and, in particular, in the context of self-similar processes (see e.g., [12, 23, 34, 66, 151, 153]).

We will proceed in two different ways. The first, already outlined in Example 8.5.4, involves isonormal processes and does not require the construction of a Gaussian measure (for the convenience of the reader, we will rewrite the content of Example 8.5.4 in more detail). The second involves building the integrals in the usual way, starting with elementary functions and a complex Gaussian measure.

9.1 The isonormal approach

Consider first the "isonormal" approach. As before, we shall define the isonormal Gaussian process as

$$X_\beta = \{X_\beta(\psi) : \psi \in \mathfrak{H}_{E,\beta}\},$$

where β is a real non-atomic symmetric measure on $\mathbb{R}$, where the symmetry consists in the property that

$$\beta(dx) = \beta(-dx), \tag{9.1.1}$$

and where

$$\mathfrak{H}_{E,\beta} = L^2_E(\mathbb{R}, d\beta) \tag{9.1.2}$$

stands for the collection of linear combinations with *real-valued* coefficients of *complex-valued* and *Hermitian* functions that are square-integrable with respect to β. Recall that a function ψ is *Hermitian*, or (equivalently) *even*, if

$$\overline{\psi(x)} = \psi(-x),$$

G. Peccati, M.S. Taqqu: Wiener Chaos: Moments, Cumulants and Diagrams –
A survey with computer implementation.
© Springer-Verlag Italia 2011

where the "bar" indicates complex conjugation. The fact that we consider a class of Hermitian functions ensures that $\mathfrak{H}_{E,\beta}$ is a real Hilbert space, endowed with the scalar product

$$(\psi_1, \psi_2)_\beta = \int_\mathbb{R} \psi_1(x)\, \overline{\psi_2(x)}\, \beta(dx) = \int_\mathbb{R} \psi_1(x)\, \psi_2(-x)\, \beta(dx). \qquad (9.1.3)$$

This scalar product is real-valued because $\overline{(\psi_1, \psi_2)_\beta} = (\psi_1, \psi_2)_\beta$. It follows from Example 8.5.4 that the Gaussian integral $I_1(\psi)$ is well-defined and is also real-valued since $\mathfrak{H}_{E,\beta}$ is a real Hilbert space. We shall denote it here

$$\widehat{I}_1(\psi)$$

to indicate that its argument ψ belongs to $\mathfrak{H}_{E,\beta} = L^2_E(\mathbb{R}, d\beta)$. In addition the chaos of any order q is defined as well, via Hermite polynomials, as in (8.4.17), as well as the full chaotic expansion as in Theorem 8.4.6. Now consider a chaos of order $q \geq 2$. By adapting the last point in the statement of Proposition 8.4.6, one sees that the symmetric power $\mathfrak{H}_{E,\beta}^{\odot q}$ can be identified with the real Hilbert space of those functions ψ_q that are symmetric on $\mathbb{R}^q$, square integrable with respect to β^q but here, in addition, such that

$$\overline{\psi_q(x_1, ..., x_q)} = \psi_q(-x_1, ..., -x_q).$$

For every $n_1, ..., n_k \geq 1$, one can write explicitly diagram formulae as follows:

$$\chi\left(\widehat{I}_{n_1}(\psi_1), \cdots, \widehat{I}_{n_k}(\psi_k)\right) = \sum_{\sigma \in \mathcal{M}_2([n], \pi^*)} \int_{\mathbb{R}^{n/2}} \psi_\sigma d\beta^{n/2}, \qquad (9.1.4)$$

and

$$\mathbb{E}\left(\widehat{I}_{n_1}(\psi_1) \cdots \widehat{I}_{n_k}(\psi_k)\right) = \sum_{\sigma \in \mathcal{M}_2^0([n], \pi^*)} \int_{\mathbb{R}^{n/2}} \psi_\sigma d\beta^{n/2}, \qquad (9.1.5)$$

where $\mathcal{M}_2([n], \pi^*)$ and $\mathcal{M}_2^0([n], \pi^*)$ are defined in (7.2.21) and (7.2.21) respectively, and ψ_σ is the function in $n/2$ variables, where

$$n = n_1 + \cdots + n_k,$$

obtained by setting $x_i = -x_j$ in

$$(\psi_1 \otimes_0 \cdots \otimes_0 \psi_k)(x_1, ..., x_n)$$
$$= \psi_1(x_1, ..., x_{n_1}) \times \psi_2(x_{n_1+1}, ..., x_{n_1+n_2}) \times \cdots \times \psi_k(x_{n_1+\cdots+n_{k-1}+1}, ..., x_n) \qquad (9.1.6)$$

if and only if $i \sim_\sigma j$. Compare these diagrams formulae with those in Corollary 7.3.1 and equation (9.1.6) with (7.3.31).

One can also define contractions as in Relation 6.2.10. Let's focus for example on the operator $\otimes_r$, which we denote $\overline{\otimes}_r$ in this context. If $f \in L^2{}_E(\mathbb{R}^p, d^p\beta)$ and $g \in L^2{}_E(\mathbb{R}'', d^q\beta)$, then

$$f\overline{\otimes}_r g\, (x_1, ..., x_{p+q-2r}) \tag{9.1.7}$$
$$= \left(f(\cdot, x_1, ..., x_{p-r}),\, g(\cdot, x_{p-r+1}, ..., x_{p+q-2r}) \right)_\beta$$
$$= \int_{\mathbb{R}^r} f\,(z_1, ..., z_r, x_1, ..., x_{p-r})\, g\,(-z_1, ..., -z_r, x_{p-r+1}, ..., x_{p+q-2r})$$
$$\beta\,(dz_1) \cdots \beta\,(dz_r).$$

We can now state the formula for the product of two integrals corresponding to the one in Proposition 6.4.1. Let $q, p \geq 1$, $f \in L^2{}_E(\mathbb{R}^p, d^p\beta)$ and $g \in L^2{}_E(\mathbb{R}^q, d^q\beta)$. Then

$$\widehat{I}_p\,(f)\,\widehat{I}_q\,(g) = \sum_{r=0}^{p\wedge q} r! \binom{p}{r}\binom{q}{r} \widehat{I}_{p+q-2r}\,(f\overline{\otimes}_r g), \tag{9.1.8}$$

where the contraction $f\overline{\otimes}_r g$ is defined in (9.1.7), and for $p = q = r$, we write

$$\widehat{I}_0\,(f\overline{\otimes}_r g) = (f,g)_{L^2_E(\mathbb{R}^r,\beta^r)}$$
$$= f\overline{\otimes}_r g$$
$$= \int_{\mathbb{R}^r} f\,(z_1, ..., z_r)\, g\,(-z_1, ..., -z_r)\, \beta\,(dz_1) \cdots \beta\,(dz_r).$$

There is nothing to prove. One just has to adapt to the scalar product of the Hilbert space.

9.2 The Gaussian measure approach

The field X_β appearing in the previous paragraph is often defined in terms of a complex Gaussian measure. In what follows, we shall briefly sketch this construction; the reader is referred to [66, Ch. 3] for an exhaustive discussion (see also the classic references [12, 34]). We start with a definition.

Definition 9.2.1 *Let β be a Borel non-atomic symmetric measure (see (9.1.1)) on the real line. A **Hermitian (complex) Gaussian measure** with control measure β is a collection of complex-valued random variables $\{\widehat{W}(A) : \beta(A) < \infty\}$ defined on some probability space $(\Omega, \mathcal{F}, \mathbb{P})$, and such that:*

(i) *The random variables $\widehat{W}(A) = \widehat{W}_1(A) + i\widehat{W}_2(A)$, $\beta(A) < \infty$, are jointly complex Gaussian.*

(ii) *$\mathbb{E}(\widehat{W}(A)) = 0$.*

(iii) *The following identity takes place for every A, B of finite measure β:*

$$\mathbb{E}\widehat{W}(A)\overline{\widehat{W}(B)} = \beta(A \cap B),$$ (9.2.9)

(iv) *The following symmetry property is satisfied:*

$$\widehat{W}(dx) = \overline{\widehat{W}(-dx)},$$ (9.2.10)

or, equivalently, $\widehat{W}(A) = \overline{\widehat{W}(-A)}$ for every set A of finite measure β, where the set $-A$ is defined by the relation: for every real x, one has that $-x \in -A$ if and only if $x \in A$.

Remark.

(a) To explicitly build a Hermitian complex Gaussian measure $\widehat{W}$ with symmetric control β, one can proceed as follows. Consider first two independent, real-valued, centered completely random Gaussian measures (say W_1^0 and W_2^0) on $\mathbb{R}_+$, each one with control measure given by the restriction of $\beta/2$ on $\mathbb{R}_+$. Then, for $A \subseteq \mathbb{R}$ such that $\beta(A) < +\infty$, define $\widehat{W}_1(A) = W_1^0(A \cap \mathbb{R}_+) + W_1^0((-A) \cap \mathbb{R}_+)$, and $\widehat{W}_2(A) = W_2^0(A \cap \mathbb{R}_+) - W_2^0((-A) \cap \mathbb{R}_+)$. It is immediate to verify that the complex-valued random measure $\widehat{W} = \widehat{W}_1 + i\widehat{W}_2$ satisfies properties (i)-(iv) of Definition 9.2.1.

(b) Reasoning as in formula (5.1.7), one also deduces from Point (iii) of Definition 9.2.1 that $\widehat{W}$ is σ-additive. Further useful properties of Hermitian Gaussian measures are collected in Section 9.6.

If ψ is simple and Hermitian as above, namely if

$$\psi = \sum_{j_i = \pm 1, \cdots, \pm N} c_{j_1}, \cdots, c_{j_q} 1_{A_{j_1}} \times \cdots 1_{A_{j_q}}$$

where the c's are real-valued, $A_{j_i} = -A_{-j_i}$ and $A_{j_i} \cap A_{j_\ell} = \emptyset$ if $i \neq \ell$, one sets

$$\widehat{I}_q(\psi) = \sum_{j_i = \pm 1, \cdots, \pm N}'' c_{j_1}, \cdots, c_{j_q} \widehat{W}(A_{j_1}) \cdots \widehat{W}(A_{j_q}).$$

The double prime in $\sum''$ indicates that one does not sum over the hyperdiagonals, that is, one excludes from the summation terms with $j_i = \pm j_\ell$ for $i \neq \ell$. The following integral notation is then used:

$$\widehat{I}_q(\psi) = \int_{\mathbb{R}^q}'' \psi(x_1, \cdots, x_q)\widehat{W}(dx_1) \cdots W(dx_q),$$

where $\int''$ refers to the fact that diagonals are excluded. One notes that the integral is real-valued because

$$\overline{\widehat{I_q}(\psi)} = \int_{\mathbb{R}^q}'' \overline{\psi(x_1, \cdots, x_q)} \; \overline{\widehat{W}(dx_1)} \cdots \overline{\widehat{W}(dx_q)}$$

$$= \int_{\mathbb{R}^q}'' \psi(-x_1, \cdots, -x_q) \; \widehat{W}(-dx_1) \cdots \widehat{W}(-dx_q) = \widehat{I_q}(\psi).$$

and also that

$$\mathbb{E}\widehat{I_p}(\psi_1)\widehat{I_q}(\psi_2) = \delta_{p,q} q! \, (\psi_1, \psi_2)_{L^2_E(\mathbb{R}^q, d^q\beta)}.$$

One then defines $\widehat{I_q}(\psi)$ for $\psi \in L^2{}_E(\mathbb{R}^q, d^q\beta)$ by the usual completion argument (see again [66, Ch. 4]).

Remark. We described two methods for defining the integral $\widehat{I_q}$. The first one, which uses the isonormal approach is the most direct and elegant. The second, which starts with defining the integral through simple functions is more explicit. It can be made simpler if one only defines the Gaussian integral $\widehat{I_1}$ and then defines $\widehat{I_q}$ through Hermite polynomials.

9.3 Spectral representation

The preceding representations are used extensively in times series analysis, where one relates a "time" representation to a "spectral" representation. This terminology will make more sense below, when we consider stochastic processes. Our aim, at this point, is to relate the Gaussian "time integral" $I_1(h)$ where $h \in L^2(\mathbb{R}, d\xi)$ to the "spectral integral" $\widehat{I_1}(\psi)$ where $\psi \in L^2_E(\mathbb{R}, d\beta)$, in such a way that

$$I_1(h) \stackrel{law}{=} \widehat{I_1}(\psi), \tag{9.3.11}$$

with equality in distribution. Both integrals are normal with mean 0 and common variance. Here

$$I_1(h) = \int_{\mathbb{R}} h(\xi) W(d\xi)$$

where h is real-valued, dW is a real-valued Gaussian measure with Lebesgue control measure $d\nu(\xi) = d\xi$, and

$$\widehat{I_1}(\psi) = \int_{\mathbb{R}} \psi(\lambda)\widehat{W}(d\lambda)$$

where ψ is Hermitian and $d\widehat{W}$ is complex Hermitian Gaussian measure with Lebesgue control measure $d\beta(\lambda) = d\lambda$ (see Definition 9.2.1). We stress that the measure $d\widehat{W}$

verifies (9.2.10) and the control measure $d\beta$ satisfies (9.1.1). What is special here is that the function ψ is taken to be the Fourier transform of h, namely

$$\psi(\lambda) = \frac{1}{\sqrt{2\pi}} \int_{\mathbb{R}} e^{i\xi\lambda} h(\xi) d\xi. \tag{9.3.12}$$

Since h is real, ψ is Hermitian. Parseval's relation gives:

$$||h||^2_{L^2(\nu)} = \int_{\mathbb{R}} h(\xi)^2 \nu(d\xi) = \int_{\mathbb{R}} |\psi(\lambda)|^2 \beta(d\lambda) = ||\psi||^2_{L^2(\beta)}. \tag{9.3.13}$$

One relates multiple integrals in a similar way:

Proposition 9.3.1 *Let $h \in L^2(\mathbb{R}^q, d^q\xi)$ and $\psi \in L^2_E(\mathbb{R}^q, d^q\lambda)$, defined by*

$$\psi(\lambda_1, \cdots, \lambda_q) = \frac{1}{(2\pi)^{q/2}} \int_{\mathbb{R}^q} e^{i(\xi_1\lambda_1 + \cdots + \xi_q\lambda_q)} h(\xi_1, \cdots, \xi_q) d\xi_1 \cdots d\xi_q. \tag{9.3.14}$$

Then

$$I_q(h) \overset{law}{=} \widehat{I}_q(\psi).$$

More generally, if $h_j \in L^2(\mathbb{R}^j, d^j\xi)$ has Fourier transform $\psi_j \in L^2_E(\mathbb{R}^j, d^j\lambda)$, $j = 1, \cdots, q$, then

$$\left(I_1(h_1), \cdots, I_q(h_q)\right) \overset{law}{=} \left(\widehat{I}_1(\psi_1), \cdots, \widehat{I}_q(\psi_q)\right).$$

To prove this proposition, start with (9.3.11) and then proceed as in Proposition 8.4.6 (see also [153]).

9.4 Stochastic processes

In the context of time series analysis, a process can be both defined in "the time domain" and in "the spectral domain". One considers then two Hilbert spaces $L^2(\mathbb{R}, d\xi)$ and $L^2_E(\mathbb{R}, d\lambda)$, both with Lebesgue control measure. Let T be some set, typically $T = \mathbb{R}$, $T = \mathbb{R}_+$ or $T = \mathbb{Z}$. Consider the following stochastic process labeled by $t \in \mathbb{R}$:

$$I_q(h_t) = \int_{\mathbb{R}}' h_t(\xi_1, \cdots, \xi_q) \, dW(\xi_1) \cdots dW(\xi_q), \ t \in T,$$

where W is a real-valued Gaussian random measure with Lebesgue control measure on $\mathbb{R}$ and $h_t \in L^2(\mathbb{R}^q, d^q\xi)$. The prime indicates that one does not integrate over hyperdiagonals. Let ψ_t denote the Fourier transform of h_t defined as in (9.3.14). It follows from Proposition 9.3.1 and the linearity of the integral that for any $n \geq 1$ and $t_1, \cdots, t_n \in T$,

$$\left(I_q(h_{t_1}), \cdots, I_q(h_{t_n})\right) \overset{law}{=} \left(\widehat{I}_q(\psi_{t_1}), \cdots, \widehat{I}_q(\psi_{t_n})\right),$$

where

$$\widehat{I}_q(\psi_t) = \int_{\mathbb{R}}^{''} \psi_t(\xi_1, \cdots, \xi_q) \, d\widehat{W}(\xi_1) \cdots d\widehat{W}(\xi_q), \ t \in T,$$

where $\widehat{W}$ has Lebesgue control measure: $d\beta(\lambda) = d\lambda$ and satisfies (9.1.1). Both $I_q(h_t)$, $t \in T$ and $\widehat{I}_q(\psi_t)$, $t \in T$ are representations of the same stochastic process. The representation $I_q(h_t)$ which involves a real-valued integrand h is said to be in the time domain, whereas the representation $\widehat{I}_q(\psi_t)$, which involves Fourier transforms, is said to be defined in the spectral domain.

9.5 Hermite processes

We shall use Hermite processes to illustrate the preceding discussion. The Hermite processes $\{X_{q,H}(t), \ t \in \mathbb{R}\}$ are self-similar processes with parameter

$$1/2 < H < 1,$$

that is, for any $c > 0$,

$$\{X_{q,H}(ct), \ t \in \mathbb{R}\} \overset{law}{=} \{c^H X_{q,H}(t), \ t \in \mathbb{R}\}, \tag{9.5.15}$$

where $\overset{law}{=}$ indicates here equality of the finite-dimensional distributions. They have stationary increments and are represented by multiple integrals of order q and hence are non-Gaussian when $q \geq 2$. They have representations both in the time domain and in the spectral domain, which are as follows (see [153]):

(a) Time domain representation: for $t \in \mathbb{R}$,

$$X_{q,H}(t) = a_{q,H_0} \int_{\mathbb{R}^q}^{'} \left\{ \int_0^t \prod_{j=1}^q (s - \xi_j)_+^{H_0 - \frac{3}{2}} ds \right\} dW(\xi_1) \dots dW(\xi_q). \tag{9.5.16}$$

(b) Spectral domain representation: for $t \in \mathbb{R}$,

$$X_{q,H}(t) \overset{law}{=} b_{q,H_0} \int_{\mathbb{R}^q}^{''} \frac{e^{i(\lambda_1 + \dots + \lambda_q)t} - 1}{i(\lambda_1 + \dots + \lambda_q)} \prod_{j=1}^q |\lambda_j|^{\frac{1}{2} - H_0} d\widehat{W}(\lambda_1) \dots d\widehat{W}(\lambda_q). \tag{9.5.17}$$

Here

$$H_0 = 1 - \frac{1 - H}{q} \in \left(1 - \frac{1}{2q}, 1\right) \tag{9.5.18}$$

ensures that the integrands are indeed square-integrable. If the Hermite process $X_{q,H}$

is normalized so that $\mathbb{E}(X_{q,H}(1)^2 = 1$, then

$$
a_{q,H_0} = \left(\frac{(q(H_0 - 1) + 1)(2q(H_0 - 1) + 1)\Gamma(\frac{3}{2} - H_0)^q}{q![\Gamma(H_0 - \frac{1}{2})\Gamma(2 - 2H_0)]^q} \right)^{1/2} \tag{9.5.19}
$$

$$
= \left(\frac{H(2H - 1)\Gamma(\frac{1}{2} + \frac{1-H}{q})^q}{q![\Gamma(\frac{1}{2} - \frac{1-H}{q})\Gamma(\frac{2(1-H)}{q})]^q} \right)^{1/2} = A_{q,H}, \tag{9.5.20}
$$

and

$$
b_{q,H_0} = \left(\frac{(q(H_0 - 1) + 1)(2q(H_0 - 1) + 1)}{q![2\Gamma(2 - 2H_0)\sin((H_0 - \frac{1}{2})\pi)]^q} \right)^{1/2} \tag{9.5.21}
$$

$$
= \left(\frac{(qH(2H - 1)}{q![2\Gamma(\frac{2(1-H)}{q})\sin((\frac{1}{2} - \frac{1-H}{q})\pi)]^q} \right)^{1/2} = B_{q,H}. \tag{9.5.22}
$$

For more details see [117, 153, 154] and for additional representations, see [117].

The Hermite processes have the same covariance as fractional Brownian motion, namely (8.3.13). In fact, the Hermite process of order $q = 1$ is nothing else than fractional Brownian motion, introduced in Example 8.3.4 and in that case $H_0 = H$. One thus obtains the following representations of fractional Brownian motion:

(a) Time domain representation of fractional Brownian motion: for $t \in \mathbb{R}$,

$$
X_{1,H}(t) = A_{1,H} \int_{\mathbb{R}} \left\{ \int_0^t (s - \xi)_+^{H-\frac{3}{2}} ds \right\} dW(\xi)
$$

$$
= \frac{A_{1,H}}{H - 1/2} \int_{\mathbb{R}} \left\{ (t - \xi)_+^{H-1/2} - (-\xi)_+^{H-1/2} \right\} dW(\xi). \tag{9.5.23}
$$

(b) Spectral domain representation fractional Brownian motion: for $t \in \mathbb{R}$,

$$
X_{1,H}(t) \overset{law}{=} B_{1,H} \int_{\mathbb{R}} \frac{e^{i\lambda t} - 1}{i\lambda} |\lambda|^{\frac{1}{2}-H} d\widehat{W}(\lambda). \tag{9.5.24}
$$

These representations are not unique. Representations (9.5.23) and (9.5.24) are valid for $0 < H < 1$, $H \neq 1/2$. For more details see [25, 135, 154].

9.6 Further properties of Hermitian Gaussian measures

Here are some of the properties of the complex Hermitian Gaussian measure $\widehat{W} = \widehat{W}_1 + i\widehat{W}_2$, where $\widehat{W}_1$ and $\widehat{W}_2$ denote respectively the real and imaginary parts $\widehat{W}$. Recall that β denotes its symmetric control measure (see Definition 9.2.1), and relation (9.2.9)).

Proposition 9.6.1 (a) $\mathbb{E}(\widehat{W}(A))^2 = 0$ if $A \cap (-A) = \emptyset$.
(b) $\mathbb{E}\widehat{W}_1(A)\widehat{W}_2(B) = 0$ (that is, $\widehat{W}_1$ and $\widehat{W}_2$ are independent).
(c) $\mathbb{E}\widehat{W}_1^2(A) = \mathbb{E}\widehat{W}_2^2(A) = \dfrac{1}{2}\mathbb{E}|\widehat{W}(A)|^2 = \dfrac{1}{2}\beta(A)$ if $A \cap (-A) = \emptyset$.
(d) $\widehat{W}(A_1), \widehat{W}(A_2), \ldots, \widehat{W}(A_n)$ are independent if $A_1 \cup (-A_1), \ldots, A_n \cup (-A_n)$ are disjoint.

Proof.
(a) $\mathbb{E}(\widehat{W}(A))^2 = \mathbb{E}\widehat{W}(A)\overline{\widehat{W}(-A)} = \beta(A \cap (-A)) = \emptyset$ if $A \cap (-A) = \emptyset$.
(b) $\mathbb{E}\widehat{W}_1(A)\widehat{W}_2(B)$

$$= \frac{1}{4}[\mathbb{E}(\widehat{W}(A) + \overline{\widehat{W}(A)})(\widehat{W}(B) + \overline{\widehat{W}(B)})]$$

$$= \frac{1}{4}[\mathbb{E}(\widehat{W}(A) + \widehat{W}(-A))(\widehat{W}(B) - \widehat{W}(-B))]$$

$$= \frac{1}{4}[\beta(A \cap B) - \beta((A \cap (-B)) + \beta((-A) \cap B) - \beta((-A) \cap (-B))] = 0$$

by (9.1.1).
(c) If $A \cap (-A) = \emptyset$, then by (a) and (b),

$$0 = \mathbb{E}(\widehat{W}(A))^2 = \mathbb{E}(\widehat{W}_1(A))^2 + 2i\widehat{W}_1(A)\widehat{W}_2(A) - (\widehat{W}_2(A))^2]$$
$$= \mathbb{E}(\widehat{W}_1(A))^2 - \mathbb{E}(\widehat{W}_2(A))^2.$$

On the other hand, by (9.2.9),

$$\beta(A) = \mathbb{E}|\widehat{W}(A)|^2 = \mathbb{E}(\widehat{W}_1(A))^2 + \mathbb{E}(\widehat{W}_2(A))^2.$$

(d) If A and B are such that $A \cup (-A)$ are $B \cup (-B)$ are disjoint, then

$$\mathbb{E}\widehat{W}((A \cup (-A))\overline{\widehat{W}(B \cup (-B))} = \beta((A \cup (-A)) \cap (B \cup (-B))) = 0.$$

∎

9.7 Caveat about normalizations

Let $d\nu(\xi) = d\xi$, $d\beta(\lambda) = d\lambda$ and set

$$\psi(\lambda) = \frac{1}{\sqrt{2\pi}}\int_{\mathbb{R}} e^{i\xi\lambda}h(\xi)d\xi \tag{9.7.25}$$

with $h \in L^2(\mathbb{R}, d\nu)$ and $\psi \in L^2_E(\mathbb{R}, d\beta)$. There are, unfortunately, different conventions for choosing the multiplicative constant in front of the Fourier transform integral. To illustrate this, let $a > 0$ and define

$$\psi_a(\lambda) = \frac{a}{\sqrt{2\pi}}\int_{\mathbb{R}} e^{i\xi\lambda}h(\xi)d\xi, \tag{9.7.26}$$

that is

$$\psi_a(\lambda) = a\psi(\lambda). \tag{9.7.27}$$

The normalization constant a is typically chosen to be either

$$2\pi, \quad \sqrt{2\pi}, \quad 1, \quad \text{or} \quad 1/\sqrt{2\pi}.$$

If $a = 1$, then $\psi_a(\lambda) = \psi(\lambda)$. If $a \neq 1$, then for this choice not to have an effect on Parseval's relation (9.3.13), namely

$$\|h\|^2_{L^2(\nu)} = \|\psi\|^2_{L^2(\beta)},$$

one should choose as control measure

$$\beta_a(d\lambda) = a^{-2}\beta(d\lambda),$$

so that

$$\|\psi\|^2_{L^2(\beta)} = \int_{\mathbb{R}} |\psi(\lambda)|^2 \beta(d\lambda) = \int_{\mathbb{R}} |a^{-1}\psi_a(\lambda)|^2 \beta(d\lambda)$$

$$= \int_{\mathbb{R}} |\psi_a(\lambda)|^2 \beta_a(d\lambda) = \|\psi_a\|^2_{L^2(\beta_a)}.$$

To shed light on the potential problems define the Fourier transform as (9.7.26) with some $a \neq 1$ and let $\zeta = \psi_a$. We shall examine the different ways of defining $\widehat{I}(\zeta)$.

(i) One can define $\widehat{I}(\zeta)$ as $\int_{\mathbb{R}} \zeta d\widehat{W}_a$, where $\widehat{W}_a$ has a control measure β_a. This is the best perspective since $E|I(h)|^2 = \|h\|^2_\nu = \|\psi\|^2_\beta = \|\zeta\|^2_{\beta_a}$. This means that one works with $L^2_E(\mathbb{R}, d\beta_a)$ rather than $L^2_E(\mathbb{R}, d\beta)$.

(ii) If the measure β has a special significance, for example if it is Lebesgue measure, then one often wants to continue using it. One can then define $\widehat{I}(\zeta)$ as usual, namely as $\int_{\mathbb{R}} \zeta d\widehat{W}$ where $\widehat{W}$ has control measure β. One then loses the isometry with the time domain since $EI(h)^2 = \|h\|^2_{L^2(\nu)} = \|\psi\|^2_{L^2(\beta)} = a^{-2}\|\zeta\|_{L^2(\beta)} \neq \|\zeta\|_{L^2(\beta)}$ but this point of view is acceptable, specially if one tends not to refer to a "time representation" of the type $I(h)$, where h is real-valued. It is, however, necessary to keep constantly track of the factor a.

The following example provides an illustration.

Example 9.7.1 Suppose $\nu(d\xi) = d\xi$, $\beta(d\lambda) = d\lambda$ and define the Fourier transform of $h(\xi) \in L^2(\mathbb{R}, d\xi)$ as

$$\zeta(\lambda) = \int_{\mathbb{R}} e^{i\zeta\lambda}h(\xi)d\xi, \tag{9.7.28}$$

that is, ζ is our ψ_a with $a = 1/\sqrt{2\pi}$, a common choice in time series analysis. If we take the *point of view (i)*, and define

$$\widehat{I}(\zeta) = \int_{\mathbb{R}} \zeta(\lambda)d\widehat{W}_a(\lambda)$$

with

$$E|d\widehat{W}_a(\lambda)|^2 = \beta_a(d\lambda) = a^{-2}d\lambda = 2\pi \, d\lambda,$$

then we have

$$I(h) \overset{law}{=} \widehat{I}(\zeta), \qquad \text{that is,} \qquad \|h\|^2_{L^2(\nu)} = \|\zeta\|^2_{L^2(\beta_a)}. \tag{9.7.29}$$

But if we take *point of view (ii)* instead and define

$$\widehat{I}(\zeta) = \int_{\mathbb{R}} \zeta(\lambda)d\widehat{W}(\lambda) \tag{9.7.30}$$

with $E|d\widehat{W}(\lambda)|^2 = d\lambda$, then by (9.7.27),

$$I(h) \overset{law}{=} a^{-1}\widehat{I}(\zeta) = (2\pi)^{1/2}\widehat{I}(\zeta) \tag{9.7.31}$$

and the isometry is lost.

Now extend all these definitions to many variables and focus on the formula (6.4.17) for $\widehat{I}_p(f)\widehat{I}_q(g)$. The best is to take *point of view (i)* and suppose that the integrals $\widehat{I}$ are defined with respect to $\widehat{W}_a$, in which case, one needs to replace each $\beta(d\lambda)$ in (9.1.7) by $\beta_a(d\lambda) = a^{-2}d\lambda = (2\pi)d\lambda$. In this case, by (9.1.7),

$$f\overline{\otimes}_r g(x_1, \ldots, x_{p+q-2r}) =$$
$$(2\pi)^r \int_{\mathbb{R}^r} f(\lambda_1, \ldots, \lambda_r, x_1, \ldots, x_{p-r})$$
$$\cdot g(-\lambda_1, \ldots, -\lambda_r, x_{p-r+1}, \ldots, x_{p+q-2r})d\lambda_1 \ldots d\lambda_r,$$

and (6.4.17) continues to hold.

Alternatively, if we take *point of view (ii)*, namely that the integrals $\widehat{I}$ are defined with respect to dW, then relation (6.4.17) should be replaced by

$$\widehat{I}_p(f)\,\widehat{I}_q(g) = \sum_{r=0}^{p\wedge q}(2\pi)^r r!\binom{p}{r}\binom{q}{r}\widehat{I}_{p+q-2r}(f\overline{\otimes}_r g), \tag{9.7.32}$$

where

$$f\overline{\otimes}_r g(x_1, \ldots, x_{p+q-2r}) = \tag{9.7.33}$$
$$\int_{\mathbb{R}^r} f(\lambda_1, \ldots, \lambda_r, x_1, \ldots, x_{p-r})g(-\lambda_1, \ldots, -\lambda_r, x_{p-r+1}, \ldots, x_{p+q-2r})d\lambda_1 \ldots d\lambda_r.$$

A direct way to check (9.7.32), under *point of view (ii)*, is to replace in formula (6.4.17), $f \in L^2_E(\mathbb{R}^p)$ and $g \in L^2_E(\mathbb{R}^q)$ by $a^p f$ and $a^q g$ respectively, and $f\overline{\otimes}_r g$ by $a^{p+q-2r}f\overline{\otimes}_r g$, where $a = (2\pi)^{-1/2}$ and $f\overline{\otimes}_r g$ is as in (9.7.33).

While the resulting formula for $\widehat{I}_p(f)\widehat{I}_q(g)$ is the same as if one had considered integrals with respect to $\widehat{W}_a$, the perspective is different.

10

Some facts about Charlier polynomials

In this short chapter, we present some explicit connections between Poisson random measures and Charlier polynomials.

There is not a unique accepted definition of Charlier polynomials. We shall consider here two related sets of polynomials, namely $\{c_n(x, a) : n \geq 0\}$ and $\{C_n(x, a) : n \geq 0\}$. The first is considered, for example, by Ogura [101] and Roman [130]. The second appears in Kabanov [49] and Surgailis [150].

The Charlier polynomials $c_n(x, a)$, $x = 0, 1, 2, ...,$ $a > 0$, are orthogonal with respect to the Poisson distribution with mean a, that is,

$$\sum_{x=0}^{\infty} c_n(x, a) c_m(x, a) e^{-a} \frac{a^x}{x!} = \frac{n!}{a^n} \delta_{nm}$$

(δ_{nm} is the Kronecker delta), and have generating function

$$\sum_{n=0}^{\infty} \frac{c_n(x, a)}{n!} t^n = e^{-t} \left(1 + \frac{t}{a} \right)^x . \tag{10.0.1}$$

They satisfy the recurrence relation: $c_0(x, a) = 1$,

$$c_{n+1}(x, a) = a^{-1} x c_n(x - 1, a) - c_n(x, a) \tag{10.0.2}$$

(see [101, Eq. (5)] or [130, Eq. 14.3.0]). There are several other recurrence equations (see [130, p. 121]). Following again [101], these polynomials can be nicely described in terms of the backward shift operator $Bf(x) = f(x - 1)$ and the Poisson probability function

$$\Pi(x, a) = e^{-a} \frac{a^x}{x!}, \quad x = 0, 1, ...,$$

as

$$c_n(x, a) = \left(\frac{x}{a} B - 1 \right) c_{n-1}(x),$$

G. Peccati, M.S. Taqqu: Wiener Chaos: Moments, Cumulants and Diagrams –
A survey with computer implementation.
© Springer-Verlag Italia 2011

which gives (10.0.2), and also

$$c_n(x, a) = \left(\frac{x}{a}B - 1\right)^n 1,$$

(compare with formula (8.1.1) for Hermite polynomials). Expanding the right-hand side of the previous expression, using the binomial expansion, yields

$$c_n(x, a) = \sum_{r=0}^{n}(-1)^{n-r}\binom{n}{r}\left(\frac{xB}{a}\right)^r 1$$

$$= \sum_{r=0}^{n}(-1)^{n-r}\binom{n}{r}\frac{x^{(r)}}{a^r},$$

where $x^{(0)} = 1$, $x^{(1)} = xB1 = x$, $x^{(2)} = (xB)^2 1 = xB(xB)1 = xBx = x(x-1)$, and in general

$$x^{(r)} = (xB)^r 1 = x(x-1)\cdots(x-r+1).$$

Thus, the first few Charlier polynomials are

$$c_0(x, a) = 1$$

$$c_1(x, a) = -a^{-1}(a - x) = \frac{x}{a} - 1$$

$$c_2(x, a) = a^{-2}(ah2 - x - 2ax + x^2) = \frac{x^{(2)}}{a^2} - 2\frac{x}{a} + 1$$

$$c_3(x, a) = -a^3(a^3 - 2x - 3ax - 3a^2x + 3x^2 + 3ax^2 - x^3)$$

$$= \frac{x^{(3)}}{a^3} - 3\frac{x^{(2)}}{a^2} + 3\frac{x}{a} - 1.$$

Remark. A slight modification of $c_n(x, a)$ is given by Engel [26], who defines $K_n(x, a) = (-a)^n c_n(x, a)/n!$.

The second set of Charlier polynomials is defined from the first by setting

$$C_n(x, a) = a^n c_n(x + a, a), \; n \geq 0, \; x = -a, -a + 1, \ldots \qquad (10.0.3)$$

Its generating function is

$$\sum_{n=0}^{\infty}\frac{C_n(x, a)}{n!}t^n = e^{-ta}(1 + t)^{x+a}. \qquad (10.0.4)$$

The first few are $C_0(x, a) = 1$, $C_1(x, a) = x$, $C_2(x, a) = x^2 - x - a$. We call the polynomials C_n "centered Charlier". The polynomials C_n are particularly suitable for centered Poisson variables. Thus, if N is Poisson with mean a and $\widehat{N} = N - a$, then the orthogonality relation becomes

$$E[C_n(\widehat{N}, a)C_m(\widehat{N}, a)] = E[C_n(N - a, a)C_m(n - a, a)]$$

$$= a^{2n}E[c_n(N, a)c_m(N, a)]$$

$$= n!a^n\delta_{nm}$$

(compare with (8.1.5)). Call $J(t)$ the generating function (10.0.4), and observe that its right-hand side verifies

$$(1+t)\frac{\partial J}{\partial t} = (x-at)J.$$

Applying this relation to the left-hand side yields the recursion relation

$$C_{n+1}(x,a) = (x-n)C_n(x,a) - anC_{n-1}(x,a), \ n \geq 1. \tag{10.0.5}$$

Remark. The polynomials $G_n(t,x,\lambda)$ considered by Kabanov in [49] equal $G_n(t,x,\lambda) = C_n(x,\lambda t)/n!$.

We now present a statement relating Charlier polynomials and multiple integrals.

Proposition 10.0.2 *Let $\widehat{N}$ be a centered Poisson measure over $(Z,\mathcal{Z})$, with σ-finite and non atomic control measure ν. Then, for every $B \in \mathcal{Z}$ such that $\nu(B) < \infty$, and every $n \geq 1$, one has that*

$$C_n(\widehat{N}(B),\nu(B)) = I_n^{\widehat{N}}(1_B^{\otimes n}) = \int_{Z_0^n} 1_B(z_1) \cdots 1_B(z_n)\widehat{N}(dz_1)\ldots\widehat{N}(dz_n).$$
$$\tag{10.0.6}$$

Proof. The result is true for $n = 1$, since $I_1^{\widehat{N}}(1_B) = \widehat{N}(B) = C_1(\widehat{N}(B),\nu(B))$. It is also true for $n = 0$ since, in this case, both sides of (10.0.6) are equal to 1 (by convention). The proof is completed by induction on n, as an application of the multiplication formula (6.5.21). This formula applied to $f(z_1,...,z_n)$ and $g(z_1)$ yields

$$I_n^{\widehat{N}}(f)I_1^{\widehat{N}}(g) = I_{n+1}^{\widehat{N}}(f \star_0^0 g) + n\big[I_n^{\widehat{N}}(f \star_1^0 g) + I_{n-1}^{\widehat{N}}(f \star_1^1 g)\big].$$

If $f(z_1,...,z_n) = 1_B^{\otimes n}(z_1,...,z_n) = \prod_{j=1}^n 1_B(z_j)$ and $g(z_1) = 1_B(z_1)$, we get $f \star_0^0 g = 1_B^{\otimes(n+1)}$, $f\star_1^0 g = 1_B^{\otimes n}$, and $f\star_1^1 g = 1_B^{\otimes(n-1)} \times \nu(B)$. Plugging these expressions into the previous computations, yields therefore

$$\begin{aligned}
I_{n+1}^{\widehat{N}}(1_B^{\otimes(n+1)}) &= I_n^{\widehat{N}}(1_B^{\otimes n}))I_1^{\widehat{N}}(1_B) - n\big[I_n^{\widehat{N}}(1_B^{\otimes n}) + \nu(B)I_{n-1}^{\widehat{N}}(1_B^{\otimes(n-1)})\big] \\
&= (\widehat{N}(B) - n)I_n^{\widehat{N}}(1_B^{\otimes n}) - \nu(B)nI_{n-1}^{\widehat{N}}(1_B^{\otimes n-1}) \\
&= (\widehat{N}(B) - n)C_n(\widehat{N}(1_B),\nu(B)) - \nu(B)nC_{n-1}(\widehat{N}(1_B),\nu(B)) \\
&= C_{n+1}(\widehat{N}(1_B),\nu(B)),
\end{aligned}$$

by using the induction hypothesis and the recurrence relation (10.0.5). ∎

Example 10.0.3 One has that

$$I^{\widehat{N}}(1_B \otimes 1_B) = C_2(\widehat{N}(B)) = \widehat{N}(B)^2 - \widehat{N}(B) - \nu(B).$$

Consider now the set

$$B_1^{i_1} \times \ldots \times B_k^{i_k} = \underbrace{(B_1 \times \cdots \times B_1)}_{i_1 \text{ times}} \times \cdots \times \underbrace{(B_k \times \ldots \times B_k)}_{i_k \text{ times}},$$

where $i_1 + \ldots + i_k = n$. It is the tensor product of B_1 repeated i_1 times, B_2 repeated i_2 times,..., B_k repeated i_k times. As shown in the next statement, if the sets B_i are disjoint, then the multiple Poisson integral of the indicator of $B_1^{i_1} \times \ldots \times B_k^{i_k}$ can be expressed as a product of Charlier polynomials.

Proposition 10.0.4 *Assume that the sets $B_1, \ldots, B_k$ are disjoint and of finite ν measure. Then,*

$$I_n^{\widehat{N}}\left(1_{B_1^{i_1} \times \cdots \times B_k^{i_k}}\right) = \prod_{a=1}^{k} C_{i_a}(\widehat{N}(B_a), \nu(B_a)). \tag{10.0.7}$$

Proof. One has clearly that $1_{B_1^{i_1} \times \cdots \times B_k^{i_k}} = 1_{B_1}^{\otimes i_1} \otimes \cdots \otimes 1_{B_k}^{\otimes i_k}$. In view of Proposition 10.0.2, it is therefore sufficient to show that

$$I_n^{\widehat{N}}\left(1_{B_1}^{\otimes i_1} \otimes \cdots \otimes 1_{B_k}^{\otimes i_k}\right) = \prod_{a=1}^{k} I_{i_a}^{\widehat{N}}(1_{B_a}^{\otimes i_a}). \tag{10.0.8}$$

To prove relation (10.0.8), observe that the functions 1_{B_a}, $a = 1, \ldots, k$, have mutually disjoint supports, so that, for $b \neq a$, contractions of the type $1_{B_a}^{\otimes i_a} \star_r^l 1_{B_b}^{\otimes i_b}$ equal zero unless $r = l = 0$ (in which case one has $1_{B_a}^{\otimes i_a} \star_0^0 1_{B_b}^{\otimes i_b} = 1_{B_a}^{\otimes i_a} \otimes 1_{B_b}^{\otimes i_b}$). An iterated application of the multiplication formula (6.5.21) yields therefore that

$$\prod_{a=1}^{k} I_{i_a}^{\widehat{N}}(1_{B_a}^{\otimes i_a}) = I_n^{\widehat{N}}\left(1_{B_1}^{\otimes i_1} \star_0^0 \cdots \star_0^0 1_{B_k}^{\otimes i_k}\right) = I_n^{\widehat{N}}\left(1_{B_1}^{\otimes i_1} \otimes \cdots \otimes 1_{B_k}^{\otimes i_k}\right),$$

thus concluding the proof.   ∎

We conclude the chapter with a proof of the fact that Poisson measures enjoy a chaotic representation property.

Corollary 10.0.5 *Let the notation and assumptions of this chapter prevail. Then, the completely random measure $\widehat{N}$ enjoys the chaotic representation property (5.9.78).*

Proof. Proposition 10.0.4 and formula (10.0.4) imply that every random variable of the type $F = \prod_{k=1,\ldots,N}(1 + t_k)^{\widehat{N}(C_k) + \nu(C_k)}$, where the C_k are disjoint, can be represented as a series of multiple integrals. Since the linear span of random variables such as F is dense in $L^2(\sigma(\widehat{N}), \mathbb{P})$, the proof is concluded by using the fact that random variables enjoying the chaotic decomposition (5.9.78) form a Hilbert space.   ∎

Remark. Hence any random variable in $L^2(\sigma(\widehat{N}), \mathbb{P})$ has the chaotic expansion (5.9.78). Contrary to the Gaussian case, however, where each term $I_n^G(f_n))$ can be expressed as a sum if products of Hermite polynomials (see, for example, (8.4.17)), here $I_n^{\widehat{N}}(f_n)$) cannot, in general, be expressed as a sum of products of Charlier polynomials, unless, for example, when $f_n = 1_{B_1^{i_1} \times \cdots \times B_k^{i_k}}$ as in Proposition 10.0.4. This is because $\int_Z f(z)G(dz)$ is Gaussian when G is a Gaussian measure, but $\int_{\mathbb{R}} f(z)\widehat{N}(dz)$ is, in general, not Poisson, and hence its powers cannot be readily expressed in terms of Charlier polynomials.

11

Limit theorems on the Gaussian Wiener chaos

In a recent series of papers (see [69, 82, 86, 87, 90, 95, 98, 103, 110, 111] for the Gaussian case, and [106, 107, 108, 109, 114] for the Poisson case) a set of new results has been established, allowing to obtain neat Central Limit Theorems (CLTs) for sequences of random variables belonging to a fixed Wiener chaos of some Gaussian or Poisson field. The techniques adopted in the above references are quite varied, as they involve stochastic calculus (see [98, 103, 111]), Malliavin calculus (see [82, 95, 110]), Stein's method (see [86, 87, 90, 106]) and decoupling (see [108, 107, 109]). However, all these contributions may be described as "drastic simplifications" of the method of *moments and cumulants* (see, for example, [12, 66], as well as the discussion below) which is a common tool for proving weak convergence results for non-linear functionals of random fields.

The aim of this chapter (and of the following) is to draw the connection between the above quoted CLTs and the method of moments and cumulants into further light, by providing a detailed discussion of the combinatorial implications of the former. This discussion will involve the algebraic formalism introduced in Chapters 2–4, as well as the diagram formulae proved in Chapter 7.

This chapter mainly focuses on random variables belonging to a Gaussian Wiener chaos, whereas Chapter 12 deals with random objects in a Poisson chaos.

In order to appreciate the subtlety of the issues faced in this chapter, in the next section we list some well-known properties of the laws of chaotic random variables.

11.1 Some features of the laws of chaotic random variables (Gaussian case)

We now consider an isonormal Gaussian process $X = \{X(h) : h \in \mathfrak{H}\}$ over some real separable Hilbert space $\mathfrak{H}$. Recall (see Chapter 8) that the notion of isonormal Gaussian process is more general than the one of Gaussian measure: indeed, as shown in Example 8.3.2, the linear space generated by a Gaussian measure can be always

G. Peccati, M.S. Taqqu: Wiener Chaos: Moments, Cumulants and Diagrams –
A survey with computer implementation.
© Springer-Verlag Italia 2011

embedded into an isonormal Gaussian process. For every $q \geq 1$, random variables of the type $I_q^X(f)$, $f \in \mathfrak{H}^{\odot q}$, are defined via formula (8.4.17) (these objects compose the qth Wiener chaos associated with X). We also recall that, if $\mathfrak{H} = L^2(Z, \mathcal{Z}, \nu)$, where $(Z, \mathcal{Z})$ is a Polish space endowed with its Borel σ-field, and the measure ν is non-atomic and σ-finite, then, for every $q \geq 2$, the symmetric tensor product $\mathfrak{H}^{\odot q}$ can be identified with the space $L_s^2(\nu^q)$ of symmetric and square-integrable functions on Z^q. Also, in this case one has that $I_q^X(f) = I_q^G(f)$, where G is the Gaussian measure (with control ν) $A \mapsto G(A) = X(\mathbf{1}_A)$, where $A \in \mathcal{Z}$ is such that $\nu(A) < \infty$.

The following remarkable facts are in order.

- For every integer $q \geq 1$ and every kernel $f \in \mathfrak{H}^{\odot q}$ such that $f \neq 0$, the law of the random variable $I_q^X(f)$ is absolutely continuous with respect to the Lebesgue measure. This fact is proved by Shigekawa in [140] by means of Malliavin-type operators. Actually, in [140] something more general is actually shown, namely: if the random variable Y is not a.s. constant and it is equal to a finite linear combination of random variables of the type $I_q^X(f)$, then the law of Y is absolutely continuous with respect to the Lebesgue measure.

- If $q = 2$, then there exists a sequence $\{\xi_i : i \geq 1\}$ of i.i.d. centered standard Gaussian random variables such that

$$I_2^X(f) = \sum_{i=1}^{\infty} \lambda_i \left(\xi_i^2 - 1\right), \qquad (11.1.1)$$

where the series converges in $L^2(\mathbb{P})$, and $\{\lambda_i : i \geq 1\}$ is the sequence of eigenvalues of the Hilbert-Schmidt operator (from $\mathfrak{H}$ into $\mathfrak{H}$) given by

$$h \mapsto f \otimes_1 h,$$

where $\otimes_1$ indicates a contraction of order 1. In particular, $I_2^X(f)$ admits some finite exponential moment, and the law of $I_2^X(f)$ is determined by its moments. A proof of these facts is given, for example, in Janson [46, Ch. VI], see in particular Theorem 6.1 in [46].

- If $q \geq 3$, the law of $I_q^X(f)$ may not be determined by its moments. A complete discussion of this point can be found in the fundamental paper by Slud [142].

- For every $q \geq 2$, a non-zero random variable of the form $I_q^X(f)$ *cannot be Gaussian*. A quick proof of this fact can be deduced from formula (11.2.7). Indeed, if $I_q^X(f)$ was non-zero and Gaussian, then one would have that $\chi_4(I_q^X(f)) = 0$ and therefore, thanks to (11.2.7), $f \otimes_p f = 0$ for every $p = 1, ..., q-1$. A little inspection shows that this last condition implies that $f = 0$, thus yielding a contradiction. Another way of proving the non-Gaussianity of multiple integrals consists in proving that a non-zero random variable of the type $I_q^X(f)$, $q \geq 2$, has some infinite exponential moments. This approach is detailed, for example, in [46, Ch. VI].

- For $q \geq 3$, and except for trivial cases (for instance, $f = 0$), there does not exist a general explicit formula for the characteristic function of $I_q^X(f)$.

11.2 A useful estimate

The main result discussed in this chapter is the forthcoming Theorem 11.4.1, providing neat necessary and sufficient conditions ensuring that a sequence of random variables inside a fixed Gaussian Wiener chaos converges to a normal distribution. We shall provide a new combinatorial proof of this result, which is based on a simple estimate for deterministic integrals built from connected and non-flat diagrams. The formal statement of such an estimate is the object of the present section.

Fix integers $q, m \geq 2$. We recall some notation from Section 7.2. The partition $\pi^* \in \mathcal{P}([mq])$ is given by

$$\pi^* = \{b_1, b_2, ..., b_m\} = \{\{1, ..., q\}, \{q + 1, ..., 2q\}, ..., \{(m - 1)q + 1, ..., mq\}\} \tag{11.2.2}$$

(compare with (6.1.1)). The class $\mathcal{M}_2([mq], \pi^*)$ is the collection of all those $\sigma \in \mathcal{P}([mq])$ such that $\sigma \wedge \pi^* = \hat{0}$, $\sigma \vee \pi^* = \hat{1}$, and every block of σ contains exactly two elements (compare with (7.2.21)). Recall that, according to (7.2.27), a partition $\sigma \in \mathcal{P}([mq])$ is an element of $\mathcal{M}_2([mq], \pi^*)$ if and only if the diagram $\Gamma(\pi^*, \sigma)$ (see Section 4.1) is Gaussian, connected and non-flat, which is equivalent to saying that the multigraph $\hat{\Gamma}(\pi^*, \sigma)$ (see Section 4.3) is connected and has no loops. Observe that $\mathcal{M}_2([mq], \pi^*) = \emptyset$ if the integer mq is odd.

Now consider a function $f \in L_s^2(\nu^q)$, where ν is some non-atomic σ-finite measure over a Polish space $(Z, \mathcal{Z})$. Fix $m \geq 3$. Recall that we defined the m-tensor product function

$$\underbrace{f \otimes_0 \cdots \otimes_0 f}_{m \text{ times}} (x_1, ..., x_{mq}) = \prod_{j=1}^{m} f(x_{(j-1)q+1}, ..., x_{jq}). \tag{11.2.3}$$

According to the conventions of Chapter 6 and Chapter 7, for every $m \geq 2$ such that mq is even and every $\sigma \in \mathcal{M}_2([mq], \pi^*)$, one can define a function

$$(x_1, ..., x_{mq/2}) \mapsto f_{\sigma,m}(x_1, ..., x_{mq/2}),$$

in $mq/2$ variables by using the following procedure:

–   build the tensor product $f \otimes_0 \cdots \otimes_0 f$ appearing in formula (11.2.3);
–   identify two variables x_i and x_j in the argument of $f \otimes_0 \cdots \otimes_0 f$ if and only if i and j are in the same block of σ.

See Example 7.3.2.

The main findings of this chapter are based on the following results.

Proposition 11.2.1 *Let the above notation prevail. Fix integers $m \geq 2$ and $q \geq 3$ such that mq is even, and consider a function $f \in L_s^2(\nu^q)$. Then, there exists $r = 1, ..., q-1$ such that*

$$\left| \int_{Z^{mq/2}} f_{\sigma,m} d\nu^{mq/2} \right| \leq \|f \otimes_r f\|_{L^2(\nu^{2q-2r})} \times \|f\|_{L^2(\nu^q)}^{m-2}, \tag{11.2.4}$$

where $f \otimes_r f$ indicates the contraction of f of order r, given by (6.2.16).

The proof of Proposition 11.2.1 is deferred to Section 11.6.

We now state the following formula (proved in [98]) which gives an explicit expression for the fourth cumulant of a random variable of the type $I_q^X(f)$, $f \in \mathfrak{H}^{\odot q}$, $q \geq 2$:

Proposition 11.2.2 Let $I_q^X(f)$, $f \in \mathfrak{H}^{\odot q}$, $q \geq 2$. Then

$$\chi_4\left(I_q^X(f)\right)$$

$$= \mathbb{E}\left[I_q^X(f)^4\right] - 3\left(\mathbb{E}\left[I_q^X(f)^2\right]\right)^2 \tag{11.2.5}$$

$$= \mathbb{E}\left[I_q^X(f)^4\right] - 3(q!)^2 \|f\|_{\mathfrak{H}^{\otimes q}}^4 \tag{11.2.6}$$

$$= \sum_{p=1}^{q-1} \frac{(q!)^4}{(p!(q-p)!)^2}\left\{\|f \otimes_p f\|_{\mathfrak{H}^{\otimes 2(q-p)}}^2 + \binom{2q-2p}{q-p}\left\|\widetilde{f \otimes_p f}\right\|_{\mathfrak{H}^{\otimes 2(q-p)}}^2\right\}. \tag{11.2.7}$$

The proof of Proposition 11.2.2 can also be found in Section 11.6.

As pointed out in Section 11.1, formula (11.2.7) can be used in order to prove that, for every isonormal Gaussian process X, every $q \geq 2$ and every $f \in \mathfrak{H}^{\odot q}$, the random variable $I_q^X(f)$ *cannot be Gaussian* (see also [46, Ch. 6]).

Corollary 11.2.3 Fix $q \geq 2$ and $f \in \mathfrak{H}^{\odot q}$ such that $\mathbb{E}\left[I_q^X(f)^2\right] = 1$. Then neither can the distribution of $I_q^X(f)$ be normal nor can we have $\mathbb{E}\left[I_q^X(f)^4\right] = 3$.

Proof. If either $I_q^X(f)$ had a normal distribution or if $\mathbb{E}\left[I_q^X(f)^4\right] = 3$, then, according to Proposition 11.2.2, for every $p = 1, ..., q-1$, we would have $f \otimes_p f = 0$. Thus, for each $v \in \mathfrak{H}^{\otimes(q-p)}$ we obtain

$$0 = (f \otimes_p f, v \otimes v)_{\mathfrak{H}^{\otimes 2(q-p)}} = \|(f, v)_{\mathfrak{H}^{\otimes(q-p)}}\|_{\mathfrak{H}^{\otimes p}}^2,$$

which implies $f = 0$, contradicting $\mathbb{E}\left[I_q^X(f)^2\right] = 1$. ∎

11.3 A general problem

In the present chapter and the next, we will be interested in several variations of the following problem.

Problem A. *Let φ be a completely random Gaussian or Poisson measure, with control measure ν, where ν is a σ-finite and non-atomic measure over some Polish space $(Z, \mathcal{Z})$. For $d \geq 1$ and $q_1, ..., q_d \geq 1$, let $\{f_j^{(n)} : j = 1, ..., d, n \geq 1\}$ be a collection* *of kernels such that* $f_j^{(n)} \in L_s^2\left(Z^{q_j}, \mathcal{Z}^{q_j}, \nu^{q_j}\right)$ *(the vector* $(q_1, ..., q_d)$ *does not depend on* n*), and*

$$\lim_{n\to\infty} \mathbb{E}\left[I_{q_i}^\varphi\left(f_i^{(n)}\right) I_{q_j}^\varphi\left(f_j^{(n)}\right)\right] = C\,(i,j)\,, \quad 1 \le i,j \le d, \qquad (11.3.8)$$

where the integrals $I_{q_i}^\varphi\left(f_i^{(n)}\right)$ *are defined via (5.5.61) and* $\mathbf{C} = \{C\,(i,j)\}$ *is a* $d \times d$ *positive definite matrix. We denote by* $\mathbf{N}_d\,(0, \mathbf{C})$ *a d-dimensional centered Gaussian vector with covariance matrix* $\mathbf{C}$*. Find conditions on the sequence* $\left(f_1^{(n)}, ..., f_d^{(n)}\right)$*,* $n \ge 1$*, in order to have the CLT*

$$\mathbf{F}_n \triangleq \left(I_{q_1}^\varphi\left(f_1^{(n)}\right), ..., I_{q_d}^\varphi\left(f_d^{(n)}\right)\right) \xrightarrow{law} \mathbf{N}_d\,(0, \mathbf{C})\,, \quad n \to \infty. \qquad (11.3.9)$$

We observe that, if $q_i \ne q_j$ in (11.3.8), then necessarily $C\,(i,j) = 0$ by Point 2 in Proposition 8.4.6. The relevance of Problem A comes from the chaotic representation (5.9.78), implying that a result such as (11.3.9) may be a key tool in order to establish CLTs for more general functionals of the random measure φ. As noted above, if φ is Gaussian and $q \ge 2$, then a random variable of the type $I_q^\varphi\,(f)$, $f \ne 0$, cannot be Gaussian.

Plainly, when φ is Gaussian, a solution of Problem A can be immediately deduced from the results discussed in Section 7.3. Indeed, if the normalization condition (11.3.8) is satisfied, then the moments of the sequence $\{\mathbf{F}_n\}$ are uniformly bounded (to see this, one can use (5.9.79)), and the CLT (11.3.9) takes place if and only if every cumulant of order ≥ 3 associated with $\mathbf{F}_n$ converges to zero when $n \to \infty$. Moreover, an explicit expression for the cumulants can be deduced from (7.3.29). This method of proving the CLT (11.3.9) (which is known as the *method of cumulants*) has been used e.g. in the references [10, 12, 34, 68, 71, 72], where the authors proved CLTs for non-linear functionals of Gaussian fields with a non trivial covariance structure (for instance, sequences with long memory or isotropic spherical fields). However, such an approach (e.g. in the study of fractional Gaussian processes) may be technically quite demanding, since it involves an infinity of asymptotic relations (one for every cumulant of order ≥ 3). If one uses the diagram formulae (7.3.29), the method of cumulants requires that one explicitly computes and controls an infinity of expressions of the type $\int_{Z^{mq/2}} f_{\sigma,m} d\nu^{mq/2}$, where the partition σ is associated with a non-flat, Gaussian and connected diagram (see Section 4.1).

Remarks. (i) We already observed that, except for trivial cases, when φ is Gaussian the explicit expression of the characteristic function of a random variable of the type $I_q^\varphi\,(f)$, $q \ge 3$, is unknown. On the other hand, explicit expressions in the case $q = 2$ can be deduced from (11.1.1); see e.g. [46, Ch. VI] [98, p. 185].

(ii) Thanks to the results discussed in Section 7.4 (in particular, formula (7.4.40)), the method of cumulants and diagrams can be also used when φ is a completely random Poisson measure. Clearly, since (7.4.40) also involves non-Gaussian diagrams, the use of this approach in the Poisson case is even more technically demanding.

In the forthcoming sections we will show how one can successfully bypass the method of moments and cumulants when dealing with CLTs on a fixed Wiener chaos.

11.4 One-dimensional CLTs in the Gaussian case

The following two results involve one-dimensional sequences of chaotic random variables, and are based on the main findings of [98] (for Theorem 11.4.1) and [86, 89] (for Theorem 11.4.3). Observe that the forthcoming statement is expressed in the general language of isonormal Gaussian processes – and therefore applies in particular to Gaussian measures.

Theorem 11.4.1 (The fourth cumulant condition – see [98]) *Fix an integer $q \geq 2$, and define random variables of the type $I_q^X(f)$ according to (8.4.17). Then, for every sequence $\{f^{(n)} : n \geq 1\}$ such that $f^{(n)} \in \mathfrak{H}^{\odot q}$ for every n, and*

$$\lim_{n \to \infty} q! \left\| f^{(n)} \right\|_{\mathfrak{H}^{\otimes q}}^2 = \lim_{n \to \infty} \mathbb{E}\left[I_q^X\left(f^{(n)}\right)^2 \right] = 1, \qquad (11.4.10)$$

the following three conditions are equivalent

1. $\lim_{n \to \infty} \chi_4\left(I_q^X\left(f^{(n)}\right)\right) = 0$;
2. for every $r = 1, ..., q - 1$,

$$\lim_{n \to \infty} \left\| f^{(n)} \otimes_r f^{(n)} \right\|_{\mathfrak{H}^{\otimes 2(q-r)}} = 0 , \qquad (11.4.11)$$

where the contraction $f^{(n)} \otimes_r f^{(n)}$ is defined according to (8.5.19);
3. as $n \to \infty$, the sequence $\left\{ I_q^X\left(f^{(n)}\right) : n \geq 1 \right\}$ converges in distribution towards a centered standard Gaussian random variable $Z \sim N(0, 1)$.

Remark. The normalization condition (11.4.10) ensures that the limit distribution of $I_q^X\left(f^{(n)}\right)$ is not degenerate.

Proof. We shall prove the implications:

$$(1.) \Rightarrow (2.) \Rightarrow (3.) \Rightarrow (1.).$$

[(1.) $\Rightarrow$ (2.)] The desired implication follows immediately from the identity (11.2.7).
[(2.) $\Rightarrow$ (3.)] By virtue of Lemma 3.2.5, we know that the centered standard Gaussian distribution is determined by its moments (and therefore by its cumulants). Since multiple stochastic integrals are centered by construction and the normalization condition (11.4.10) is in order, to prove the desired implication it is sufficient to show that, for every $m \geq 3$, the sequence $\chi_m(I_q^X(f^{(n)}))$, $n \geq 1$, converges to zero. We will write down the proof in the special case $\mathfrak{H} = L^2(Z, \mathcal{Z}, \nu)$, where ν is σ-finite and non-

atomic, so that one has that $I_q^X(f^{(n)}) = I_q^G(f^{(n)})$, where G is a Gaussian measure with control ν. Also, the contraction $f^{(n)} \otimes_r f^{(n)}$ is obtained according to (6.2.16). In this case, we may use the explicit expression (7.3.29) and deduce that, for $n \geq 1$ and mq even,

$$\chi_m(I_q^G(f^{(n)})) = \sum_{\sigma \in \mathcal{M}_2([mq], \pi^*)} \int_{Z^{mq/2}} f_{\sigma,m}^{(n)} d\nu^{mq/2},$$

where the functions $f_{\sigma,m}^{(n)}$ are constructed following the procedure described in Section 11.2. According to Proposition 11.2.1, one has that, for every $\sigma \in \mathcal{M}_2([mq], \pi^*)$, the estimate

$$\left| \int_{Z^{mq/2}} f_{\sigma,m}^{(n)} d\nu^{mq/2} \right| \leq \|f^{(n)} \otimes_r f^{(n)}\|_{L^2(\nu^{2q-2r})} \times \|f^{(n)}\|_{L^2(\nu^q)}^{m-2}$$

holds for some $r = 1, ..., q - 1$. Again by virtue of the normalization condition (11.4.10), this yields the desired conclusion. The case of a general Hilbert space $\mathfrak{H}$ can be dealt with by means of an isomorphism argument: the details are left to the reader (see e.g. [98, Sect. 2.2] for a full discussion of this point).

[(3.) $\Rightarrow$ (1.)] In view of (8.4.18) and of the normalization condition (11.4.10), one has that, for every $m \geq 3$, the sequence $E(|I_q^X(f^{(n)})|^m)$ is bounded. It follows in particular, given (11.2.5), that if (3.) is in order, then $\mathbb{E}(I_q^X(f^{(n)})^4)$ must converge to $\mathbb{E}(Z^4) = 3$, readily yielding the desired conclusion.   ■

The equivalence of (1.), (2.) and (3.) in the previous statement has been first proved in [98] by means of stochastic calculus techniques. The paper [95] contains an alternate proof with additional necessary and sufficient conditions, as well as several crucial connections with Malliavin calculus operators (see e.g. [94]). As already observed, by (11.2.6), one has that

$$\mathbb{E}\left[I_q^X\left(f^{(n)}\right)^4\right] > 3\mathbb{E}\left[I_q^X\left(f^{(n)}\right)^2\right]^2,$$

unless $f^{(n)} = 0$ since the expression in (11.2.7) is non-negative. Observe that condition (1.) in the previous statement holds if and only if

$$\lim_{n \to \infty} \mathbb{E}\left[I_q^X\left(f^{(n)}\right)^4\right] = 3.$$

Remark. It is the condition (11.4.11) which is most often used in practice. It is therefore desirable to state explicit expressions for $\|f \otimes_r f\|_{\mathfrak{H}^{2(q-r)}}$, whenever $\mathfrak{H} = L^2(\nu)$, where ν is a positive, non-atomic and σ-finite measure on some Polish space $(Z, \mathcal{Z})$. In this case, one has indeed that

$$f \otimes_r f(\mathbf{a}_{q-r}, \mathbf{a}'_{q-r}) = \int_{Z^r} f(\mathbf{a}_{q-r}, \mathbf{b}_r) f(\mathbf{a}'_{q-r}, \mathbf{b}_r) \nu^r(d\mathbf{b}_r),$$

and hence since f is symmetric,

$$\|f \otimes_r f\|^2_{L^2(\nu^{2(q-r)})}$$

$$= \int_{Z^{q-r}} \int_{Z^{q-r}} \left(\int_{Z^r} f(\mathbf{a}_{q-r}, \mathbf{b}_r) f(\mathbf{b}_r, \mathbf{a}'_{q-r}) \nu^r(d\mathbf{b}_r) \right)^2 \nu^{q-r}(d\mathbf{a}_{q-r}) \nu^{q-r}(d\mathbf{a}'_{q-r})$$

$$= \int_{Z^{q-r}} \int_{Z^{q-r}} \int_{Z^r} \int_{Z^r} f(\mathbf{a}_{q-r}, \mathbf{b}_r) f(\mathbf{b}_r, \mathbf{a}'_{q-r}) f(\mathbf{a}'_{q-r}, \mathbf{b}'_r) f(\mathbf{b}'_r, \mathbf{a}_{q-r})$$

$$\hspace{11.4.12}$$

$$\nu^{q-r}(d\mathbf{a}_{q-r}) \nu^{q-r}(d\mathbf{a}'_{q-r}) \nu^r(d\mathbf{b}_r) \nu^r(d\mathbf{b}'_r)$$

$$= \int_{Z^r} \int_{Z^r} \left(\int_{Z^{q-r}} f(\mathbf{a}_{q-r}, \mathbf{b}_r) f(\mathbf{b}'_r, \mathbf{a}_{q-r}) \nu^{q-r}(d\mathbf{a}_{q-r}) \right)^2 \nu^r(d\mathbf{b}_r) \nu^r(d\mathbf{b}'_r)$$

$$= \|f \otimes_{q-r} f\|^2_{L^2(\nu^{2r})}. \hspace{2cm} (11.4.13)$$

Other characterizations of the quantity $\|f \otimes_r f\|^2_{L^2(\nu^{2(q-r)})}$ are provided in Lemma 11.7.3 below.

Example 11.4.2 Suppose $q = 4$, then for $r = 1$,

$$\|f \otimes_1 f\|^2_{L^2(\nu^6)}$$

$$= \int_{Z^6} \left[\int_Z f(x_1, x_2, x_3, y) f(x_4, x_5, x_6, y) \nu(dy) \right]^2 \prod_{i=1}^{6} \nu(dx_i)$$

$$= \int_{Z^8} f(x_1, x_2, x_3, y) f(x_4, x_5, x_6, y) \times$$

$$\times f(x_1, x_2, x_3, y') f(x_4, x_5, x_6, y') \nu(dy) \nu(dy') \prod_{i=1}^{6} \nu(dx_i)$$

$$= \int_Z \left[\int_{Z^3} f(x_1, x_2, x_3, y) f(x_1, x_2, x_3, y') \prod_{i=1}^{3} \nu(dx_i) \right]^2 \nu(dy) \nu(dy')$$

$$= \|f \otimes_3 f\|^2_{L^2(\nu^2)};$$

and for $r = 2$,

$$\|f \otimes_2 f\|^2_{L^2(\nu^4)}$$

$$= \int_{Z^8} f(x_1, x_2, y_1, y_2) f(x_3, x_4, y_1, y_2) \times$$

$$\times f(x_1, x_2, y'_1, y'_2) f(x_3, x_4, y'_1, y'_2) \prod_{j=1}^{2} [\nu(dy_j) \nu(dy'_j)] \prod_{i=1}^{4} \nu(dx_i).$$

Moreover, $\|f \otimes_3 f\|_{L^2(\nu^2)} = \|f \otimes_1 f\|_{L^2(\nu^6)}$.

Remark. Theorem 11.4.1, as well as its multidimensional generalizations (see Section 11.8 below), has been applied to a variety of frameworks, such as: quadratic functionals of bivariate Gaussian processes (see [19]), quadratic functionals of fractional processes (see [98]), high-frequency limit theorems on homogeneous spaces (see [69, 70]), self-intersection local times of fractional Brownian motion (see [39, 95]), Berry-Esséen bounds in CLTs for Gaussian subordinated sequences (see [86, 87, 90]), needlets analysis on the sphere (see [5]), power variations of iterated processes (see [85]), weighted variations of fractional processes (see [83, 91]) and of related random functions (see [7, 17]).

We now recall that the *total variation distance* between the law of two general real-valued random variables Y and Z, is given by

$$d_{TV}(Y, Z) = \sup |\mathbb{P}(Y \in B) - \mathbb{P}(Z \in B)|,$$

where the supremum is taken over all Borel sets $B \in \mathcal{B}(\mathbb{R})$. Observe that the topology induced by d_{TV}, on the class of probability measures on $\mathbb{R}$, is strictly stronger than the topology of weak convergence, and thus $\lim_{n\to\infty} d_{TV}(Y_n, Y) = 0$ is a stronger result than $Y_n \overset{law}{\to} Y$. Indeed, taking $B = (-\infty, x]$ shows that $\lim_{n\to\infty} d_{TV}(Y_n, Y) = 0$ implies $Y_n \overset{law}{\to} Y$, but the converse is in general not true. For example, take $\mathbb{P}(Y_n = 1/n) = 1$. Then, Y_n converges in law to 0, but

$$d_{TV}(Y_n, 0) \geq |\mathbb{P}(Y_n \leq 0) - \mathbb{P}(0 \leq 0)| = |0 - 1| = 1.$$

See e.g. Dudley [24, Ch. 11] for a discussion of other relevant properties of d_{TV}.

The following statement provides an explicit upper bound in the total variation distance for the normal approximations studied in Theorem 11.4.1. Note that, for the sake of simplicity, we only deal with chaotic random variables with variance 1 (the extension to a general variance is left to the reader as an exercise).

Theorem 11.4.3 (see [86, 89]) *Let Z be a standard centered normal random variable. Let $q \geq 2$ and $f \in \mathfrak{H}^{\odot q}$ be such that $q!\|f\|^2_{\mathfrak{H}^{\otimes q}} = \mathbb{E}[(I_q^X(f))^2] = 1$. Then, the following estimate in total variation holds:*

$$d_{TV}(I_q^X(f), Z)$$

$$\leq 2q \sqrt{\sum_{r=1}^{d-1} (2q - 2r)! \, (r-1)!^2 \binom{q-1}{r-1}^4 \left\| \widetilde{f \otimes_r f} \right\|^2_{\mathfrak{H}^{\otimes 2(d-r)}}} \tag{11.4.14}$$

$$\leq 2\sqrt{\frac{q-1}{3q}} \times \sqrt{\mathbb{E}\left[I_q^X(f)^4 \right] - 3}. \tag{11.4.15}$$

As usual, in the previous equation the symbol $\widetilde{f \otimes_r f}$ stands for the symmetrization of the contraction $f \otimes_r f$.

Remark on Stein's method. The upper bound (11.4.14) is proved in [86], by means of Malliavin calculus (see e.g. [94]) and the so-called *Stein's method* for normal approximations (see e.g. [14, 147]). The inequality (11.4.15) is proved through a direct computation in [89]. The main idea behind Stein's method (for normal approximations in the total variation distance) is the following. Fix a standard Gaussian random variable Z and, for every Borel set $A \subset \mathbb{R}$, consider the so-called *Stein's equation*

$$\mathbf{1}_A(x) - \mathbb{P}(Z \in A) = h'(x) - xh(x), \quad x \in \mathbb{R}. \tag{11.4.16}$$

A solution to (11.4.16) is a function h that is almost everywhere differentiable and such that there exists a version of h' verifying (11.4.16) for every real x. It is not difficult to prove that, for every A, (11.4.16) admits a solution, say h_A, such that $|h_A| \le \sqrt{\pi/2}$ and $|h'_A| \le 2$. Now replace in x in (11.4.16) by a random variable W and take expectations on both sides:

$$\mathbf{1}_A(W) - \mathbb{P}(Z \in A) = h'(W) - Wh(W),$$
$$\mathbb{P}(W \in A) - \mathbb{P}(Z \in A) = \mathbb{E}h'(W) - \mathbb{E}Wh(W).$$

Then recalling the definition of the total variation distance d_{TV} given above, yields the following remarkable estimate:

Stein's bound. *For every real-valued random variable W*

$$d_{TV}(W, Z) \le \sup_h |\mathbb{E}(h'(W) - Wh(W))|, \tag{11.4.17}$$

where the supremum is taken over the class of smooth bounded functions h such that $|h'| \le 2$.

Observe that the bound (11.4.17) provides an alternate proof of Stein's Lemma 3.2.4 in the case $\sigma = 1$. Indeed, if W is such that $\mathbb{E}(h'(W) - Wh(W)) = 0$ for every smooth h, then $d_{TV}(W, Z) = 0$, meaning that W has a centered standard Gaussian distribution. The bound (11.4.14) is obtained in [86] by means of the so called 'integration by parts formula' of Malliavin calculus, namely by using the fact that, if $W = I_q^X(f)$, then

$$\mathbb{E}(Wh(W)) = \frac{1}{q}\mathbb{E}(h'(W)\|DW\|_{\mathfrak{H}}^2),$$

where the Malliavin derivative of W, written DW, has to be regarded a random element with values in $\mathfrak{H}$. Applying the previous identity to (11.4.17), and recalling that the supremum is taken over smooth bounded functions h such that $|h'| \le 2$, one infers that

$$d_{TV}(I_q^X(f), Z) \le 2\mathbb{E}\left|1 - \frac{1}{q}\|DI_q^X(f)\|_{\mathfrak{H}}^2\right|. \tag{11.4.18}$$

The estimate (11.4.14) is then deduced by explicitly assessing the expectation on the RHS of (11.4.14).

11.5 An application to Brownian quadratic functionals

Let W_t, $t \in [0, 1]$ be a standard Brownian motion on $[0, 1]$ starting at zero. One has

$$\mathbb{E} \int_0^1 \frac{W_t^2}{t^2} dt = \int_0^1 \frac{dt}{t} = \infty,$$

but since standard Brownian motion satisfies the following law of the iterated logarithm (see [52]):

$$\underline{\lim}_{t \downarrow 0} \frac{W_t}{\sqrt{2t \log \log(1/t)}} = -1, \qquad \overline{\lim}_{t \downarrow 0} \frac{W_t}{\sqrt{2t \log \log(1/t)}} = 1, \quad \text{a.s,}$$

the following stronger relation holds:

$$\int_0^1 \frac{W_t^2}{t^2} dt = \infty \quad \text{a.s.}$$

This phenomenon has important consequences in the theory of the so-called "enlargement of filtrations". See e.g. Lemma 3.22, p. 44 in [47] for a discussion of this fact and for alternate proofs.

In what follows we shall use Theorem 11.4.1 to establish the following CLT which states that the deviation of $\int_\epsilon^1 t^{-2} W_t^2 dt$ from its mean $\log \frac{1}{\epsilon}$, adequately normalized, becomes Gaussian as ϵ converges to zero.

Proposition 11.5.1 *As $\epsilon \downarrow 0$,*

$$\frac{1}{2\sqrt{\log\left(\frac{1}{\epsilon}\right)}} \left[\int_\epsilon^1 \frac{W_t^2}{t^2} dt - \log\left(\frac{1}{\epsilon}\right) \right] \xrightarrow{law} Z \sim N(0, 1). \qquad (11.5.19)$$

Proof. Recall that $W_t = I_1^X(\mathbf{1}_{[0,t]})$, where X is the isonormal Gaussian process (on $\mathfrak{H} = L^2([0, 1], dx)$) generated by W. Using the multiplication formula in Proposition 8.5.3, we deduce

$$W_t^2 = I_1^X(\mathbf{1}_{[0,t]}) I_1^X(\mathbf{1}_{[0,t]}) = I_0^X(\mathbf{1}_{[0,t]} \otimes_1 \mathbf{1}_{[0,t]}) + I_2^X(\mathbf{1}_{[0,t]} \otimes_0 \mathbf{1}_{[0,t]})$$
$$= t + I_2^X(\mathbf{1}_{[0,t]} \otimes_0 \mathbf{1}_{[0,t]}).$$

As a consequence

$$\int_\epsilon^1 \frac{W_t^2}{t^2} dt - \log \frac{1}{\epsilon} = \int_\epsilon^1 I_2^X(\mathbf{1}_{[0,t]} \otimes_0 \mathbf{1}_{[0,t]}) \frac{dt}{t^2}. \qquad (11.5.20)$$

We now want to apply the stochastic Fubini's Theorem 5.13.1 in order to change the order of integration. This is possible because

$$\int_\epsilon^1 \left[\int_0^1 \int_0^1 \mathbf{1}_{[0,t]}(x) \mathbf{1}_{[0,t]}(y) dx dy \right]^{1/2} \frac{dt}{t^2} = \int_\epsilon^1 \left[\int_0^1 \mathbf{1}_{[0,t]}(x) dx \right] \frac{dt}{t^2}$$
$$= \int_\epsilon^1 \frac{dt}{t} < \infty.$$

Thus relation (11.5.20) becomes

$$\int_\epsilon^1 I_2^X(1_{[0,t]} \otimes_0 1_{[0,t]}) \frac{dt}{t^2} = I_2^X(f_\epsilon),$$

where

$$f_\epsilon(x,y) = \int_\epsilon^1 1_{[0,t]}(x) 1_{[0,t]}(y) \frac{dt}{t^2} = \frac{1}{x \vee y \vee \epsilon} - 1,$$

because $\int_a^1 t^{-2} dt = \frac{1}{a} - 1$. Here "$\vee$" denotes the usual maximum: $x \vee y \vee \epsilon = \max(x, y, \epsilon)$. Therefore

$$\frac{1}{2\sqrt{\log \frac{1}{\epsilon}}} \left[\int_\epsilon^1 \frac{W_t^2}{t^2} dt - \log \frac{1}{\epsilon} \right] = I_2^X(\tilde{f}_\epsilon),$$

where

$$\tilde{f}_\epsilon(x,y) = \frac{f_\epsilon(x,y)}{2\sqrt{\log \frac{1}{\epsilon}}}.$$

In view of Theorem 11.4.1 it is now sufficient to show that

$$
\begin{aligned}
2\|\tilde{f}_\epsilon\|_{L^2([0,1]^2, dxdy)}^2 &= \mathbb{E}[I_2^X(\tilde{f}_\epsilon)^2] \\
&= 2 \int_0^1 \int_0^1 dxdy\, \tilde{f}_\epsilon(x,y)^2 \\
&\xrightarrow[\epsilon \to 0]{} 1,
\end{aligned}
\tag{11.5.21}
$$

and

$$\|\tilde{f}_\epsilon \otimes_1 \tilde{f}_\epsilon\|_{L^2([0,t]^2, dxdy)}^2 = \int_0^1 \int_0^1 dxdy \left(\int_0^1 dz\, \tilde{f}_\epsilon(x,z) \tilde{f}_\epsilon(z,y) \right)^2 \xrightarrow[\epsilon \to 0]{} 0. \tag{11.5.22}$$

We first show that relation (11.5.21) holds:

$$
\begin{aligned}
\|f_\epsilon\|^2 &= 2 \int_0^1 \int_0^x \left(\frac{1}{x \vee \epsilon} - 1 \right)^2 dydx \quad \text{(by symmetry)} \\
&= 2 \int_0^1 \left(\frac{1}{x \vee \epsilon} - 1 \right)^2 xdx \tag{11.5.23} \\
&= 2 \int_0^1 \left(\frac{x}{(x \vee \epsilon)^2} + x - 2\frac{x}{x \vee \epsilon} \right) dx \\
&= A_\epsilon + B_\epsilon + C_\epsilon
\end{aligned}
$$

where

$$A_\epsilon = 2 \int_\epsilon^1 \frac{dx}{x} + 2 \int_0^\epsilon \frac{xdx}{\epsilon^2} = -2 \log \epsilon + 1 = 2 \log \frac{1}{\epsilon} + 1$$

$$B_\epsilon = 2 \int_0^1 x \, dx = 1$$

$$C_\epsilon = -4 \left[\int_\epsilon^1 dx + \int_0^\epsilon \frac{x}{\epsilon} dx \right] = -4(1 - \epsilon) - 2\epsilon.$$

Therefore

$$2\|\widetilde{f}_\epsilon\|^2 = \frac{2}{4 \log \frac{1}{\epsilon}} [A_\epsilon + B_\epsilon + C_\epsilon] \sim \frac{2}{4 \log \frac{1}{\epsilon}} \left(2 \log \frac{1}{\epsilon} \right) \xrightarrow[\epsilon \to 0]{} 1.$$

We need now to prove Relation (11.5.22).

$$\|\widetilde{f}_\epsilon \otimes_1 \widetilde{f}_\epsilon\|^2_{L^2([0,t]^2, dx dy)}$$

$$= \int_0^1 \int_0^1 dx dy \left(\int_0^1 dz \widetilde{f}_\epsilon(x, z) \widetilde{f}_\epsilon(z, y) \right)^2$$

$$= \frac{1}{16 \left(\log \frac{1}{\epsilon} \right)^2} \int_0^1 \int_0^1 dx dy \left(\int_0^1 dz \widetilde{f}_\epsilon(x, z) \widetilde{f}_\epsilon(y, z) \right)^2$$

$$= \frac{1}{16 \left(\log \frac{1}{\epsilon} \right)^2} \int_0^1 \int_0^1 dx dy \int_0^1 dz \left(\frac{1}{z \vee x \vee \epsilon} - 1 \right) \left(\frac{1}{z \vee y \vee \epsilon} - 1 \right)$$

$$\times \int_0^1 dz' \left(\frac{1}{z' \vee x \vee \epsilon} - 1 \right) \left(\frac{1}{z' \vee y \vee \epsilon} - 1 \right)$$

$$= \frac{1}{16 \left(\log \frac{1}{\epsilon} \right)^2} \int_{[0,1]^4} dx_1 dx_2 dx_3 dx_4 \left(\frac{1}{x_1 \vee x_3 \vee \epsilon} - 1 \right) \left(\frac{1}{x_2 \vee x_3 \vee \epsilon} - 1 \right)$$

$$\times \left(\frac{1}{x_1 \vee x_4 \vee \epsilon} - 1 \right) \left(\frac{1}{x_2 \vee x_4 \vee \epsilon} - 1 \right)$$

$$= D_\epsilon.$$

To prove that $D_\epsilon \to 0$ as $\epsilon \to 0$, one has to write

$$\int_{[0,1]^4} = \sum_{\pi \in \mathfrak{S}_4} \int_{S_\pi}$$

where the sum sums over the set $\mathfrak{S}_4$ of all 4! permutations of $[4] = \{1, 2, 3, 4\}$ and

$$S_\pi = \{x_{\pi(1)} > x_{\pi(2)} > x_{\pi(3)} > x_{\pi(4)}\},$$

and show that, for each $\pi \in \mathfrak{S}_4$,

$$\frac{1}{\left(\log \frac{1}{\epsilon} \right)} \int_{S_\pi} (\dots) \to 0.$$

We shall only do this for π equal to the identity permutation, i.e. $\pi(i) = i, i = 1, 2, 3, 4$. In this case $\int_{S_\pi} = \int_{x_1 > x_2 > x_3 > x_4}$ so that

$$\int_{S_\pi} dx_1 dx_2 dx_3 dx_4 (\ldots)(\ldots)(\ldots)(\ldots)$$

$$= \int_0^1 dx_1 \int_0^{x_1} dx_2 \int_0^{x_2} dx_3 \int_0^{x_3} dx_4 \left(\frac{1}{x_1 \vee \epsilon} - 1\right)^2 \left(\frac{1}{x_2 \vee \epsilon} - 1\right)^2$$

$$= \int_0^1 dx_1 \left(\frac{1}{x_1 \vee \epsilon} - 1\right)^2 \int_0^{x_1} \left(\frac{1}{x_2 \vee \epsilon} - 1\right)^2 \frac{x_2^2}{2} dx_2$$

$$\leq \frac{1}{2} \int_0^1 dx_1 \left(\frac{1}{x_1 \vee \epsilon} - 1\right)^2 x_1 \sim \log \frac{1}{\epsilon}$$

using the computations for (11.5.23).   ∎

Proposition 11.5.1 can be extended to fractional Brownian motion, denoted here B_t^H (see Nualart and Peccati [98, Proposition 7]).

Proposition 11.5.2 *If* $1/2 < H < 1$, *then*

$$\frac{1}{\sqrt{\log \frac{1}{\epsilon}}} \left(\int_\epsilon^1 (B_t^H)^2 \frac{dt}{t^{2H+1}} - \log \frac{1}{\epsilon}\right) \xrightarrow{law} k_H Z,$$

where $Z \sim N(0, 1)$ *and* $k_H > 00$ *is a constant depending only on* H.

11.6 Proof of the propositions

We prove here the two key propositions used in the proof of Theorem 11.4.1, namely Proposition 11.2.1 and Proposition 11.2.2.

We start with a detailed proof of Proposition 11.2.1 for a specific choice of m, q and σ. We will then provide an argument allowing to show Proposition 11.2.1 in the general case. Note that the general proof given at the end of the section is only sketched, since all the necessary ideas are already contained in the first part. The reader is invited to fill in the missing details as an exercise.

Proof of Proposition 11.2.1 in a specific case. We shall provide a proof of Proposition 11.2.1 in the specific case $q = 4$, $m = 5$ and

$$\sigma = \{\{1, 19\}, \{2, 20\}, \{3, 7\}, \{4, 8\}, \{5, 10\}, \{6, 11\}, \{9, 15\}, \{12, 16\},$$
$$\{13, 17\}, \{14, 18\}\}$$

(this situation corresponds to the connected diagram in Fig. 11.1). In particular, our

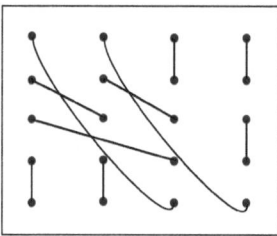

Fig. 11.1. A connected non-flat diagram

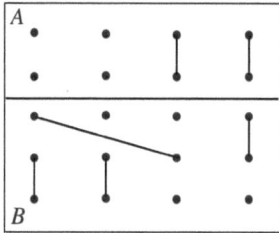

Fig. 11.2. Dividing a diagram

aim is to show that, for every $f \in L^2_s(\nu^q)$,

$$\left| \int_{Z^{10}} f_{\sigma,5} d\nu^{10} \right| \leq \|f \otimes_2 f\|_{L^2(\nu^4)} \times \|f\|^3_{L^2(\nu^4)},$$

which proves the statement for $r = 2$.

To do this, first divide the five blocks of π^* into two groups, namely $A = \{b_1, b_2\}$ (the first two rows of the diagram) and $B = \{b_3, b_4, b_5\}$ (the last three rows of the diagram). Then, remove from the diagram every segment corresponding to a block of sigma linking a block of A and a block of B. This procedure is illustrated in Fig. 11.2.

Now consider the two blocks of π^* in A (resp. the three blocks of π^* in B), and build a function F_A (resp. F_B) in 4 variables according to the following procedure:

– build the 2-tensor (resp. 3-tensor) product $f \otimes_0 f(x_1, ..., x_8)$ (resp. $f \otimes_0 f \otimes_0 f(x_1, ..., x_{12})$);
– identify two variables x_i and x_j in the argument of $f \otimes_0 f$ (resp. $f \otimes_0 f \otimes_0 f$) if and only if $\{i, j\}$ is a block of σ that has not been removed;
– integrate the identified variables with respect to ν and let the remaining variables free to move.

The two functions F_A and F_B thus obtained are:

$$F_A(z_1, z_2, z_3, z_4) = \int_Z \int_Z f(z_1, z_2, x, y) f(z_3, z_4, x, y) \nu(dx) \nu(dy)$$
$$= f \otimes_2 f(z_1, z_2, z_3, z_4),$$

$$F_B(z_1, z_2, z_3, z_4) =$$

$$\int_Z \int_Z \int_Z \int_Z f(x, z_1, z_2, y) f(v, w, x, y) f(v, w, z_3, z_4)\nu(dx)\nu(dy)\nu(dv)\nu(dw).$$

An iterated use of the Cauchy-Schwarz inequality, implies in particular that

$$\|F_B\|_{L^2(\nu^4)} \leq \|f\|^3_{L^2(\nu^4)}. \tag{11.6.24}$$

To see this, note that

$$F_B^2(z_1, z_2, z_3, z_4)$$
$$\leq \left(\int_{Z^4} f^2(v, w, x, y)\nu(dv)\nu(dw)\nu(dx)\nu(dy) \right) g_1(z_1, z_2) g_2(z_3, z_4)$$
$$= \|f\|^2_{L^2(\nu^4)} g_1(z_1, z_2) g_2(z_3, z_4),$$

where

$$g_1(z_1, z_2) = \int_{Z^2} f^2(x, y, z_1, z_2)\nu(dx)\nu(dy)$$
$$g_2(z_3, z_4) = \int_{Z^2} f^2(v, w, z_3, z_4)\nu(dv)\nu(dw),$$

so that

$$\|F_B\|^2_{L^2(\nu^4)} = \int_{Z^4} F_B^2(z_1, z_2, z_3, z_4)\nu(dz_1)\nu(dz_2)\nu(dz_3)\nu(dz_4)$$
$$\leq \|f\|^2_{L^2(\nu^4)} \int_{Z^2} g_1(z_1, z_2)\nu(dz_1)\nu(dz_2) \int_{Z^2} g_2(z_3, z_4)\nu(dz_3)\nu(dz_4)$$
$$= \|f\|^6_{L^2(\nu^4)}.$$

Now label the four dots in the upper and lower parts of the diagram by means of the integers $1, 2, 3, 4$ (as done in Fig. 11.3), and draw again the four blocks of σ linking

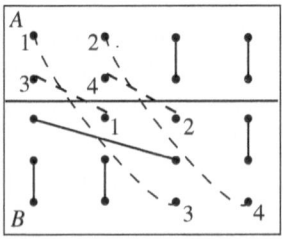

Fig. 11.3. Linking A and B

A and B, represented as dashed segments in Fig. 11.3. These dashed segments define a permutation β of the set $[4] = \{1, 2, 3, 4\}$, given by

$$\beta(1) = 3, \quad \beta(2) = 4, \quad \beta(3) = 1, \quad \beta(4) = 2,$$

and one has that

$$\int_{Z^{mq/2}} f_{\sigma,m} \, d\nu^{mq/2} = \int_{Z^{10}} f_{\sigma,5} \, d\nu^{10}$$

$$= \int_Z \int_Z \int_Z \int_Z F_A(z_1, z_2, z_3, z_4) F_B(z_{\beta(1)}, z_{\beta(2)}, z_{\beta(3)}, z_{\beta(4)})$$

$$\nu(dz_1)\nu(dz_2)\nu(dz_3)\nu(dz_4).$$

Since $F_A = f \otimes_2 f$, we can use the Cauchy-Schwarz inequality and the estimate (11.6.24) to deduce that

$$\left| \int_{Z^{10}} f_{\sigma,5} d\nu^{10} \right| \le \left(\|f \otimes_2 f\|_{L^2(\nu^4)}^2 \times \|F_B\|_{L^2(\nu^4)}^2 \right)^{1/2}$$

$$\le \|f \otimes_2 f\|_{L^2(\nu^4)} \times \|f\|_{L^2(\nu^4)}^3,$$

yielding the desired conclusion. ∎

Proof of Proposition 11.2.1 in the general case. Let mq be even. Since the diagram $\Gamma(\pi^*, \sigma)$ is non-flat and connected, there exist two blocks of π^*, say b_i and b_j, that are connected exactly by r blocks of σ, for some $r = 1, ..., q-1$ (note that, if two blocks of π^* were connected by q blocks of σ, then $\Gamma(\pi^*, \sigma)$ would not be connected). By partitioning the blocks of π^* (or, equivalently, the rows of $\Gamma(\pi^*, \sigma)$) into two subsets $A = \{b_i, b_j\}$ and $B = \{b_k : k \ne i, j\}$, and by reasoning as in the first part of this section (in particular, by repeatedly applying the Cauchy-Schwarz inequality), one sees that

$$\int_{Z^{mq/2}} f_{\sigma,m} \, d\nu^{mq/2}$$

$$= \int_{Z^{2q-2r}} f \otimes_r f(z_1, ..., z_{2q-2r}) F_B(z_1, ..., z_{2q-2r}) \nu(dz_1) \cdots \nu(dz_{2q-2r}),$$

for some function $F_B : Z^{2q-2r} \to \mathbb{R}$ verifying

$$\|F_B\|_{L^2(\nu^{2q-2r})} \le \|f\|_{L^2(\nu^q)}^{m-2}.$$

The conclusion follows from the Cauchy-Schwarz inequality. ∎

Proof of Proposition 11.2.2. [1] Our starting point is the formula for $\mathbb{E}\left[I_q^X(f)\right]^4$ given in (6.4.20), which we rewrite as

$$\mathbb{E}\left[I_q^X(f)\right]^4 = (q!)^2 \|f\|_{L^2(\nu^q)}^4 + (2q!) \left\|f \widetilde{\otimes}_0 f\right\|_{L^2(\nu^{2q})}^2$$

$$+ \sum_{p=1}^{q-1} \left[p!\binom{q}{p}^2\right]^2 (2q-2p)! \left\|f \widetilde{\otimes}_p f\right\|_{L^2(\nu^{2q-2p})}^2 . \qquad (11.6.25)$$

The key is to evaluate $\left\|f \widetilde{\otimes}_0 f\right\|_{L^2(\nu^{2q})}^2$. Call Π_{2q} the set of the $(2q)!$ permutations of $\{1, ..., 2q\}$, whose generic elements will be denoted by π, or π'. On such a set, we introduce the following notation: for $p = 0, ..., q$ we write

$$\pi \overset{p}{\sim} \pi'$$

if the set

$$(\pi(1), ..., \pi(q)) \cap (\pi'(1), ..., \pi'(q)) \qquad (11.6.26)$$

contains exactly p elements. Note that, for a given $\pi \in \Pi_{2q}$, there are $(q!)^2 \binom{q}{p}^2$ permutations π' such that

$$\pi' \overset{p}{\sim} \pi.$$

This fact is clear for $p = 0, q$. To prove it for $p = 1, ..., q - 1$, fix $\pi \in \Pi_{2q}$ and observe that, in order to build a permutation π' such that $\pi' \overset{p}{\sim} \pi$, one has first to specify the following: (i) the p elements of $\{\pi(1), ..., \pi(q)\}$ that will be in common with $\{\pi'(1), ..., \pi'(q)\}$ (this can be done in $\binom{q}{p}$ different ways), (ii) the $q - p$ elements of $\{\pi(q+1), ..., \pi(2q)\}$ that will be compose the remaining part of $\{\pi'(1), ..., \pi'(q)\}$ (this can be done in $\binom{q}{q-p} = \binom{q}{p}$ different ways). Operations (i) and (ii) determine which integers will be in the first q places and the last q places of π': in order to completely specify π' it is the necessary to chose (a) the order in which the first q integers will appear (this can be done in $q!$ different ways), and (b) the order in which the last q integers will appear (that can be again done in $q!$ different ways). Putting (i), (ii), (a) and (b) together, we obtain the coefficient $(q!)^2 \binom{q}{p}^2$.

An important observation is that if $\pi' \overset{p}{\sim} \pi$, that is, if the first q elements of π and π' have exactly p elements in common, then necessarily the last q elements of π and π' will also have exactly p elements in common. For example, if $q = 4$ and $p = 2$, then a particular outcome for π and π' is:

$$\begin{array}{c|c|c}\pi & 1234 & 5678 \\ \pi' & 5634 & 1278\end{array}$$

---

[1] We focus on the case $\mathfrak{H}^{\otimes q} = L^2(\nu^q)$, where ν is non-atomic, for convenience.

where the last first four elements have 3,4 in common and the last four have 7,8 in common.

For convenience set $g = f \otimes_0 f$, that is,

$$g(z_1, \cdots, z_{2q}) = f(z_1, \cdots, z_q) f(z_{q+1}, \cdots, z_{2q}),$$

and observe that

$$\widetilde{g}(z_1, \cdots, z_{2q}) = \frac{1}{(2q)!} \sum_{\pi \in \Pi_{2q}} g(z_{\pi(1)}, \cdots, z_{\pi(2q)}).$$

In order to compute

$$\|\widetilde{g}\|^2 = \frac{1}{((2q)!)^2} \sum_{\pi \in \Pi_{2q}} \sum_{\pi' \in \Pi_{2q}} \int_{Z^{2q}} g(z_{\pi(1)}, \cdots, z_{\pi(2q)}) g(z_{\pi'(1)},$$

$$\cdots, z_{\pi'(2q)}) \nu(dz_1) \cdots \nu(dz_{2q}),$$

we need to evaluate the above integral. In view of the observation made above, this integral, *when* $\pi \overset{p}{\sim} \pi'$ $(p = 1, ..., q - 1)$, equals

$$\int_{Z^{2q}} f(z_{\pi(1)}, \cdots, z_{\pi(q)}) f(z_{\pi(q+1)}, \cdots, z_{\pi(2q)}) f(z_{\pi'(1)}, \cdots, z_{\pi'(q)})$$

$$\times f(z_{\pi'(q+1)}, \cdots, z_{\pi'(2q)}) \nu(dz_1) \cdots \nu(dz_{2q})$$

$$= \int_{Z^{2q-2p}} \left[\int_{Z^p} f(z_{\pi(1)}, \cdots, z_{\pi(q)}) f(z_{\pi'(1)}, \cdots, z_{\pi'(q)}) \right]$$

$$\times \left[\int_{Z^p} f(z_{\pi(q+1)}, \cdots, z_{\pi(2q)}) f(z_{\pi'(q+1)}, \cdots, z_{\pi'(2q)}) \right] \nu(dz_1) \cdots \nu(dz_{2q})$$

$$= \int_{Z^{2q-2p}} f \otimes_p f(z_1, \cdots, z_{2q-2p}) \, f \otimes_p f(z_1, \cdots, z_{2q-2p}) \, \nu(dz_1) \cdots \nu(dz_{2q-2p})$$

$$= \|f \otimes_p f\|^2_{L^2(\nu^{2q-2p})}. \tag{11.6.27}$$

This is easy to evaluate if $\pi' \overset{0}{\sim} \pi$ or $\pi' \overset{q}{\sim} \pi$, because in these cases, $\|f \otimes_p f\|^2_{L^2(\nu^{2q-2p})} = \|f\|^4_{L^2(\nu^q)}$. For example, if $p = 0$,

$$(f \otimes_0 f)^2(z_1, \cdots, z_{2q}) = f^2(z_1, \cdots, z_q) \, f^2(z_{q+1}, \cdots, z_{2q}).$$

Since

$$\frac{1}{((2q)!)^2} \sum_{\pi \in \Pi_{2q}} \sum_{\pi' \in \Pi_{2q}} = \frac{1}{((2q)!)^2} \sum_{\pi \in \Pi_{2q}} \left(\sum_{\pi' \overset{0}{\sim} \pi} + \sum_{\pi' \overset{q}{\sim} \pi} + \sum_{p=1}^{q-1} \sum_{\pi' \overset{p}{\sim} \pi} \right), \tag{11.6.28}$$

we get that the first two terms in the parenthesis contribute

$$\frac{1}{((2q)!)^2} \sum_{\pi \in \Pi_{2q}} \left((q!)^2 + (q!)^2\right) \|f\|^4_{L^2(\nu^q)} = \frac{1}{((2q!))^2}(2q)!2(q!)^2\|f\|^4_{L^2(\nu^q)}$$

$$= \frac{2(q!)^2}{(2q)!}\|f\|^4_{L^2(\nu^q)}, \qquad (11.6.29)$$

since $(q!)^2\binom{q}{p}^2$ equals $(q!)^2$ when $p = 0$ or q.

We now focus on the sum $\sum_{\pi' \overset{p}{\sim} \pi}$ where $p = 1, \ldots, q - 1$. By (11.6.27), for such a p,

$$\sum_{\pi' \overset{p}{\sim} \pi} \int_{Z^{2q}} g(z_{\pi(1)}, \cdots, z_{\pi(2q)}) g(z_{\pi'(1)}, \cdots, z_{\pi'(2q)}) \nu(dz_1) \cdots \nu(dz_{2q})$$

$$= \sum_{\pi' \overset{p}{\sim} \pi} \|f \otimes_p f\|^2_{L^2(\nu^{2q-2p})} = (q!)^2\binom{q}{p}^2 \|f \otimes_p f\|^2_{L^2(\nu^{2q-2p})}.$$

Hence, the second term in (11.6.25) equals

$$(2q)!\|f \widetilde{\otimes}_0 f\|^2_{L^2(\nu^{2q})} = (2q)!\|\widetilde{g}\|^2_{L^2(\nu^{2q})} \qquad (11.6.30)$$

$$= 2(q!)^2\|f\|^4_{L^2(\nu^q)} + \frac{1}{(2q)!} \sum_{\pi \in \Pi_{2q}} \sum_{p=1}^{q-1} (q!)^2\binom{q}{p}^2 \|f \widetilde{\otimes}_p f\|^2_{L^2(\nu^{2q-2p})}$$

$$= 2(q!)^2\|f\|^4_{L^2(\nu^q)} + \sum_{p=1}^{q-1} \frac{(q!)^4}{(p!(q-p)!)^2}\|f \widetilde{\otimes}_0 f\|^2_{L^2(\nu^{2q-2p})}.$$

We now consider the last term in (11.6.25) which will be left essentially unchanged. Its coefficient is

$$\left[p!\binom{q}{p}^2\right]^2 (2q-2p)! = \left[\frac{(q!)^4}{(p!(q-p)!)^2}\right] \frac{(2q-2p)!}{(q-p)^2} = \left[\frac{(q!)^4}{(p!(q-p)!)^2}\right]\binom{2q-2p}{q-p}.$$

We therefore get

$$\mathbb{E}\left[I_q^X(f)^4\right] - 3(q!)^2\|f\|^4_{L^2(\nu^q)}$$

$$= \sum_{p=1}^{q-1} \frac{(q!)^4}{(p!(q-p)!)^2}\left\{\|f \otimes_p f\|^2_{L^2(\nu^{2q-2p})} + \binom{2q-2p}{q-p}\left\|f \widetilde{\otimes}_p f\right\|^2_{L^2(\nu^{2q-2p})}\right\}.$$

∎

Remark. What is remarkable is that in (11.6.30) we get a factor of $2(q!)^2\|f\|^4_{L^2(\nu^q)}$, which adds to the factor $(q!)^2\|f\|^4_{L^2(\nu^q)}$ in (11.6.25) to give $3(q!)^2\|f\|^4_{L^2(\nu^q)} = 3(\mathbb{E}\left[I_q^X(f)\right]^2)^2$, which is exactly what we need for the 4th cumulant of the normal distribution.

11.7 Further combinatorial implications

We focus here on some interesting combinatorial implications of Theorem 11.4.1, in the special case $\mathfrak{H} = L^2(Z, \mathcal{Z}, \nu)$, with ν a σ-finite and non-atomic measure over a Polish space $(Z, \mathcal{Z})$. According to Proposition 8.4.6, in this case the random variable $I_q^X(f)$, where $f \in L_s^2(Z^q, \mathcal{Z}^q, \nu^q) = L_s^2(\nu^q)$, is the multiple Wiener-Itô integral of f with respect to the Gaussian measure $A \to G(A) = X(1_A)$ – see Definition 5.5.4. The implication (1.) $\Longrightarrow$ (3.) in Theorem 11.4.1 provides the announced "drastic simplification" of the methods of moments and cumulants. However, as demonstrated by the applications of Theorem 11.4.1 listed above, condition (11.4.11) is often much easier to verify. Indeed, it turns out that the implication (2.) $\Longrightarrow$ (3.) has an interesting combinatorial interpretation.

To see this, we shall fix $q \geq 2$ and adopt the notation of Section 11.2. One should note, in particular, that the definition of π^* given in (11.2.2) depends on the choice of the integers m, q, and that this dependence will be always clear from the context. Given $m \geq 2$, we will also denote by

$$\mathcal{M}_2^c([mq], \pi^*)$$

the subset of $\mathcal{M}_2([mq], \pi^*)$ composed of those partitions σ such that the diagram

$$\Gamma(\pi^*, \sigma)$$

is *circular* (see Section 4.1). We also say that a partition $\sigma \in \mathcal{M}_2^c([mq], \pi^*)$ has *rank* r ($r = 1, ..., q - 1$) if the diagram $\Gamma(\pi^*, \sigma)$ has exactly r edges linking the first and the second row.

Example 11.7.1   (i) The partition whose diagram is given in Fig. 7.6 is an element of

$$\mathcal{M}_2^c([8], \pi^*)$$

and has rank $r = 1$ (there is only on edge linking the first and second rows).

(ii) Consider the case $q = 3$ and $m = 4$, as well as the partition $\sigma \in \mathcal{M}_2([12], \pi^*)$ given by

$$\sigma = \{\{1, 4\}, \{2, 5\}, \{3, 12\}, \{6, 9\}, \{7, 10\}, \{8, 11\}\}.$$

Then, the diagram $\Gamma(\pi^*, \sigma)$ is the one in Fig. 11.4, and therefore $\sigma \in \mathcal{M}_2^c([12], \pi^*)$ and σ has rank $r = 2$.

The following technical result links the notions of circular diagram, rank and contraction. For $q \geq 2$ and $\sigma \in \mathcal{M}_2^c([4q], \pi^*)$, let $f_{\sigma,4}$ be the function in $2q$ variables obtained by identifying x_i and x_j in the argument of the tensor product (11.2.3) (with $m = 4$) if and only if $i \sim_\sigma j$.

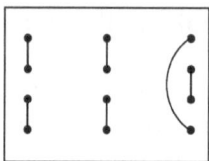

Fig. 11.4. A circular diagram

Example 11.7.2 If $q = 3$ and $\sigma \in \mathcal{M}_2^c([12], \pi^*)$ is associated with the diagram in Fig. 11.4, then

$$f_{\sigma,4}(x_1, x_2, x_3, x_6, x_7, x_8) = f(x_1, x_2, x_3)f(x_1, x_2, x_6)f(x_7, x_8, x_6)f(x_7, x_8, x_3).$$

Relabeling the variables, one obtains

$$f_{\sigma,4}(x_1, x_2, x_3, x_4, x_5, x_6) = f(x_1, x_2, x_3)f(x_1, x_2, x_4)f(x_5, x_6, x_4)f(x_5, x_6, x_3).$$

Lemma 11.7.3 *Fix $f \in L_s^2(\nu^q)$, $q \geq 2$, and, for $r = 1, ..., q - 1$, define the contraction $f \otimes_r f$ according to (6.2.16). Then, for every $\sigma \in \mathcal{M}_2^c([4q], \pi^*)$ with rank $r \in \{1, ..., q - 1\}$,*

$$\int_{Z^{2q}} f_{\sigma,4} d\nu^{2q} = \|f \otimes_r f\|_{L^2(\nu^{2(q-r)})}^2 = \|f \otimes_{q-r} f\|_{L^2(\nu^{2r})}^2. \tag{11.7.31}$$

Proof. It is sufficient to observe that f is symmetric by definition, and then to use the relation (11.4.13), as well as a standard version of the Fubini theorem. ∎

Remark. Formula (11.7.31) implies that, for a fixed f and for every $\sigma \in \mathcal{M}_2^c([4q], \pi^*)$, the value of the integral $\int_{Z^{2q}} f_{\sigma,4} d\nu^{2q}$ depends on σ uniquely through r (or $q - r$), where r is the rank of σ.

By using Lemma 11.7.3, one obtains immediately the following result, which provides a combinatorial description of the implication (2.) $\Longrightarrow$ (3.) in Theorem 11.4.1.

Proposition 11.7.4 *For every $q \geq 2$ and every sequence $\{f^{(n)} : n \geq 1\} \subset L_s^2(\nu^q)$ such that $q! \|f^{(n)}\|_{L^2(\nu^q)}^2 \to 1$ $(n \to \infty)$, the following relations are equivalent:*

1. as $n \to \infty$

$$\sum_{\sigma \in \mathcal{M}_2([mq], \pi^*)} \int_{Z^{mq/2}} f_{\sigma,m}^{(n)} d\nu^{mq/2} \to 0, \quad \forall m \geq 3; \tag{11.7.32}$$

2. for every partition $\sigma \in \mathcal{M}_2^c([4q], \pi^*)$, as $n \to \infty$,

$$\int_{Z^{2q}} f_{\sigma,4}^{(n)} d\nu^{2q} \to 0. \tag{11.7.33}$$

Proof. Thanks to formula 7.3.29, one deduces that

$$\sum_{\sigma \in \mathcal{M}_2([mq],\pi^*)} \int_{Z^{mq/2}} f_{\sigma,m}^{(n)} d\nu^{mq/2} = \chi_m \left(I_q^X \left(f^{(n)} \right) \right),$$

where $I_q^X \left(f^{(n)} \right)$ is the multiple Wiener-Itô integral of $f^{(n)}$ with respect to the Gaussian measure induced by X, and χ_m indicates the mth cumulant. It follows that, since $q! \left\| f^{(n)} \right\|_{L^2(\nu^q)}^2 = \mathbb{E}\left[I_q^X \left(f^{(n)} \right)^2 \right] \to 1$, relation (11.7.32) is equivalent (by virtue of Lemma 3.2.5) to $I_q^X \left(f^{(n)} \right) \overset{law}{\to} Z \sim N(0,1)$. On the other hand, one deduces from Lemma 11.7.3 that (11.7.33) takes place if and only if (11.4.11) holds. Since, according to Theorem 11.4.1, condition (11.4.11) is necessary and sufficient in order to have $I_q^X \left(f^{(n)} \right) \overset{law}{\to} Z$, we immediately obtain the desired conclusion. ∎

Corollary 11.7.5 *Fix $q \geq 2$ and suppose that the sequence $\left\{ f^{(n)} : n \geq 1 \right\} \subset L_s^2 (\nu^q)$ is such that $q! \left\| f^{(n)} \right\|_{L^2(\nu^q)}^2 \to 1$ ($n \to \infty$). Then, $I_q^X (f^{(n)}) \overset{law}{\to} Z \sim N(0,1)$ if and only if (11.7.33) takes place.*

Proof. As pointed out in the proof of Proposition 11.7.4, since the normalization condition $q! \left\| f^{(n)} \right\|_{L^2(\nu^q)}^2 \to 1$ is in order, relation (11.7.32) is equivalent to the fact that the sequence $I_q^X \left(f^{(n)} \right), n \geq 1$, converges in law to a standard Gaussian random variables. The implication (2.) $\Longrightarrow$ (1.) in the statement of Proposition 11.7.4 yields the desired result. ∎

Remarks. (1) Corollary 11.7.5 implies that, in order to prove a CLT on a fixed Wiener chaos, *it is sufficient to compute and control a finite number of expressions of the type* $\int_{Z^{2q}} f_{\sigma}^{(n)} d\nu^{2q}$, *where σ is associated with a connected Gaussian circular diagram with four rows.* Moreover, these expressions determine the speed of convergence in total variation, via the upper bound given in (11.4.14).

(2) In view of Lemma 11.7.3, relation (11.7.31) also implies that: (i) for q even, (11.7.33) takes place for every

$$\sigma \in \mathcal{M}_2^c \left([4q], \pi^* \right)$$

if and only if for every $r = 1, ..., q/2$, there exists a partition $\sigma \in \mathcal{M}_2^c \left([4q], \pi^* \right)$ with rank r and such that (11.7.33) holds; (ii) for q odd, (11.7.33) takes place for every $\sigma \in \mathcal{M}_2^c \left([4q], \pi^* \right)$ if and only if for every $r = 1, ..., (q+1)/2$, there exists a partition $\sigma \in \mathcal{M}_2^c \left([4q], \pi^* \right)$ with rank r and such that (11.7.33) holds.

11.8 A multidimensional CLT

The paper [111] (but see also [90, 96, 103]) contains a complete solution of Problem A in the Gaussian case, for every vector of dimension $d \geq 2$. For such an index d, we denote by V_d the set of all vectors $(i_1, i_2, i_3, i_4) \in (1, ..., d)^4$ such that at least one of the following three properties is verified:

(a) the last three variables are equal and different from the first;
(b) the second and third variables are equal and different from the others (that are also distinguished);
(c) all variables are different.

Formally:

(a) $i_1 \neq i_2 = i_3 = i_4$;
(b) $i_1 \neq i_2 = i_3 \neq i_4$ and $i_4 \neq i_1$;
(c) the elements of $(i_1, ..., i_4)$ are all different.

In what follows, $X = \{X(h) : h \in \mathfrak{H}\}$ indicates an isonormal Gaussian process over a real separable Hilbert space $\mathfrak{H}$.

Theorem 11.8.1 *Let $d \geq 2$ and $q_1, ..., q_d \geq 1$ be fixed and let*

$$\left\{ f_j^{(n)} : j = 1, ..., d, \ n \geq 1 \right\}$$

be a collection of kernels such that $f_j^{(n)} \in \mathfrak{H}^{\odot q_j}$ and the normalization condition (11.3.8) is verified. Then, the following conditions are equivalent:

1. *as $n \to \infty$, the vector $\mathbf{F}_n = \left(I_{q_1}^X \left(f_1^{(n)} \right), ..., I_{q_d}^X \left(f_d^{(n)} \right) \right)$ converges in law towards a d-dimensional Gaussian vector $\mathbf{N}_d(0, \mathbf{C}) = (N_1, ..., N_d)$ with covariance matrix $\mathbf{C} = \{C(i, j)\}$;*

2.

$$\lim_{n \to \infty} \mathbb{E}\left[\left(\sum_{i=1,...,d} I_{q_i}^X \left(f_i^{(n)} \right) \right)^4 \right]$$

$$= 3 \left(\sum_{i=1}^{d} C(i, i) + 2 \sum_{1 \leq i < j \leq d} C(i, j) \right)^2 = \mathbb{E}\left[\left(\sum_{i=1}^{d} N_i \right)^4 \right],$$

and

$$\lim_{k \to \infty} \mathbb{E}\left[\prod_{l=1}^{4} I_{q_{i_l}}^X \left(f_{i_l}^{(n)} \right) \right] = \mathbb{E}\left[\prod_{l=1}^{4} N_{i_l} \right]$$

$\forall (i_1, i_2, i_3, i_4) \in V_d$;

3. *for every $j = 1, ..., d$, the sequence $I_{q_j}^X \left(f_j^{(n)} \right) \ n \geq 1$, converges in law towards N_j, that is, towards a centered Gaussian variable with variance $C(j, j)$;*

4. $\forall j = 1, ..., d, \ \lim_{n \to \infty} \chi_4 \left(I_{q_j}^X \left(f_j^{(n)} \right) \right) = 0$;

5. $\forall j = 1, ..., d$

$$\lim_{n \to \infty} \left\| f_j^{(n)} \otimes_r f_n^{(n)} \right\|_{\mathfrak{H}^{\otimes 2(q_j - r)}} = 0, \tag{11.8.34}$$

$\forall r = 1, ..., q_j - 1$.

The original proof of Theorem 11.8.1 uses arguments from stochastic calculus. See [90] and [96], respectively, for alternate proofs based on Malliavin calculus and Stein's method. In particular, in [90] one can find bounds analogous to (11.4.14), concerning the multidimensional Gaussian approximation of $\mathbf{F}_n$ in the Wasserstein distance. The crucial element in the statement of Theorem 11.8.1 is the implication $(3.) \Rightarrow (1.)$, which yields the following result.

Corollary 11.8.2 *Let the vectors $\mathbf{F}_n$, $n \geq 1$, be as in the statement of Theorem 11.8.1, and suppose that (11.3.8) is satisfied. Then, the convergence in law of each component of the vectors $\mathbf{F}_n$, towards a Gaussian random variable, always implies the joint convergence of $\mathbf{F}_n$ towards a Gaussian vector with covariance $\mathbf{C}$.*

Thanks to Theorem 11.4.1, it follows that a CLT such as (11.3.9) can be uniquely deduced from (11.3.8) and from the relations (11.8.34), involving the contractions of the kernels $f_j^{(n)}$.

When $\mathfrak{H} = L^2(Z, \mathcal{Z}, \nu)$ (with ν non atomic), the combinatorial implications of Theorem 11.8.1 are similar to those of Theorem 11.4.1. Indeed, thanks to the implication $(5.) \Rightarrow (1.)$, one deduces that, for a sequence $\left(f_1^{(n)}, ..., f_d^{(n)} \right)$, $n \geq 1$, as in (11.3.8), if

$$\int_{Z^{2q_j}} \left(f_j^{(n)} \right)_\sigma d\nu^{2q_j} \to 0, \quad \forall \sigma \in \mathcal{M}_2^c \left([4q_j], \pi^* \right),$$

then, the joint cumulants of order ≥ 3 converge to zero, namely

$$\sum_{\sigma \in \mathcal{M}_2([M], \pi^*)} \int_{Z^{M/2}} f_{\sigma, \ell}^{(n)} d\nu^{M/2} \to 0,$$

for every integer M which is the sum of $\ell \geq 3$ components $(q_{i_1}, q_{i_2}, ..., q_{i_\ell})$ of the vector $(q_1, ..., q_m)$ (with possible repetitions of the indices $i_1, ..., i_\ell$), with

$$\pi^* = \left\{ \{1, ..., q_{i_1}\}, ..., \{q_1 + ... + q_{i_{\ell-1}} + 1, ..., M\} \right\} \in \mathcal{P}([M]),$$

and every function $f_{\sigma, \ell}^{(n)}$, in $M/2$ variables, is obtained by identifying two variables x_a and x_b in the argument of $f_{i_1}^{(n)} \otimes_0 \cdots \otimes_0 f_{i_\ell}^{(n)}$ if and only if $a \sim_\sigma b$.

As already pointed out, the chaotic representation property (5.9.78) allows to use Theorem 11.8.1 in order to obtain CLTs for general functionals of an isonormal Gaussian process X. We now present a result in this direction, obtained in [39], whose proof can be deduced from Theorem 11.8.1.

Theorem 11.8.3 (see [39]) *We consider a sequence $\{F_n : n \geq 1\}$ of centered and square-integrable functionals of an isonormal Gaussian process X, admitting the chaotic decomposition*

$$F_n = \sum_{q=1}^{\infty} I_q^X \left(f_q^{(n)} \right), \quad n \geq 1.$$

Assume that

- $\lim_{N \to \infty} \lim \sup_{k \to \infty} \sum_{q \geq N+1} q! \left\| f_q^{(k)} \right\|_{\mathfrak{H}^{\otimes q}}^2 \to 0,$

- *for every $q \geq 1$, $\lim_{n \to \infty} q! \left\| f_q^{(n)} \right\|_{\mathfrak{H}^{\otimes q}}^2 = \sigma_q^2,$*

- $\sum_{q=1}^{\infty} \sigma_q^2 \triangleq \sigma^2 < \infty,$

- *for every $q \geq 1$, $\lim_{n \to \infty} \left\| f_q^{(n)} \otimes_r f_q^{(n)} \right\|_{\mathfrak{H}^{\otimes 2(q-r)}} = 0, \forall r = 1, ..., q-1.$*

Then, as $n \to \infty$, $F_n \overset{law}{\to} N\left(0, \sigma^2\right)$, where $N\left(0, \sigma^2\right)$ is a centered Gaussian random variable with variance σ^2.

12

CLTs in the Poisson case:
the case of double integrals

We conclude this monograph by discussing a simplified CLT for sequences of double integrals with respect to a Poisson random measure. Note that this result (originally obtained in [109]) has been generalized in [106], where one can find CLTs for sequences of multiple integrals of arbitrary orders – with explicit Berry-Esséen bounds in the Wasserstein distance obtained once again via Stein's method.

In this chapter, $(Z, \mathcal{Z}, \nu)$ is a Polish measure space, with ν σ-finite and non-atomic. Also, $\widehat{N} = \{\widehat{N}(B) : B \in \mathcal{Z}_\nu\}$ is a compensated Poisson measure with control measure given by ν. In [109], we have used some decoupling techniques developed in [107] in order to prove CLTs for sequences of random variables of the type:

$$F_n = I_2^{\widehat{N}}\left(f^{(n)}\right), \quad n \geq 1, \tag{12.0.1}$$

where $f^{(n)} \in L_s^2(\nu^2)$. In particular, we focus on sequences $\{F_n\}$ satisfying the following assumption

Assumption N. The sequence $f^{(n)}$, $n \geq 1$, in (12.0.1) verifies :

N.i (*integrability*) $\forall n \geq 1$,

$$\int_Z f^{(n)}(z, \cdot)^2 \nu(dz) \in L^2(\nu) \quad \text{and} \quad \left\{\int_Z f^{(n)}(z, \cdot)^4 \nu(dz)\right\}^{\frac{1}{2}} \in L^1(\nu); \tag{12.0.2}$$

N.ii (*normalization*) As $n \to \infty$,

$$2 \int_Z \int_Z f^{(n)}(z, z')^2 \nu(dz) \nu(dz') \to 1; \tag{12.0.3}$$

N.iii (*fourth power*) As $n \to \infty$,

$$\int_Z \int_Z f^{(n)}(z, z')^4 \nu(dz) \nu(dz') \to 0 \tag{12.0.4}$$

G. Peccati, M.S. Taqqu: Wiener Chaos: Moments, Cumulants and Diagrams –
A survey with computer implementation.
© Springer-Verlag Italia 2011

(in particular, this implies that $f^{(n)} \in L^4(\nu^2)$).

Remarks. (1) The conditions in (12.0.2) are technical : the first ensures the existence of the stochastic integral of $\int_Z f^{(n)}(z, \cdot)^2 \nu(dz)$ with respect to $\widehat{N}$; the second allows to use some Fubini arguments in the proof of the results to follow.

(2) Suppose that there exists a set B, independent of n, such that $\nu(B) < \infty$ and $f^{(n)} = f^{(n)} 1_B$, a.e.–$d\nu^2$, $\forall n \geq 1$ (this holds, in particular, when ν is finite). Then, by the Cauchy-Schwarz inequality, if (12.0.4) is true, then $(f^{(n)})$ converges necessarily to zero. Therefore, in order to study more general sequences $(f^{(n)})$, we must assume that $\nu(Z) = +\infty$.

The next theorem is the main result of [109]. It concerns convergence of the second Wiener chaos.

Theorem 12.0.4 *Let* $F_n = I_2^{\widehat{N}}(f^{(n)})$ *with* $f^{(n)} \in L_s^2(\nu^2)$, $n \geq 1$, *and suppose that Assumption N is verified. Then, for every* $n \geq 1$,

$$f^{(n)} \star_1^0 f^{(n)} \in L^2(\nu^3),$$

and

$$f^{(n)} \star_1^1 f^{(n)} \in L_s^2(\nu^2),$$

and also :

1. *if*

$$\left\| f^{(n)} \star_2^1 f^{(n)} \right\|_{L^2(\nu)} \to 0 \text{ and } \left\| f^{(n)} \star_1^1 f^{(n)} \right\|_{L^2(\nu^2)} \to 0, \tag{12.0.5}$$

then

$$F_n \overset{law}{\to} N(0, 1), \tag{12.0.6}$$

where $N(0, 1)$ *is a centered Gaussian random variable with unitary variance;*

2. *if* $F_n \in L^4(\mathbb{P})$, $\forall n$, *a sufficient condition in order to have (12.0.5) is*

$$\chi_4(F_n) \to 0; \tag{12.0.7}$$

3. *if the sequence* $\{F_n^4 : n \geq 1\}$ *is uniformly integrable, then the three conditions (12.0.5), (12.0.6) and (12.0.7) are equivalent.*

Remark. See [18] and [105] for several applications of Theorem 12.0.4 to Bayesian non-parametric survival analysis.

We now give a combinatorial interpretation (in terms of diagrams) of the three asymptotic conditions appearing in formulae (12.0.4) and (12.0.5). To do this, consider the set $[8] = \{1, ..., 8\}$, as well as the partition $\pi^* = \{\{1, 2\}, \{3, 4\}, \{5, 6\}, \{7, 8\}\} \in$

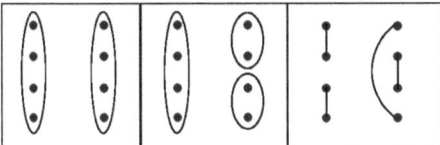

Fig. 12.1. Three diagrams associated with contractions

$\mathcal{P}([8])$. We define the set of partitions $\mathcal{M}_{\geq 2}([8],\pi^*) \subset \mathcal{P}([8])$ according to (7.2.23). Given an element $\sigma \in \mathcal{M}_{\geq 2}([8],\pi^*)$ and given $f \in L^2_s(\nu^2)$, the function $f_{\sigma,4}$, in $|\sigma|$ variables, is obtained by identifying the variables x_i and x_j in the argument of $f \otimes f \otimes f \otimes f$ (as defined in (6.1.3)) if and only if $i \sim_\sigma j$. We define three partitions $\sigma_1, \sigma_2, \sigma_3 \in \mathcal{M}_{\geq 2}([8],\pi^*)$ as follows:

$$\sigma_1 = \{\{1,3,5,7\},\{2,4,6,8\}\}$$
$$\sigma_2 = \{\{1,3,5,7\},\{2,4\},\{6,8\}\}$$
$$\sigma_3 = \{\{1,3\},\{4,6\},\{5,7\},\{2,8\}\}.$$

The diagrams $\Gamma(\pi^*,\sigma_1)$, $\Gamma(\pi^*,\sigma_2)$ and $\Gamma(\pi^*,\sigma_3)$ are represented (in order) in Fig. 12.1.

One has therefore the following combinatorial representation of the three norms appearing in formulae (12.0.4) and (12.0.5) (the proof is elementary, and left to the reader).

Proposition 12.0.5 *For every* $f \in L^2_s(\nu^2)$, *one has that*

$$\int_Z \int_Z f(z,z')^4 \, \nu(dz)\nu(dz') = \|f\|^4_{L^4(\nu^2)} = \int_{Z^2} f_{\sigma_1,4}(z,z')\nu(dz)\nu(dz')$$

$$\int_Z \left[\int_Z f(z,z')^2 \, \nu(dz)\right]^2 \nu(dz') = \|f \star^1_2 f\|^2_{L^2(\nu)} = \int_Z f_{\sigma_2,4}(z)\nu(dz)$$

$$\int_{Z^2} \left[\int_Z f(z,z')f(z,z'')\nu(dz)\right]^2 \nu(dz')\nu(dz'') = \|f \star^1_1 f\|^2_{L^2(\nu)}$$

$$= \int_Z f_{\sigma_3,4}(z)\nu(dz).$$

In particular, Proposition 12.0.5 implies that, on the second Wiener chaos of a Poisson measure, one can establish CLTs by focusing uniquely on expressions related to three connected diagrams with four rows. Similar characterizations for sequences belonging to chaoses of higher orders can be deduced from the main findings of [106].

Appendix A

Practical implementation using Mathematica

The purpose of this appendix is to illustrate some of the set partitions and diagram formulae we have encountered. We do this through a practical implementation of these formulae by using the software program *Mathematica*[1][2]. The source code and more detailed examples are contained in a companion *Mathematica* notebook. This notebook can be downloaded by using the link:

http://extras.springer.com, Password: 978-88-470-1678-1

To view the notebook you may use the free *Mathematica Reader* which you can download from the Internet or you can use the *Mathematica* program itself. The *Mathematica* program (version 7.0 or later) must be installed on your computer or remotely through your system if you want to execute the source code.

A.1 General introduction

The filename of a *Mathematica* notebook ends with a ".nb". Place the notebook and the style file *StyleFile978-88-470-1678-1.nb* in the same folder. This style file formats the notebook nicely but it is not absolutely necessary to have it.

In Windows, double-clicking on the notebook will open the notebook and start *Mathematica* (or the *Mathematica Reader*). It is best to begin by executing the entire notebook because some functions depend on previously defined functions. To execute the entire notebook (on Windows), press Control-A (to select all the cells in the notebook), followed by Shift-Enter. (On a Mac, press Command-A followed by Shift-Enter.) These actions will execute all of the function definitions and examples in this notebook. This may take a little time. You will see "Running" on the top left of the

---

[1] *Mathematica* is a computational software program developed by Wolfram Research.
[2] For further references on the use of *Mathematica* in the context of multiple stochastic integrals, see for example, [54], [155].

G. Peccati, M.S. Taqqu: Wiener Chaos: Moments, Cumulants and Diagrams –
A survey with computer implementation.
© Springer-Verlag Italia 2011

notebook. After the execution has completed, you will be able to apply the functions created in this notebook to new set partitions or to create additional examples of your own. You may type, paste, or outline a command and then press Shift-Enter to execute it. You may do this in the present notebook or, preferably, in a new notebook that you would open.

Please note that the execution of any of these functions requires the use of the *Combinatorica* package in *Mathematica* (which can be activated by executing the line Needs["Combinatorica"]. We automatically execute this package in this notebook.

To type a command, simply type the appropriate function and the desired argument and press Shift-Enter to execute the cell in which the command is located. *Mathematica* will label the command as a numbered input and the output will be numbered correspondingly. One can recall previous outputs with commands like %5, which would recover the fifth output in the current *Mathematica* session. Unexecuted cells have no input label.

Warning. Some of the functions in this notebook use a brute-force methodology. The number of set partitions associated to a positive integer n grows very rapidly, and so any brute-force methodology may suffer from combinatorial explosion. Unless you have lots of time and computational resources at your disposal, it may be better to investigate examples for which n is small.

Comment about In[ ] and Out[ ]. *Mathematica*, after execution of a command, automatically numbers it, for example, "In[1]:=". The output lines are also numbered, for example, the corresponding output line will start with: "Out[1]=". Here, we symbolically start a command input line with "*In:=*" and the corresponding output starts with "*Out=*" . Thus, in the sequel do not type, for example, "In:=" nor "In[269]:=", as this is automatically generated by *Mathematica* after you execute the command by pressing Shift-Enter.

Comments about the time it takes to evaluate a function. To evaluate the time it takes to simulate 1,000,000 realizations of a standard normal random variable, type:

In :=   RandomReal[NormalDistribution[0, 1], {1 000 000}]; // AbsoluteTiming
Out=   0.2187612, Null

The semicolon suppresses the display of the output (except for the timing result), which is generated by AbsoluteTiming. The first part of the output is the number of seconds it took *Mathematica* to generate 1,000,000 realizations of a standard normal random variable. The output was suppressed, which is why the symbol Null appears in the second position.

Comments about the function names. We typically give the functions long, informative names. In practical use, one may sometimes want to use a shorter name. For example, to have the function Chi do the same thing as the function CumulantVector-ToMoments, set:

In :=   Chi = CumulantVectorToMoments

A few similar functions may have different names because they are used in different contexts. This is the case, for example, of the functions MeetSolve, PiStarMeet Solve and MZeroSets.

Comments about typing Greek symbols. To type symbols, such as μ and κ, either copy and paste them with the mouse from somewhere else in the notebook or type respectively \[Mu] and \[Kappa]. *Mathematica* converts these immediately to greek.

Comments about *Mathematica* notation. In *Mathematica*, a set partition is denoted in a particular way, for example, $\{\{1,2,4\},\{3\},\{5\}\}$ is a set partition of the set $\{1,2,3,4,5\}$. The trivial, or maximal, partition consisting of the entire set is denoted $\{\{1,2,3,4,5\}\}$ and the set partition consisting entirely of singletons is denoted $\{\{1\},\{2\},\{3\},\{4\},\{5\}\}$.

Comments about hard returns. *Mathematica* ignores hard returns. Therefore, if the input line is long, one can press "Enter" in order to subdivide it.

Comments about stopping the execution of a function. Hold down the "Alt" key and then press the "." key, or go to the menu at the top, choose "Evaluation" and then choose "Abort Evaluation."

Comments about the examples below. The examples are particularly simple. Their only purpose is to illustrate the *Mathematica* commands. The *Mathematica* output will sometimes not be listed below, because it is either too long or would require complicated formatting.

A.2 When Starting

Type first the following two commands:

In :=   Needs ["Combinatorica"] ; Off[First::first]

The first command loads the package *Combinatorica*. The second command shuts off an error message that does not impact any of the function outputs. These commands will be executed in sequence.

Note: These two commands are listed at the beginning of the notebook. They will be automatically executed if you execute the whole notebook.

A.3 Integer Partitions

A.3.1 IntegerPartitions

Find all nonnegative integers that sum up to an integer n, ignoring order; this is a built-in *Mathematica* function.

▶ Please refer to Section 2.1.

Example. Find all integer partitions of the number 5. The output $\{3, 1, 1\}$ should be read as the decomposition $5 = 3 + 1 + 1$.

$In :=$ IntegerPartitions[5]
$Out=$ $\{\{5\}, \{4, 1\}, \{3, 2\}, \{3, 1, 1\}, \{2, 2, 1\}, \{2, 1, 1, 1\}, \{1, 1, 1, 1, 1\}\}$

Example. Count the number of integer partitions of the number 5.

$In :=$ Length[IntegerPartitions[5]]
$Out=$ 7

A.3.2 IntegerPartitionsExponentRepresentation

This function represents the output of IntegerPartitions in exponent notation; namely, $1^{r_1} 2^{r_2} \cdots n^{r_n}$, where r_i is the number of times that integer i appears in the decomposition, $i = 1, ..., n$.

Example. Find all integer partitions of the number 5, and express them in exponent notation. Thus $5 = 3 + 1 + 1$ is expressed as $\{1^2 2^0 3^1 4^0 5^0\}$ since there are 2 ones and 1 three.

$In :=$ IntegerPartitionsExponentRepresentation[5]
$Out=$ $\{\{1^0 2^0 3^0 4^0 5^1\}, \{1^1 2^0 3^0 4^1 5^0\}, \{1^0 2^1 3^1 4^0 5^0\}, \{1^2 2^0 3^1 4^0 5^0\},$
      $\{1^1 2^2 3^0 4^0 5^0\}, \{1^3 2^1 3^0 4^0 5^0\}, \{1^5 2^0 3^0 4^0 5^0\}\}$

A.4 Basic Set Partitions

▶ Please refer to Section 2.2.

A.4.1 SingletonPartition

Generates the singleton set partition of order n, namely $\hat{0} = \{\{1\}, \{2\}, \cdots, \{n\}\}$. This is a partition made up of singletons.

Example. Generate the singleton set partition of order 12.

In := SingletonPartition[12]
Out= $\{\{1\}, \{2\}, \{3\}, \{4\}, \{5\}, \{6\}, \{7\}, \{8\}, \{9\}, \{10\}, \{11\}, \{12\}\}$

Example. Same example with grid display.

In := Grid[SetPartitions[3], Frame $\rightarrow$ All]

Out=

$\{1,2,3\}$		
$\{1\}$	$\{2,3\}$	
$\{1,2\}$	$\{3\}$	
$\{1,3\}$	$\{2\}$	
$\{1\}$	$\{2\}$	$\{3\}$

A.4.2 MaximalPartition

Generates the maximal set partition of order n, namely $\hat{1} = \{\{1, 2, ..., n\}\}$.

Example. Generate the maximal set partition of order 12.

In := MaximalPartition[12]
Out= $\{\{1, 2, 3, 4, 5, 6, 7, 8, 9, 10, 11, 12\}\}$

A.4.3 SetPartitions

Enumerates all of the set partitions associated with a particular integer n, namely all set partitions of $\{1, 2, \cdots, n\}$; this is a built-in *Mathematica* function.

▶ Please refer to Section 2.4.

Example. Enumerate all set partitions of $\{1, 2, 3\}$.

In := SetPartitions[3]
Out= $\{\{\{1, 2, 3\}\}, \{\{1\}, \{2, 3\}\}, \{\{1, 2\}, \{3\}\}, \{\{1, 3\}, \{2\}\}, \{\{1\}, \{2\}, \{3\}\}\}$

A.4.4 PiStar

Generates a set partition of $\{1, 2, \cdots, n_1 + \cdots + n_k\}$, with k blocks, the first of size n_1 and the last of size n_k, that is,

$$\pi^* = \{\{1, \cdots, n_1\}, \{n_1 + 1, \cdots, n_1 + n_2\}, \cdots, \{n_1 + \cdots + n_{k-1} + 1, \cdots, n\}\}.$$

▶ Please refer to Formula (6.1.1).

Example. Create a set partition π^* with blocks of size 1, 4, 7, and 9.

In := PiStar[{1, 2, 4, 7, 9}]
Out= $\{\{1\}, \{2, 3\}, \{4, 5, 6, 7\}, \{8, 9, 10, 11, 12, 13, 14\},$
        $\{15, 16, 17, 18, 19, 20, 21, 22, 23\}\}$

A.5 Operations with Set Partitions

A.5.1 PartitionIntersection

Finds the meet $\sigma \wedge \pi$ between two set partitions σ and π; each block of $\sigma \wedge \pi$ is a non-empty intersection between one block of σ and one block of π.

Example. Find the meet between the set partitions $\{\{1, 4, 5\}, \{2\}, \{3\}\}$ and $\{\{1, 2, 5\}, \{3\}, \{4\}\}$.

In := PartitionIntersection $[\{\{1, 4, 5\}, \{2\}, \{3\}\}, \{\{1, 2, 5\}, \{3\}, \{4\}\}]$
Out= $\{\{1, 5\}, \{2\}, \{3\}, \{4\}\}$

Example. If the partitions σ and π are not of the same order, then PartitionIntersection returns a message.

In := PartitionIntersection[SingletonPartition[8], $\{\{1, 2, 5, 6, 7\}, \{3\}, \{4\}\}]$
Out= Error

A.5.2 PartitionUnion

Finds the join $\sigma \vee \pi$ between two set partitions σ and π; that is, the union of blocks of σ and π that have at least one element in common.

Example. Find the join between the set partitions $\{\{1, 4, 5\}, \{2\}, \{3\}\}$ and $\{\{1, 2, 5\}, \{3\}, \{4\}\}$.

In := PartitionUnion [{{1,4,5},{2},{3}} , {{1,2,5},{3},{4}}]
Out= $\{\{1, 2, 4, 5\}, \{3\}\}$

A.5.3 MeetSolve

For a fixed set partition π, find all set partitions σ such that the meet of π and σ is the singleton partition $\hat{0}$, that is, $\sigma \wedge \pi = \hat{0}$.

▶ Please refer to Section 4.2.

Example. Find all set partitions σ such that the meet of σ and $\{\{1,4,5\},\{2\},\{3\}\}$ is the singleton partition $\hat{0}$.

In := MeetSolve[$\{\{1,4,5\},\{2\},\{3\}\}$]
Out= $\{\{\{1\},\{2,3,4\},\{5\}\},\ \{\{1\},\{2,5\},\{3,4\}\},\ \{\{1\},\{2,3,5\},\{4\}\},$
    $\{\{1\},\ \{2,4\},\{3,5\}\},\{\{1,2\},\{3,4\},\{5\}\},\ \{\{1,2\},\{3,5\},\{4\}\},$
    $\{\{1,2,3\},\{4\},\{5\}\},\ \{\{1,3\},\{2,4\},\{5\}\},\ \{\{1,3\},\{2,5\},\{4\}\},$
    $\{\{1\},\{2\},\{3,4\},\{5\}\},\ \{\{1\},\{2\},\{3,5\},\{4\}\},\ \{\{1\},\{2,3\},\{4\},\{5\}\},$
    $\{\{1\},\{2,4\},\{3\},\{5\}\},\ \{\{1\},\{2,5\},\{3\},\{4\}\},\ \{\{1,2\},\{3\},\{4\},\{5\}\},$
    $\{\{1,3\},\{2\},\{4\},\{5\}\},\ \{\{1\},\{2\},\{3\},\{4\},\{5\}\}\}$

The output also displays the number of solutions (namely 17).

A.5.4 JoinSolve

For a fixed set partition π, find all set partitions σ such that the join of π and σ is the maximal partition $\hat{1}$, that is, $\sigma \vee \pi = \hat{1}$.

Example. Find all set partitions σ such that the join of σ and $\{\{1,4,5\},\{2\},\{3\}\}$ is the maximal partition $\hat{1}$.

In := JoinSolve[$\{\{1,4,5\},\{2\},\{3\}\}$]
Out= $\{\{\{1,2,3,4,5\}\},\ \{\{1\},\{2,3,4,5\}\},\ \{\{1,2\},\{3,4,5\}\},$
    $\{\{1,2,3\},\{4,5\}\},\ \{\{1,3\},\{2,4,5\}\},\ \{\{1,2,3,4\},\{5\}\},$
    $\{\{1,5\},\{2,3,4\}\},\ \{\{1,2,5\},\{3,4\}\},\{\{1,3,4\},\{2,5\}\},$
    $\{\{1,2,3,5\},\{4\}\},\ \{\{1,4\},\{2,3,5\}\},\ \{\{1,2,4\},\{3,5\}\},$
    $\{\{1,3,5\},\{2,4\}\},\ \{\{1\},\{2,3,4\},\{5\}\},\ \{\{1\},\{2,5\},\{3,4\}\},$
    $\{\{1\},\{2,3,5\},\{4\}\},\ \{\{1\},\{2,4\},\{3,5\}\},\ \{\{1,2\},\{3,4\},\{5\}\},$
    $\{\{1,2\},\{3,5\},\{4\}\},\ \{\{1,2,3\},\{4\},\{5\}\},$
    $\{\{1,3\},\{2,4\},\{5\}\},\ \{\{1,3\},\{2,5\},\{4\}\}\}$

A.5.5 JoinSolveGrid

Display the output of JoinSolve in a grid.

Example. Find all set partitions σ such that the join of σ and $\{\{1,2\},\{3,4\}\}$ is the maximal partition $\hat{1}$.

In :=   JoinSolveGrid$[\{\{1,2\},\{3,4\}\}]$

Out=

$\{1,2,3,4\}$		
$\{1\}$	$\{2,3,4\}$	
$\{1,3,4\}$	$\{2\}$	
$\{1,2,3\}$	$\{4\}$	
$\{1,4\}$	$\{2,3\}$	
$\{1,2,4\}$	$\{3\}$	
$\{1,3\}$	$\{2,4\}$	
$\{1\}$	$\{2,3\}$	$\{4\}$
$\{1\}$	$\{2,4\}$	$\{3\}$
$\{1,3\}$	$\{2\}$	$\{4\}$
$\{1,4\}$	$\{2\}$	$\{3\}$

A.5.6 MeetAndJoinSolve

For a fixed set partition π, find all set partitions σ such that the join of π and σ is the maximal partition $\hat{1}$ AND the meet of π and σ is the singleton partition $\hat{0}$. that is, $\sigma \wedge \pi = \hat{0}$ and $\sigma \vee \pi = \hat{1}$.

Example. Find all set partitions σ such that the join of σ and $\{\{1,4,5\},\{2\},\{3\}\}$ is the maximal partition $\hat{1}$ and such that the meet of σ and $\{\{1,4,5\},\{2\},\{3\}\}$ is the singleton partition $\hat{0}$. Include the timing of the computation.

In :=   MeetAndJoinSolve $[\{\{1,4,5\},\{2\},\{3\}\}] \,//$ AbsoluteTiming
Out= $\{0.2031250, \{\{\{1\},\{2,3,4\},\{5\}\}, \{\{1\},\{2,5\},\{3,4\}\},$
    $\{\{1\},\{2,3,5\},\{4\}\}, \{\{1\},\{2,4\},\{3,5\}\}, \{\{1,2\},\{3,4\},\{5\}\},$
    $\{\{1,2\},\{3,5\},\{4\}\}, \{\{1,2,3\},\{4\},\{5\}\},$
    $\{\{1,3\},\{2,4\},\{5\}\}, \{\{1,3\},\{2,5\},\{4\}\}\}\}$

The output in the first position is the number of seconds it took to generate the rest of the output. The number of solutions will also be specified (namely 9).

A.5.7 CoarserSetPartitionQ

This is a built-in *Mathematica* function; CoarserSetPartitionQ[σ,π] yields "True" if the set partition π is coarser than the set partition σ; that is, if $\sigma \leq \pi$, i.e., if every block in σ is contained in a block of π.

Example. Is the set partition $\pi = \{1,2\}, \{3,4\}, \{5\}$ coarser than the set partition $\sigma = \{\{1,2,3,4\}, \{5\}\}$?

> *In :=*   CoarserSetPartitionQ[{{1, 2, 3, 4}, {5}}, {{1, 2}, {3, 4}, {5}}]
> *Out=*   False

A.5.8 CoarserThan

For a fixed set partition π, find all set partitions σ such that $\sigma \geq \pi$, i.e., all set partitions such that every block of π is fully contained in some block of σ.

Example. Find all of the set partitions σ such that $\sigma \geq \{\{1,2\}, \{3,4\}, \{5\}\}$.

> *In :=*   CoarserThan[{{1, 2}, {3, 4}, {5}}]
> *Out=*   {{{1, 2, 3, 4, 5}}, {{1, 2}, {3, 4, 5}}, {{1, 2, 3, 4}, {5}},
>   {{1, 2, 5}, {3, 4}}, {{1, 2}, {3, 4}, {5}}}

A.6 Partition Segments

A.6.1 PartitionSegment

For set partitions σ and π, find the partition segment $[\sigma, \pi]$, that is, all set partitions ρ such that $\sigma \leq \rho \leq \pi$.

▶ Please refer to Definition 2.2.4.

Example. Find all of the set partitions ρ such that

$$\{\{1\}, \{2\}, \{3\}, \{4, 5\}\} = \sigma \leq \rho \leq \pi = \{\{1, 2, 3\}, \{4, 5\}\}.$$

> *In :=*   PartitionSegment[{{1}, {2}, {3}, {4, 5}}, {{1, 2, 3}, {4, 5}}]
> *Out=*   {{{1, 2, 3}, {4, 5}}, {{1}, {2, 3}, {4, 5}}, {{1, 2}, {3}, {4, 5}},
>   {{1, 3}, {2}, {4, 5}}, {{1}, {2}, {3}, {4, 5}}}

The output also specifies the number of partitions in the segment (namely 5).

A.6.2 ClassSegment

For set partitions σ and π, with $\sigma \leq \pi$, obtain the "class" of the segment $[\sigma, \pi]$, namely

$$\lambda(\sigma, \pi) = (1^{r_1} 2^{r_2} \cdots |\sigma|^{r_{|\sigma|}}),$$

where r_i is the number of blocks of π that contain exactly i blocks of σ (terms with exponent zero are ignored in the output).

▶ Please refer to Definition 2.3.1.

Example. Compute the class segment when $\sigma = \{\{1,2\}, \{3\}, \{4,5\}\}$ and $\pi = \{\{1,2,3\}, \{4,5\}\}$.

In :=   ClassSegment[{{1,2}, {3}, {4,5}}, {{1,2,3}, {4,5}}]
Out= $(1^1 2^1)$

This is because only one block of π (namely, $\{4,5\}$) contains one block of σ and one block of π (namely, $\{1,2,3\}$) contains two blocks of σ. The answer could be more fully written as $(1^1 2^1 3^0)$.

A.6.3 ClassSegmentSolve

Given a partition σ, find all coarser partitions π with a specific class segment $\lambda(\sigma, \pi) = (1^{r_1} 2^{r_2} \cdots |\sigma|^{r_{|\sigma|}})$ written as $\{\{1, r_1\}, \{2, r_2\}, \cdots, \{|\sigma|, r_{|\sigma|}\}\}$ with $r_i \geq 0$ or merely $r_i > 0$. We want r_i blocks of π to contain exactly i blocks of σ, for $i = 1, ..., n$.

Input format. ClassSegmentSolve[$\sigma, \lambda(\sigma, \pi)$]

Example. Find the set of all partitions π such that the class segment of $\sigma = \{\{1,2\}, \{3\}, \{4,5\}\}$ and π is equal to $\lambda(\sigma, \pi) = (1^1 2^1)$. We want $r_1 = 1$ block of π to contain one block of σ and $r_2 = 1$ block of π to contain two blocks of σ.

In :=   ClassSegmentSolve[{{1,2}, {3}, {4,5}}, {{1, 1}, {2, 1}}]
Out=   {{{1,2}, {3,4,5}},   {{1,2,3}, {4,5}},   {{1,2,4,5}, {3}}}

If $\pi = \{\{1,2\}, \{3,4,5\}\}$, then $\{1,2\}$ contains the block $\{1,2\}$ in $\sigma = \{\{1,2\}, \{3\}, \{4,5\}\}$ and $\{3,4,5\}$ contains the blocks $\{3\}$ and $\{4,5\}$ of σ.

A.6.4 SquareChoose

For a list of non-negative integers $\{r_1, r_2, ..., r_n\}$, with $1r_1 + 2r_2 + \cdots + nr_n = n$, compute

$$\begin{bmatrix} n \\ \lambda \end{bmatrix} = \begin{bmatrix} n \\ r_1, ..., r_n \end{bmatrix} = \frac{n!}{(1!)^{r_1} r_1! (2!)^{r_2} r_2! \cdots (n!)^{r_n} r_n!}.$$

If one relabels the integers and denote by $\{r_1, r_2, \cdots, r_k\}$, with $k \leq n$ and k equal to the greatest integer in $[n]$ such that $r_k > 0$, then $1r_1 + 2r_2 + \cdots + kr_k = n$, and

$$\begin{bmatrix} n \\ \lambda \end{bmatrix} = \begin{bmatrix} n \\ r_1, ..., r_k \end{bmatrix} = \frac{n!}{(1!)^{r_1} r_1! (2!)^{r_2} r_2! \cdots (k!)^{r_k} r_k!}.$$

▶ Please refer to Formula (2.3.8).

Example. For the list $\{1, 3, 0, 0, 0, 0, 0\}$ or the list $\{1, 3\}$ compute

$$\frac{7!}{[(1!)^1 (1!)][(2!)^3 (3!)]}.$$

In :=   SquareChoose[{1, 3, 0, 0, 0, 0, 0, 0}]
In :=   SquareChoose[{1, 3}]

The output is 105 in both cases.

A.7 Bell and Touchard polynomials

▶ Please refer to Section 2.4.

A.7.1 StirlingS2

Obtain the Stirling numbers $S(n, k)$ of the second kind, where $1 \leq k \leq n$. This is a built-in mathematica function. Recall that

$$S(n, k) = \sum_{r_1, ..., r_{n-k+1}} \frac{n!}{(1!)^{r_1} r_1! (2!)^{r_2} r_2! \cdots (n-k+1)!^{r_{n-k+1}} r_{n-k+1}!},$$

where the sum runs over all vectors on nonnegative integers $(r_1, ..., r_{n-k+1})$ satisfying $r_1 + \cdots + r_{n-k+1} = k$ and $1r_1 + 2r_2 \cdots + (n-k+1) r_{n-k+1} = n$.

▶ Please refer to Proposition 2.3.4.

Example. Get $S(3, 2)$.

$In :=$ StirlingS2[3, 2]
$Out=$ $= 3$

Example. Get all the Stirling numbers of the second kind for $n = 10$.

$In :=$ Table[StirlingS2[10, k], k, 10]
$Out=$ 1, 511, 9330, 34105, 42525, 22827, 5880, 750, 45, 1

Example. Make a table of the Stirling numbers of the second kind for $n = 1, ..., 10$
and $k = 1, ..., n$.

$In :=$ Table[If[n >= k, StirlingS2[n, k]], {n, 1, 10}, {k, 1, 10}] // MatrixForm

$Out=$

$$
\begin{pmatrix}
1 & \text{Null} & \text{Null} & \text{Null} & \text{Null} & \text{Null} & \text{Null} & \text{Null} & \text{Null} & \text{Null} \\
1 & 1 & \text{Null} & \text{Null} & \text{Null} & \text{Null} & \text{Null} & \text{Null} & \text{Null} & \text{Null} \\
1 & 3 & 1 & \text{Null} & \text{Null} & \text{Null} & \text{Null} & \text{Null} & \text{Null} & \text{Null} \\
1 & 7 & 6 & 1 & \text{Null} & \text{Null} & \text{Null} & \text{Null} & \text{Null} & \text{Null} \\
1 & 15 & 25 & 10 & 1 & \text{Null} & \text{Null} & \text{Null} & \text{Null} & \text{Null} \\
1 & 31 & 90 & 65 & 15 & 1 & \text{Null} & \text{Null} & \text{Null} & \text{Null} \\
1 & 63 & 301 & 350 & 140 & 21 & 1 & \text{Null} & \text{Null} & \text{Null} \\
1 & 127 & 966 & 1701 & 1050 & 266 & 28 & 1 & \text{Null} & \text{Null} \\
1 & 255 & 3025 & 7770 & 6951 & 2646 & 462 & 36 & 1 & \text{Null} \\
1 & 511 & 9330 & 34105 & 42525 & 22827 & 5880 & 750 & 45 & 1
\end{pmatrix}
$$

The output has "Null" in the upper quadrant since the $S(n, k)$ are either not defined or
are set equal to 0 when $k > n$.

A.7.2 BellPolynomialPartial

Generate the partial Bell polynomials $B_{n,k}(x_1, ..., x_{n-k+1})$, defined for $1 \le k \le n$.
Recall that

$$
B_{n,k}(x_1, ..., x_{n-k+1}) =
$$

$$
\sum_{r_1,...,r_{n-k+1}} \frac{n!}{r_1! r_2! \cdots r_{n-k+1}!} \left(\frac{x_1}{1!} \right)^{r_1} \cdots \left(\frac{x_{n-k+1}}{(n-k+1)!} \right)^{r_{n-k+1}}
$$

where the sum runs over all vectors on nonnegative integers $(r_1, ..., r_{n-k+1})$ satisfying
$r_1 + \cdots + r_{n-k+1} = k$ and $1r_1 + 2r_2 \cdots + (n-k+1) r_{n-k+1} = n$.

▶ Please refer to Definition 2.4.1.

Input format. BellPolynomialPartial$[n, k, x]$ where $1 \le k \le n$ and x is a vector of length at least $n - k + 1$ (the extra components are ignored). The index k can also be "All". This outputs $B_{n,1}, \cdots, B_{n,n}$ in a list.

Example. Generate the partial Bell polynomial $n = 4, k = 2$ and 3 variables $x1, x2, x3, x4$. These form a vector and must be inserted with curly brackets.

In := BellPolynomialPartial$[4, 2, \{x1, x2, x3\}]$
Out= $3x2^2 + 4x1x3$

The output should be read as $3x_2^2 + 4x_1x_3$.

Example. Numerical computation with $n = 70, k = 50$, and the variables taking consecutive numerical values.

In := BellPolynomialPartial$[40, 33, \{1, 2, 3, 4, 5, 6, 7, 8\}]$
Out= 794559498748278120

Example. Numerical computation with $n = 70, k = 50$, and the variables taking consecutive numerical values, defined through a table.

In := BellPolynomialPartial[70, 50, Table[i, {i, 1, 70 - 50 + 1}]]
Out= 154385188734681964874267578125000000000000000000000

A.7.3 BellPolynomialPartialAuto

Automatically generates symbolic form of BellPolynomialPartial$[n, k]$ without having to input a vector. Moreover k can be 'All', in which case the output is $B_{n,1}, \cdots, B_{n,n}$ in a list.

Example. Generate the partial Bell polynomial $n = 4, k = 2$ but with unspecified variables.

In := BellPolynomialPartialAuto[4, 2]
Out= $3x_2^2 + 4x_1x_3$

Example. Generate all the partial Bell polynomials with $n = 4$.

In := BellPolynomialPartialAuto[4,"All"]
Out= $\{x_4, \ 3x_2^2 + 4x_1x_2, \ 6x_1^2x_3, \ x_1^4\}$

A.7.4 BellPolynomialComplete

Generate the complete Bell polynomial of order n, that is,

$$B_n(x_1, x_2, \cdots, x_n) = \sum \begin{bmatrix} n \\ r_1, \ldots, r_n \end{bmatrix} x_1^{r_1} x_2^{r_2} \cdots x_n^{r_n}$$

$$= \sum \frac{n!}{(1!)^{r_1} r_1! (2!)^{r_2} r_2! \cdots (n!)^{r_n} r_n!} x_1^{r_1} x_2^{r_2} \cdots x_n^{r_n},$$

where the sum is over all non-negative $r_1, r_2, \cdots, r_n$ satisfying $1r_1 + 2r_2 + \cdots + nr_n = n$.

▶ Please refer to Definition 2.4.1.

Input format. BellPolynomialComplete[x] where x is a vector. The length of the vector determines the value of n.

   Example. Generate the complete Bell polynomial with 4 variables $x1, x2, x3, x4$. These form a vector and must be inserted with curly brackets.

   In :=   BellPolynomialComplete[$\{x1, x2, x3, x4\}$]
   Out= $x1^4 + 6x1^2 x2 + 3x2^2 + 4x1x3 + x4$

The output should be read as $x_1^4 + 6x_1^2 x_2 + 3x_2^2 + 4x_1 x_3 + x_4$.

   Example. Generate the complete Bell polynomial with 4 identical variables.

   In :=   BellPolynomialComplete[$\{x, x, x, x\}$]
   Out= $x + 7x^2 + 6x^3 + x^4$

A.7.5 BellPolynomialCompleteAuto

Automatically generates a symbolic form of complete Bell polynomial of order n. There is no need to input a vector.

Input format. BellPolynomialCompleteAuto[n].

   Example. Obtain the complete Bell polynomial of order 4.

   In :=   BellPolynomialCompleteAuto[4]
   Out= $x_1^4 + 6x_1^2 x_2 + 3x_2^2 + 4x_1 x_3 + x_4$

Example. Make a table of Complete Bell polynomials of order 1 to 6.

In :=   TableForm[Table[BellPolynomialCompleteAuto[n], n, 1, 6],
TableHeadings → "n=1", "n=2", "n=3", "n=4", "n=5", "n=6", None]

Out=
n=1 x_1
n=2 $x_1^2 + x_2$
n=3 $x_1^3 + 3x_1x_2 + x_3$
n=4 $x_1^4 + 6x_1^2x_2 + 3x_2^2 + 4x_1x_3 + x_4$
n=5 $x_1^5 + 10x_1^3x_2 + 15x_1x_2^2 + 10x_1^2x_3 + 10x_2x_3 + 5x_1x_4 + x_5$
n=6 $x_1^6 + 15x_1^4x_2 + 45x_1^2x_2^2 + 15x_2^3 + 20x_1^3x_3 + 60x_1x_2x_3 + 10x_3^2$
$$+15x_1^2x_4 + 15x_2x_4 + 6x_1x_5 + x_6$$

A.7.6 BellB

Gives the complete Bell polynomial when all the arguments are identical and computes the Bell number of order n. These are buit-in mathematica functions.

Input format. BellB$[n, x]$ where n is a non-negative integer and x is a scalar. This is equivalent to BellPolynomialComplete$[\{x, x, .., x, x\}]$ with the scalar x appearing n times.

Example. Get BellPolynomialComplete$[\{x, x, x, x\}]$ using BellB$[4, x]$.

In :=   BellB$[4, x]$
Out= $x + 7x^2 + 6x^3 + x^4$

Input format. BellB$[n]$.

This is the Bell number of order n. It is equivalent to BellPolynomialComplete$[\{1, ..., 1\}]$ with 1 appearing n times and also equivalent to $T_n(1)$, where T_n is the Touchard polynomial.

Example. Get BellPolynomialComplete$[\{1, 1, 1, 1\}]$ using BellB$[4]$.

In :=   BellB[4]
Out=   15

Example. Get the Bell numbers up to 20.

In :=   Table[BellB[k], {k, 20}]
Out=   {1, 2, 5, 15, 52, 203, 877, 4140, 21147, 115975, 678570, 4213597, 27644437, 190899322, 1382958545, 10480142147, 82864869804, 682076806159, 5832742205057, 51724158235372 }

Example. Compute the sum of *SquareChoose* of over all values of its non-negative arguments, that is:

$$\sum \begin{bmatrix} n \\ r_1, \ldots, r_n \end{bmatrix} = \sum \frac{n!}{(1!)^{r_1} \, r_1! \, (2!)^{r_2} \, r_2! \cdots (n!)^{r_n} \, r_n!}$$

This amounts to evaluate the corresponding complete Bell polynomial with the variables x all equal to 1 or BellB[n]. Do it for $n = 5$.

In :=   BellB[5]
Out=   52

A.7.7 TouchardPolynomial

Obtain the Touchard polynomial $T_n(x)$ of order n, where x is a scalar. It is, in fact, identical to BellB$[n, x]$. The Touchard polynomial of order n is equal to the moment of order n of a Poisson with parameter x.

▶ Please refer to Definition 2.4.1.

Input format. TouchardPolynomial$[n, x]$ where n is a non-negative integer and x is a scalar.

Example. Get the Touchard polynomial of order $n = 4$.

In :=   TouchardPolynomial[4, x]
Out= $x + 7x^2 + 6x^3 + x^4$

Input format. TouchardPolynomial$[n, x, \text{"All"}]$. This generates a full list of Touchard Polynomials of order 0 up to order n.

Example. Get the Touchard polynomial of order 0 up to order 5.

In :=   TouchardPolynomial[5, x, "All"]
Out= $= \{1, x, x+x^2, x+3x^2+x^3, x+7x^2+6x^3+x^4, x+15x^2+25x^3+10x^4+x^5\}$

A.7.8 TouchardPolynomialTable

Generates a table of Touchard polynomials from order 0 to n.

Example. Get a table of the Touchard polynomials of order 0 up to order 5.

In := TouchardPolynomialTable[5]

Out=

n	T_n
0	1
1	x
2	$x + x^2$
3	$x + 3x^2 + x^3$
4	$x + 7x^2 + 6x^3 + x^4$
5	$x + 15x^2 + 25x^3 + 10x^4 + x^5$

A.7.9 TouchardPolynomialCentered

Computes the centered Touchard polynomial of order n. It is equal to the moment of order n of a centered Poisson with parameter x.

▶ Please refer to Proposition 3.3.4.

Input format. TouchardPolynomialCentered$[n, x]$ where n is a non-negative integer and x is a scalar.

Example. Get the Touchard polynomial of order $n = 5$.

In := TouchardPolynomialCentered$[5, x]$
Out= $x + 10x^2$

Input format. TouchardPolynomialCentered$[n, x, \text{"All"}]$. This generates a full list of centered Touchard polynomials of order 0 up to order n.

Example. Get the centered Touchard polynomial of order 0 up to order 5.

In := TouchardPolynomialCentered$[5, x, \text{"All"}]$
Out= $\{1, 0, x, x, x + 3x^2, x + 10x^2\}$

A.7.10 TouchardPolynomialCenteredTable

Generates a table of centered Touchard polynomials from order 0 to n.

Example. Get a table of the centered Touchard polynomials of order 0 up to order 5.

In :=   TouchardPolynomialCenteredTable[5]

Out=

n	$\tilde{T}_n$
0	1
1	0
2	x
3	x
4	$x + 3x^2$
5	$x + 10x^2$

A.8 Mobius Formula

▶ Please refer to Section 2.5.

A.8.1 MobiusFunction

For set partitions $\sigma \leq \pi$, compute the Mobius function

$$\mu(\sigma, \pi) = (-1)^{n-r} (2!)^{r_3} (3!)^{r_4} \cdots ((n-1)!)^{r_n}$$
$$= (-1)^{n-r} \prod_{j=2}^{n-1} (j!)^{r_{j+1}} = (-1)^{n-r} \prod_{j=0}^{n-1} (j!)^{r_{j+1}},$$

where $n = |\sigma|$, $r = |\pi|$, and $\lambda(\sigma, \pi) = (1^{r_1} 2^{r_2} \cdots n^{r_n})$ (that is, there are exactly r_i blocks of π containing exactly i blocks of σ).

Input format. MobiusFunction[σ,π].

Example. Evaluation of the Mobius function with $\sigma = \{\{1\}, \{2\}, \{3, 4\}, \{5, 6\}, \{7\}, \{8, 9\}\}$ and $\pi = \{\{1, 2\}, \{3, 4, 5, 6, 7, 8, 9\}\}$.

In :=   MobiusFunction[{{1}, {2}, {3, 4}, {5, 6}, {7}, {8, 9}}, {{1, 2},
               {3, 4, 5, 6, 7, 8, 9}}]

Out=   6

Let's check this result with ClassSegment.

In := [{{1}, {2}, {3, 4}, {5, 6}, {7}, {8, 9}}, {{1, 2}, {3, 4, 5, 6, 7, 8, 9}}]
Out= $(2^1 4^1)$

Thus $r_2 = 1, r_4 = 1, n = 6, r = 2$ which explains why $\mu(\sigma, \pi) = (-1)^{6-2}$ $(3!)^1 = 6$.

A.8.2 MobiusRecursionCheck

Verify that the Mobius function recursion

$$\mu(\sigma, \pi) = - \sum_{\sigma \leq \rho < \pi} \mu(\sigma, \rho),$$

holds; in the output, the number in the first coordinate is $\mu(\sigma, \pi)$ and the number in the second coordinate is $-\sum_{\sigma \leq \rho < \pi} \mu(\sigma, \rho)$; they should be equal.

Example. Check the Mobius function recursion when $\sigma = \{\{1\}, \{2\}, \{3\}, \{4\}\}$ and $\pi = \{\{1, 2, 3, 4\}\}$.

In := MobiusRecursionCheck [{{{1}, {2}, {3}, {4}}, {{1, 2, 3, 4}}}]

The output is $\{-6, -6\}$, confirming equality.

A.8.3 MobiusInversionFormulaZero

Compute the Mobius inversion formula given by

$$F(\pi) = \sum_{\hat{0} \leq \sigma \leq \pi} \mu(\sigma, \pi) G(\sigma)$$

where F and G are functions taking set partitions as their arguments.

Input format. MobiusInversionFormulaZero[π].

Example. Let $\pi = \{\{1, 2, 3\}, \{4, 5\}\}$. For all $\hat{0} \leq \sigma \leq \pi$, obtain the Mobius function coefficient $\mu(\sigma, \pi)$ and print $G(\sigma)$.

In := MobiusInversionFormulaZero[{{1, 2, 3}, {4, 5}}]

Out=   Number of partitions in the segment = 10

σ	$\|\mu(\sigma,\pi)\|$	$G(\sigma)$
$\{\{1,2,3\},\{4,5\}\}$	1	$G[\{\{1,2,3\},\{4,5\}\}]$
$\{\{1\},\{2,3\},\{4,5\}\}$	-1	$G[\{\{1\},\{2,3\},\{4,5\}\}]$
$\{\{1,2\},\{3\},\{4,5\}\}$	-1	$G[\{\{1,2\},\{3\},\{4,5\}\}]$
$\{\{1,3\},\{2\},\{4,5\}\}$	-1	$G[\{\{1,3\},\{2\},\{4,5\}\}]$
$\{\{1,2,3\},\{4\},\{5\}\}$	-1	$G[\{\{1,2,3\},\{4\},\{5\}\}]$
$\{\{1\},\{2\},\{3\},\{4,5\}\}$	2	$G[\{\{1\},\{2\},\{3\},\{4,5\}\}]$
$\{\{1\},\{2,3\},\{4\},\{5\}\}$	1	$G[\{\{1\},\{2,3\},\{4\},\{5\}\}]$
$\{\{1,2\},\{3\},\{4\},\{5\}\}$	1	$G[\{\{1,2\},\{3\},\{4\},\{5\}\}]$
$\{\{1,3\},\{2\},\{4\},\{5\}\}$	1	$G[\{\{1,3\},\{2\},\{4\},\{5\}\}]$
$\{\{1\},\{2\},\{3\},\{4\},\{5\}\}$	-2	$G[\{\{1\},\{2\},\{3\},\{4\},\{5\}\}]$

Here G is not specified so the output merely lists the various terms in the sum. For example, in the output, the term corresponding to $\sigma = \{\{1\},\{2,3\},\{4,5\}\}$ is $(-1)G(\{\{1\},\{2,3\},\{4,5\}\})$.

The function MobiusInversionFormulaZero is also useful for computing

$$G\left(\pi\right) = \sum_{\hat{0}\leq\sigma\leq\pi} F\left(\sigma\right).$$

Merely ignore $\mu(\sigma,\pi)$ and replace $G(\sigma)$ by $F(\sigma)$.

A.8.4 MobiusInversionFormulaOne

Compute the Mobius inversion formula given by

$$F\left(\pi\right) = \sum_{\pi\leq\sigma\leq\hat{1}} \mu\left(\pi,\sigma\right)G\left(\sigma\right)$$

where F and G are functions taking set partitions as their arguments.

Input format. MobiusInversionFormulaOne[π].

Example. Let $\pi = \{\{1,2,3,\},\{4,5\}\}$. For all $\pi \leq \sigma \leq \hat{1}$, obtain the Mobius function coefficient $\mu\left(\pi,\sigma\right)$ and print $G(\sigma)$.

In :=   MobiusInversionFormulaOne[$\{\{1,2,3\},\{4,5\}\}$]

Out=   Number of partitions in the segment = 2

σ	$\|\mu(\sigma,\pi)\|$	$G(\sigma)$
$\{\{1,2,3,4,5\}\}$	-1	$G[\{\{1,2,3,4,5\}\}]$
$\{\{1,2,3\},\{4,5\}\}$	1	$G[\{\{1,2,3\},\{4,5\}\}]$

The output has 2 terms only in this case:

$$(-1)G[\{\{1,2,3,4,5\}\}] + (1)G[\{\{1,2,3\},\{4,5\}\}].$$

The function MobiusInversionFormulaOne is also useful for computing

$$G(\pi) = \sum_{\pi \leq \sigma \leq \hat{1}} F(\sigma).$$

Merely ignore $\mu(\sigma, \pi)$ and replace $G(\sigma)$ by $F(\sigma)$.

A.9 Moments and Cumulants

▶ Please refer to Chapter 3. Recall the notation:

For every subset $b = \{j_1, ..., j_k\} \subseteq [n] = \{1, ..., n\}$, we write

$$\boxed{\mathbf{X}_b = (X_{j_1}, ..., X_{j_k})} \quad \text{and} \quad \boxed{\mathbf{X}^b = X_{j_1} \times \cdots \times X_{j_k},}$$

where $\times$ denotes the usual product. For instance, $\forall m \leq n$,

$$\mathbf{X}_{[m]} = (X_1, .., X_m) \quad \text{and} \quad \mathbf{X}^{[m]} = X_1 \times \cdots \times X_m.$$

The function names below acquire their meaning when read from left to right. Thus, the function *MomentToCumulants* expresses a given moment in terms of cumulants. In practice, this function is used to transform cumulants into moments.

A.9.1 MomentToCumulants

Express the moment of a random variable in terms of the cumulants.

Input format. To express $\mathbb{E}[X^m]$ in terms of cumulants, use MomentToCumulants[m].

▶ Please refer to Formula (3.2.16).

  Example. Express the third moment $\mathbb{E}[X^3]$ of a random variable X in terms of cumulants.

In := MomentToCumulants[3]
Out= $\chi_1[X]^3 + 3\chi_1[X]\chi_2[X] + \chi_3[X]$

Notation: $\chi_1[X]^3$ denotes the third power of the first cumulant of X.

A.9.2 MomentToCumulantsBell

Use complete Bell polynomials to express a moment in terms of cumulants. One can use either the function *MomentToCumulantsBell* or directly *BellPolynomialComplete*.

▶ Please refer to Proposition 3.3.1.

Input format. MomentToCumulantsBell$[x]$ where x is a vector (of cumulants).

Example. Express the sixth moment of a random variable X in terms of cumulants (using Bell polynomials). Let $k1, \cdots, k6$ denote the cumulants.

In := MomentToCumulantsBell$[\{k1, k2, k3, k4, k5, k6\}]$
Out= $k1^6 + 15k1^4k2 + 45k1^2k2^2 + 15k2^3 + 20k1^3k3 + 60k1k2k3 + 10k3^2 + 15k1^2k4 + 15k2k4 + 6k1k5 + k6$

In := BellPolynomialComplete$[\{k1, k2, k3, k4, k5, k6\}]$
Out= $k1^6 + 15k1^4k2 + 45k1^2k2^2 + 15k2^3 + 20k1^3k3 + 60k1k2k3 + 10k3^2 + 15k1^2k4 + 15k2k4 + 6k1k5 + k6$

Example. Express the second moment of a random variable X in terms of cumulants (using complete Bell polynomials). Let $\chi_1[X], \chi_2[X]$ denote the cumulants. This example demonstrates that the arguments can be any expression.

In := BellPolynomialComplete$[\{\chi_1[X], \chi_2[X]\}]$
Out= $\chi_1[X]^2 + \chi_2[X]$

Example. Express the sixth central moment of a random variable X in terms of cumulants (using complete Bell polynomials). Let $k1, \cdots, k6$ denote the cumulants. This amounts to supposing that the mean, that is, the first cumulant $k1 = 0$.

In := BellPolynomialComplete$[\{0, k2, k3, k4, k5, k6\}]$
Out= $15k2^3 + 10k3^2 + 15k2k4 + k6$

Example. In the previous example, suppose that the random variable X is normalized to have mean 0 and variance 1. This amounts to setting $k1 = 0$ and $k2 = 1$.

In := BellPolynomialComplete$[\{0, 1, k3, k4, k5, k6\}]$
Out= $15 + 10k3^2 + 15k4 + k6$

Example. Obtain the sixth moment of a standard normal. This amounts to setting in addition all cumulants higher than 2 equal to 0.

In :=   BellPolynomialComplete[{0, 1, 0, 0, 0, 0}]
Out=   15

Remark. From now on, we shall sometimes adopt the notation $\chi_j[X] = \kappa_j$, $j = 1, 2, \dots$.

Example. Display the results in a table (up to order 5). For a more involved table use *MomentToCumulantsBellTable*.

In :=   Column[Table[BellPolynomialComplete[Table[Subscript[\[Kappa], i], i, 1, j]], j, 1, 5]]

The output is a table:

$$\kappa_1$$
$$\kappa_1^2 + \kappa_2$$
$$\kappa_1^3 + 3\kappa_1\kappa_2 + \kappa_3$$
$$\kappa_1^4 + 6\kappa_1^2\kappa_2 + 3\kappa_2^2 + 4\kappa_1\kappa_3 + \kappa_4$$
$$\kappa_1^5 + 10\kappa_1^3\kappa_2 + 15\kappa_1\kappa_2^2 + 10\kappa_1^2\kappa_3 + 10\kappa_2\kappa_3 + 5\kappa_1\kappa_4 + \kappa_5.$$

A.9.3 MomentToCumulantsBellTable

Express moments in terms of cumulants and display the result in a table.

Input format. MomentToCumulantsBellTable[κ] , where κ is a vector of cumulants.

Proceed interactively. How many rows do you want to see?
In :=   NumRows = 5
Out=   5

Sometimes you may need to clear the previous definitions of μ and κ. To type these symbols copy and paste them with the mouse from somewhere else in the notebook or type respectively as \[Mu] and \[Kappa].

In :=   ClearAll[μ, κ]

Input is a list of cumulants, for example:
In := κ = Table[Subscript[κ, i], i, 1, NumRows]
Out= $\{\kappa_1, \kappa_2, \kappa_3, \kappa_4, \kappa_5\}$

Then, you make a table by plugging this into *MomentToCumulantsBellTable*:

In :=   MomentToCumulantsBellTable[κ]

Out=

n	μ_n
1	κ_1
2	$\kappa_1^2 + \kappa_2$
3	$\kappa_1^3 + 3\kappa_1\kappa_2 + \kappa_3$
4	$\kappa_1^4 + 6\kappa_1^2\kappa_2 + 3\kappa_2^2 + 4\kappa_1\kappa_3 + \kappa_4$
5	$\kappa_1^5 + 10\kappa_1^3\kappa_2 + 15\kappa_1\kappa_2^2 + 10\kappa_1^2\kappa_3 + 10\kappa_2\kappa_3 + 5\kappa_1\kappa_4 + \kappa_5$

To obtain centered moments, make the first element of your input equal to 0:

In :=   \[Kappa][[1]] = 0

Out=   0

Check the value of the vector κ and generate the table:

In := κ

Out= $\{0, \kappa_2, \kappa_3, \kappa_4, \kappa_5\}$

In :=   MomentToCumulantsBellTable[κ]

Out=

n	μ_n
1	0
2	κ_2
3	κ_3
4	$3\kappa_2^2 + \kappa_4$
5	$10\kappa_2\kappa_3 + \kappa_5$

A.9.4 CumulantToMoments

Express the cumulant of a random variable in terms of moments.

▶ Please refer to Proposition 3.3.1.

Input format. To express $\chi_m[X]$ in terms of moments, use CumulantToMoments[m].

▶ Please refer to Formula (3.2.19).

Example. Express the third cumulant $\chi_3[X]$ of a random variable X in terms of moments.

In :=   CumulantToMoments[3]

Out= $2\mathbb{E}[X]^3 - 3\mathbb{E}[X]\mathbb{E}[X^2] + \mathbb{E}[X^3]$

Notation: $\mathbb{E}[X]^3$ denotes the third power of the first moment of X.

A.9.5 CumulantToMomentsBell

Use partial Bell polynomials to express a cumulant in terms of moments.

Input format. CumulantToMomentsBell$[x]$ where x is a vector (of moments).

Example. Express the moments in terms of cumulants (using partial Bell polynomials).

In := CumulantToMomentsBell$[\{m1, m2\}]$
Out= $-m1^2 + m2$

Example. Display the conversion results in a table (up to order 5). For a more involved table use *CumulantToMomentsBellTable*.

In := Column[Table[CumulantToMomentsBell[Table[Subscript[\[Mu], i], i, 1, j]], j, 1, 5]]
The output is a table:

$$\mu_1$$
$$-\mu_1^2 + \mu_2$$
$$2\mu_1^3 - 3\mu_1\mu_2 + \mu_3$$
$$-6\mu_1^4 + 12\mu_1^2\mu_2 - 3\mu_2^2 - 4\mu_1\mu_3 + \mu_4$$
$$24\mu_1^5 - 60\mu_1^3\mu_2 + 30\mu_1\mu_2^2 + 20\mu_1^2\mu_3 - 10\mu_2\mu_3 - 5\mu_1\mu_4 + \mu_5.$$

A.9.6 CumulantToMomentsBellTable

Express cumulants in terms of moments and display the result in a table.

Input format. CumulantToMomentsBellTable$[\mu]$, where μ is a vector of moments.

Proceed interactively. How many rows do you want to see?
In := NumRows = 5
Out= 5

Sometimes you may need to clear the previous definitions of μ and κ. To type these symbols copy and paste them with the mouse from somewhere else in the notebook or type respectively as \[Mu] and \[Kappa].

In := ClearAll$[\mu, \kappa]$

Input is a list of moments, for example:

In := $\mu = \text{Table}[\text{Subscript}[\mu, i], i, 1, \text{NumRows}]$

Out= $\{\mu_1, \mu_2, \mu_3, \mu_4, \mu_5\}$

Then, you make a table by plugging this into *CumulantToMomentsBellTable*:

In :=   CumulantToMomentsBellTable$[\mu]$

Out=

$$
\begin{array}{c||c}
n & \kappa_n \\
\hline
1 & \mu_1 \\
2 & -\mu_1^2 + \mu_2 \\
3 & 2\mu_1^3 - 3\mu_1\mu_2 + \mu_3 \\
4 & -6\mu_1^4 + 12\mu_1^2\mu_2 - 3\mu_2^2 - 4\mu_1\mu_3 + \mu_4 \\
5 & 24\mu_1^5 - 60\mu_1^3\mu_2 + 30\mu_1\mu_2^2 + 20\mu_1^2\mu_3 - 10\mu_2\mu_3 - 5\mu_1\mu_4 + \mu_5
\end{array}
$$

To obtain a table involving centered moments and cumulants, make the first element of your input equal to 0:

In :=   \[Mu][[1]] = 0

Out=   0

Check the value of the vector μ and generate the table:

In := μ

Out= $\{0, \mu_2, \mu_3, \mu_4, \mu_5\}$

In :=   CumulantToMomentsBellTable$[\mu]$

Out=

n	κ_n
1	0
2	μ_2
3	μ_3
4	$-3\mu_2^2 + \mu_4$
5	$-10\mu_2\mu_3 + \mu_5$

A.9.7 MomentProductToCumulants

Express the expected value of the product of random variables $\mathbf{X}^b = X_{j_1} \cdots X_{j_k}$, where $b = \{j_1, ..., j_k\} \subseteq [n] = \{1, ..., n\}$, in terms of cumulants.

▶ Please refer to Formula (3.2.7).

Example. Express the expected value $\mathbb{E}\mathbf{X}^{\{1,2\}} = \mathbb{E}X_1X_2$ in terms of cumulants.

In :=   MomentProductToCumulants $[X, \{\{1,2\}\}]$
Out= $\chi[X_1]\chi[X_2] + \chi[X_1, X_2]$

A.9.8 CumulantVectorToMoments

Express the cumulant of a random vector $\mathbf{X}_b = (X_{j_1}, ..., X_{j_k})$, where $b = \{j_1, ..., j_k\} \subseteq [n] = \{1, ..., n\}$, in terms of moments.

▶ Please refer to Formula (3.2.6).

Example. Express the cumulant of $X_{\{1,2\}} = (X_1, X_2)$ which is the covariance of X_1 and X_2, in terms of moments.

In :=   CumulantVectorToMoments $[X, \{\{1,2\}\}]$
Out= $-\mathbb{E}[X_1]\mathbb{E}[X_2] + \mathbb{E}[X_1X_2]$

A.9.9 CumulantProductVectorToCumulantVectors

Express the joint cumulant of a random vector $(\mathbf{X}_{b_1}, \mathbf{X}_{b_2}, \cdots, \mathbf{X}_{b_k})$ where $\{b_1, b_2, \cdots, b_k\} \subseteq [n]$, in terms of cumulants of random vectors whose components are not products (Malyshev's formula).

▶ Please refer to Formula (3.2.8).

Example. Express the cumulants $\chi[X_1X_2, X_3]$ in terms of cumulants of random vectors whose components are not products.

In :=   CumulantProductVectorToCumulantVectors $[X, \{\{1,2\}, \{3\}\}]$
Out= $\chi(X_2)\chi(X_1, X_3) + \chi(X_1)\chi(X_2, X_3) + \chi(X_1, X_2, X_3)$

A.9.10 GridCumulantProductVectorToCumulantVectors

Display the output of CumulantProductVectorToCumulantVectors in a grid.

Example. Display in a grid CumulantProductVectorToCumulantVectors of $\chi[X_1X_2X_3, X_4]$.

In :=   GridCumulantProductVectorToCumulantVectors $[X, \{\{1,2,3\}, \{4\}\}]$

Out=   Number of terms in summation = 10

$$
\begin{aligned}
&\chi\,[X_2]\,\chi\,[X_3]\,\chi\,[X_1,X_4] &+\\
&\chi\,[X_1,X_4]\,\chi\,[X_2,X_3] &+\\
&\chi\,[X_1]\,\chi\,[X_3]\,\chi\,[X_2,X_4] &+\\
&\chi\,[X_1,X_3]\,\chi\,[X_2,X_4] &+\\
&\chi\,[X_1]\,\chi\,[X_2]\,\chi\,[X_3,X_4] &+\\
&\chi\,[X_1,X_2]\,\chi\,[X_3,X_4] &+\\
&\chi\,[X_3]\,\chi\,[X_1,X_2,X_4] &+\\
&\chi\,[X_2]\,\chi\,[X_1,X_3,X_4] &+\\
&\chi\,[X_1]\,\chi\,[X_2,X_3,X_4] &+\\
&\chi\,[X_1,X_2,X_3,X_4] &+
\end{aligned}
$$

A.9.11 CumulantProductVectorToMoments

Express the joint cumulant of a random vector $(\mathbf{X}^{b_1},\mathbf{X}^{b_2},\cdots,\mathbf{X}^{b_k})$ where $\{b_1,b_2,\cdots,b_k\}\subseteq[n]$, in terms of moments.

Example. Express the cumulants $\chi[X_1X_2,X_3]$ in terms of moments.

In :=   CumulantProductVectorToMoments$[X,\{\{1,2\},\{3\}\}]$
Out= $-\mathbb{E}[X_1X_2]\mathbb{E}[X_3]+\mathbb{E}[X_1X_2X_3]$

A.10 Gaussian Multiple Integrals

A.10.1 GaussianIntegral

This function can be used to compute multiple Gaussian integrals

$$
\int_\sigma f(x_1,...,x_n)G(dx_1)\cdots G(dx_n)
$$

over a set partition σ and

$$
\int_{\geq\sigma} f(x_1,...,x_n)G(dx_1)\cdots G(dx_n)
$$

over all partitions at least as coarse as σ, where $f(x_1,...,x_n)$ is a symmetric function, for example, $g(x_1)\cdots g(x_n)$, and where G is a Gaussian measure with control measure ν. This function generates the set partition data necessary to compute multiple Gaussian integrals. This data is listed under two lists: "control measure ν" and "Gaussian measure G". The output is "The integral is zero" when this is the case.

▶ Please refer to Theorem 5.11.1.

Example. When computing the multiple Gaussian integral over a set partition with a block size greater than or equal to three, the integral is zero.

In :=   GaussianIntegral$[\{\{1,2,3\},\{4,5\},\{6,7\},\{8\}\}]$
Out=   The integral is zero.

Rules for interpreting the data listed in output:

Rule 1. To obtain a Gaussian multiple integral over σ, equate the variables listed under "control measure ν" and integrate them with respect to ν; the variables listed under "Gaussian measure G" are integrated with respect to G over the off-diagonals.

Rule 2. To obtain a Gaussian multiple integral over "$\geq \sigma$", equate the variables listed under "control measure ν" and integrate.

Example. Obtain the block data necessary to decompose the multiple Gaussian integral over a particular set partition.

In :=   GaussianIntegral$[\{\{1,3\},\{4,5\},\{6\},\{7\},\{2\}\}]$

Out=

Control measure ν
$\{1,3\}$
$\{4,5\}$

Gaussian measure G
$\{6\}$
$\{7\}$
$\{2\}$

The output lists $\{1,3\}, \{4,5\}$ under "control measure ν" and it lists $\{6\}, \{7\}, \{2\}$ under "Gaussian measure G".

Interpretation: Let $\sigma = \{\{1,3\},\{4,5\},\{6\},\{7\},\{2\}\}$. Integrating $g(x_1)\ldots g(x_7)$ over σ gives:

$$\left(\int g^2(x)\nu(dx)\right)^2 \int_{x_2 \neq x_6 \neq x_7} g(x_2)g(x_6)g(x_7)G(dx_2)G(dx_6)G(dx_7).$$

If one integrates it over "$\geq \sigma$" instead, one obtains

$$\left(\int g^2(x)\nu(dx)\right)^2 \left(\int g(x)G(dx)\right)^3.$$

Integrating a symmetric function $f(x_1, \cdots , x_7)$ over σ gives

$$\int_{x_2 \neq x_6 \neq x_7} f(x_3, x_3, x_5, x_5, x_2, x_6, x_7)\nu(dx_3)\nu(dx_5)G(dx_2)G(dx_6)G(dx_7).$$

If one integrates it over "$\geq \sigma$" instead, one obtains

$$\int f(x_3, x_3, x_5, x_5, x_2, x_6, x_7)\nu(dx_3)\nu(dx_5)G(dx_2)G(dx_6)G(dx_7).$$

A.11 Poisson Multiple Integrals

We first define some functions which are useful for the evaluation of Poisson multiple integrals. Please refer to the main text.

A.11.1 BOne

For a given set partition σ, find $B_1(\sigma)$, that is, isolate the blocks of σ that are singletons.

▶ Please refer to Formula (5.12.85).

  Example. Find all of the singleton blocks of $\{\{1, 2\}, \{3, 4\}, \{5\}\}$.

  In := BOne[$\{\{1, 2\}, \{3, 4\}, \{5\}\}$]
  Out= $\{\{5\}\}$

  Example. The set partition $\{\{1, 2, 3\}, \{4, 5\}\}$ has no singleton blocks.

  In := BOne[$\{\{1, 2, 3\}, \{4, 5\}\}$]
  Out= $\{\}$

A.11.2 BTwo

For a given set partition σ, find $B_2(\sigma)$, that is, isolate the blocks of σ that are not singletons.

▶ Please refer to Formula (5.12.86).

  Example. Find all the blocks of $\{\{1, 2\}, \{3, 4\}, \{5\}\}$ that are not singletons.

  In := BTwo[$\{\{1, 2\}, \{3, 4\}, \{5\}\}$]
  Out= $\{\{1, 2\}, \{3, 4\}\}$

A.11.3 PBTwo

For a fixed set partition σ, find $B_2(\sigma)$; viewing $B_2(\sigma)$ as a set, find all ordered partitions of $B_2(\sigma)$ with exactly two blocks, called R_1 and R_2.

▶ Please refer to Formula (5.12.87).

Example. For the set partition $[\{\{1,2\},\{3,4\},\{5,6\},\{7\}\}]$, find $B_2([\{\{1,2\},\{3,4\},\{5,6\},\{7\}\}])$.

In := PBTwo$[\{\{1,2\},\{3,4\},\{5,6\},\{7\}\}]$

Out=

R1	R2
$\{\{1,2\}\}$	$\{\{3,4\},\{5,6\}\}$
$\{\{3,4\}\}$	$\{\{1,2\},\{5,6\}\}$
$\{\{5,6\}\}$	$\{\{1,2\},\{3,4\}\}$
$\{\{1,2\},\{3,4\}\}$	$\{\{5,6\}\}$
$\{\{1,2\},\{5,6\}\}$	$\{\{3,4\}\}$
$\{\{3,4\},\{5,6\}\}$	$\{\{1,2\}\}$

Example. For the set partition $\{\{1,2\},\{3\}\}$, $PB_2(\{\{1,2\},\{3\}\})$ is empty.

In := PBTwo$[\{\{1,2\},\{3\}\}]$

Out=
R1	R2

A.11.4 PoissonIntegral

Let $N(dx)$ be a Poisson measure with control measure ν and let $\widehat{N}$ be the corresponding compensated Poisson measure, that is,

$$\widehat{N}(dx) = N(dx) - \mathbb{E}N(dx) = N(dx) - \nu(dx).$$

The function "PoissonIntegral" can be used to compute multiple Poisson integrals

$$\int_{\sigma} f(x_1, ..., x_n)\widehat{N}(dx_1)\cdots\widehat{N}(dx_n)$$

over a set partition σ, where $f(x_1, ..., x_n)$ is a symmetric function, for example, $g(x_1)\cdots g(x_n)$.

▶ Please refer to Theorem 5.12.2, Formula (5.12.91).

This function generates the set partition data necessary to compute multiple Poisson integrals, namely

$$R_1, R_2, B_1, B_2.$$

Rules for interpreting the output. The integral of a symmetric function over σ with respect to a product of compensated Poisson measures $\widehat{N}$ is a sum of $K + 2$ terms, where K is the number of rows in the display (R_1, R_2, B_1). They are obtained as follows:

(1) The variables with indices in R_1 are identified and integrated over ν; the variables with indices in a block of R_2 are identified. These identified variables, together with the variables with indices in B_1, are integrated over $\widehat{N}$ on the off-diagonals. This is done for each row of the display (R_1, R_2, B_1).

There are two other terms:

(2) The variables with indices in blocks of B_2 are identified. These identified variables, together with the variables in B_1, are integrated over $\widehat{N}$ on the off-diagonals.

(3) The variables with indices in each block of B_2 are identified and integrated over ν; the variables with indices in B_1 are integrated over $\widehat{N}$ on the off-diagonals.

Example. Decompose a Poisson integral on $\sigma = \{\{1,2\},\{3,4\},\{5\},\{6\}\}$.

$In :=$ PoissonIntegral$[\{\{1,2\},\{3,4\},\{5\},\{6\}\}]$

The output is a chart with two groups of outputs. The first gives (R_1, R_2, B_1). The second gives B_2, B_1. The first group (R_1, R_2, B_1) is as follows:

$$\{\{1,2\}\} \qquad \{\{3,4\}\} \qquad \{\{5\},\{6\}\}$$
$$\{\{3,4\}\} \qquad \{\{1,2\}\} \qquad \{\{5\},\{6\}\},$$

the first column corresponding to R_1, the second to R_2, the last to B_1. The second group B_2, B_1 is as follows:

$$\{\{1,2\},\{3,4\}\} \qquad \{\{5\},\{6\}\}.$$

This corresponds to the following decomposition:

$$\int_\sigma g(x_1)\cdots g(x_6)\widehat{N}(dx_1)\cdots\widehat{N}(dx_6) =$$

$$2\left(\int g^2(x_2)\nu(dx_2)\right)\int_{x_4\neq x_5\neq x_6} g^2(x_4)g(x_5)g(x_6)\widehat{N}(dx_4)\widehat{N}(dx_5)\widehat{N}(dx_6)$$

$$+\int_{x_2\neq x_4\neq x_5\neq x_6} g^2(x_2)g^2(x_4)g(x_5)g(x_6)\widehat{N}(dx_2)\widehat{N}(dx_4)\widehat{N}(dx_5)\widehat{N}(dx_6)$$

$$+\left(\int g^2(x_2)\nu(dx_2)\right)^2\int_{x_5\neq x_6} g(x_5)g(x_6)\widehat{N}(dx_5)\widehat{N}(dx_6).$$

Integrating a symmetric function $f(x_1,\cdots,x_7)$ over σ gives

$$\int_\sigma f(x_1,\cdots,x_6)\widehat{N}(dx_1)\cdots\widehat{N}(dx_6) =$$

$$2\int_{x_4\neq x_5\neq x_6} f(x_2,x_2,x_4,x_4,x_5,x_6)\nu(dx_2)\widehat{N}(dx_4)\widehat{N}(dx_5)\widehat{N}(dx_6)$$

$$+\int_{x_2\neq x_4\neq x_5\neq x_6} f(x_2,x_2,x_4,x_4,x_5,x_6)\widehat{N}(dx_2)\widehat{N}(dx_4)\widehat{N}(dx_5)\widehat{N}(dx_6)$$

$$+\int_{x_5\neq x_6} f(x_2,x_2,x_4,x_4,x_5,x_6)\nu(dx_2)\nu(dx_4)\widehat{N}(dx_5)\widehat{N}(dx_6).$$

A.11.5 PoissonIntegralExceed

This function can be used to compute multiple Poisson integrals

$$\int_{\geq\sigma} f(x_1,...,x_n)\widehat{N}(dx_1)\cdots\widehat{N}(dx_n)$$

over all partitions at least as coarse as σ, where $f(x_1,...,x_n)$ is a symmetric function, for example, $g(x_1)\cdots g(x_n)$, and where $\widehat{N}$ is a compensated Poisson measure with control measure ν.

▶ Please refer to Theorem 5.12.2, Formula (5.12.90).

This function generates the set partition data necessary to compute multiple Poisson integrals, namely B_2, B_1.

Rules for interpreting the output. Equate the variables in each block of B_2 and integrate them with respect to the uncompensated Poisson measure N; the variables listed in B_1 are integrated with respect to the compensated Poisson measure $\widehat{N}$.

Example. Decompose a Poisson integral over $\geq \sigma$ where $\sigma = \{\{1, 2, 3\}, \{4, 5\}, \{6\}, \{7\}\}$.

In :=   PoissonIntegralExceed$[\{\{1, 2, 3\}, \{4, 5\}, \{6\}, \{7\}\}]$

The output is a chart listing B_2, and B_1 as follows:

$$\{\{1, 2, 3\}, \{4, 5\}\} \qquad \{\{6\}, \{7\}\}.$$

This corresponds to the decomposition

$$\int_{\geq \sigma} g(x_1) \cdots g(x_7) \widehat{N}(dx_1) \cdots \widehat{N}(dx_7) =$$
$$\int g^3(x_3) N(dx_3) \int g^2(x_5) N(dx_5) \left(\int g(x_7) \widehat{N}(dx_7) \right)^2.$$

Integrating a symmetric function $f(x_1, \cdots, x_7)$ over $\geq \sigma$ gives

$$\int_{\geq \sigma} f(x_1, ..., x_n) \widehat{N}(dx_1) \cdots \widehat{N}(dx_n) =$$
$$\int f(x_3, x_3, x_3, x_5, x_5, x_6, x_7) N(dx_3) N(dx_5) \widehat{N}(dx_6) \widehat{N}(dx_7).$$

A.12 Contraction and Symmetrization

▶ Please refer to Section 6.2 and Section 6.3.

A.12.1 ContractionWithSymmetrization

In the product of two symmetric functions of p and q variables, respectively, identify r variables in each function and symmetrize the result.

Input format. ContractionWithSymmetrization$[p, q, r]$, where $r \leq \min\{p, q\}$.

Example.

In :=   ContractionWithSymmetrization$[2, 2, 1]$

The input function is $f(x_1, x_2) g(x_3, x_4)$. The output is a chart corresponding to

$$(1/2)(f(x_1, x_2) g(x_1, x_3) + f(x_1, x_3) g(x_1, x_2)).$$

A.12.2 ContractionIntegration

In the product of two symmetric functions of p and q variables, respectively, identify r variables in each function, and integrate l of these r variables. The integrated variables are designated by an overtilde, e.g. $\tilde{1}$.

Input format. ContractionIntegration$[p, q, r, l]$, where $l \leq r \leq \min\{p, q\}$.

Example. The input function is $f(x_1, x_2, x_3)g(x_4, x_5)$. Identify one variable and integrate it.

> $In :=$ ContractionIntegration$[3, 2, 1, 1]$
> $Out=$ $\{\{\tilde{1}, 2, 3\}, \{\tilde{1}, 4\}\}$

The output corresponds to $\int f(x_1, x_2, x_3)g(x_1, x_4)\nu(dx_1)$.

A.12.3 ContractionIntegrationWithSymmetrization

In the product of two symmetric functions of p and q variables, respectively, identify r variables in each function, integrate l of these r variables, and then symmetrize the result.

Example. The input function is $f(x_1, x_2, x_3)g(x_4, x_5)$. Identify one variable, integrate it and then symmetrize the result.

> $In :=$ ContractionIntegrationWithSymmetrization$[3, 2, 1, 1]$

The output gives the number of terms (namely 3) and displays a chart corresponding to

$$(1/3) \int f(x_1, x_2, x_3)g(x_1, x_4) + f(x_1, x_2, x_4)g(x_1, x_3) + f(x_1, x_3, x_4)g(x_1, x_2)\nu(dx_1).$$

A.13 Solving Partition Equations Involving π^*

Some of the functions here are similar to earlier ones which involved an arbitrary set partition π. Here, we are interested in partitions π of the form

$$\pi^* = \{\{1, \cdots, n_1\}, \{n_1 + 1, \cdots, n_1 + n_2\}, ..., \{n_1 + \cdots + n_{k-1} + 1, \cdots, n\}\},$$

as defined in Section A.4.4. The input is a list $(n_1, ..., n_k)$ of positive integers defining π^*.

A.13.1 PiStarMeetSolve

For a finite sequence of positive integers, generate a unique set partition π^* whose block sizes are equal to these integers, and then find all set partitions σ such that $\sigma \wedge \pi^*$ is the singleton partition $\hat{0}$. The result is used in Rota and Wallstrom's Theorem.

Input format. PiStarMeetSolve[list], where 'list" is a list of positive integers that defines the set partition π^*.

  Example. For the sequence $\{2, 2\}$, generate π^* and find the meets of π^* that give $\hat{0}$, that is, all partitions σ such that $\sigma \wedge \pi^* = \hat{0}$.

  In :=   PiStarMeetSolve[$\{2, 2\}$]

Here $\pi^* = \{\{1, 2\}, \{3, 4\}\}$. The output gives the number of solutions (namely 7) and lists them:

  Out=   {{{1, 4}, {2, 3}}, {{1, 3}, {2, 4}}, {{1}, {2, 3}, {4}},
       {{1}, {2, 4}, {3}}, {{1, 3}, {2}, {4}}, {{1, 4}, {2}, {3}},
       {{1}, {2}, {3}, {4}}}

A.13.2 PiStarJoinSolve

For a finite sequence of positive integers, generate a unique set partition π^* whose block sizes are equal to these integers, and then find all set partitions σ such that $\sigma \vee \pi^*$ is the maximal partition $\hat{1}$.

Input format. PiStarJoinSolve[list], where 'list" is a list of positive integers defining the set partition π^*.

  Example. For the sequence $\{2, 2\}$, generate π^* and find the joins of π^* that give $\hat{1}$, that is, all partitions σ such that $\sigma \vee \pi^* = \hat{1}$.

  In :=   PiStarMeetSolve[$\{2, 2\}$]

Here $\pi^* = \{\{1, 2\}, \{3, 4\}\}$. The output gives the number of solutions (namely 11) and lists them:

  Out=   213]= {{{1, 2, 3, 4}}, {{1}, {2, 3, 4}}, {{1, 3, 4}, {2}}, {{1, 2, 3}, {4}},
       {{1, 4}, {2, 3}}, {{1, 2, 4}, {3}}, {{1, 3}, {2, 4}}, {{1}, {2, 3}, {4}},
       {{1}, {2, 4}, {3}}, {{1, 3}, {2}, {4}}, {{1, 4}, {2}, {3}}}

A.13.3 PiStarMeetAndJoinSolve

For a finite sequence of positive integers, generate a unique set partition π^* whose block sizes are equal to these integers, and then find all set partitions σ such that $\sigma \wedge \pi^*$ is the singleton partition $\hat{0}$ and $\sigma \vee \pi^*$ is the maximal partition $\hat{1}$. This function is used in diagram formulae.

Example. For the sequence $\{2, 2\}$, generate π^* and find the intersection of the meets of π^* that give $\hat{0}$ and the joins of π^* that give $\hat{1}$, that is, all partitions σ such that $\sigma \wedge \pi^* = \hat{0}$ and $\sigma \vee \pi^* = \hat{1}$.

$In :=$ PiStarMeetAndJoinSolve$[\{2, 2\}]$

Here $\pi^* = \{\{1, 2\}, \{3, 4\}\}$. The output gives the number of solutions to the meet and join problem (namely 6) and lists them:

$Out=$ $\{\{\{1, 4\}, \{2, 3\}\}, \{\{1, 3\}, \{2, 4\}\}, \{\{1\}, \{2, 3\}, \{4\}\},$
$\{\{1\}, \{2, 4\}, \{3\}\}, \{\{1, 3\}, \{2\}, \{4\}\}, \{\{1, 4\}, \{2\}, \{3\}\}\}$

A.14 Product of Two Gaussian Multiple Integrals

A.14.1 ProductTwoGaussianIntegrals

Computes the product $I_p^G(f) I_q^G(g)$ of two Gaussian multiple integrals, one of order $p \geq 1$ and one of order $q \geq 1$; the function f has p variables and the function g has q variables; both f and g are symmetric functions. Namely it computes

$$I_p^G(f) I_q^G(g) = \sum_{r=0}^{p \wedge q} r! \binom{p}{r} \binom{q}{r} I_{p+q-2r}^G (f \otimes_r g),$$

In the displayed output of the function,

$$\text{coefficient} = r! \binom{p}{r} \binom{q}{r}.$$

Observe that $I_{p+q-2r}^G (f \otimes_r g)$ is equal to $I_{p+q-2r}^G \left(\widetilde{f \otimes_r g} \right)$ where $\sim$ denotes symmetrization.

▶ Please refer to Proposition 6.4.1.

Input format. ProductTwoGaussianIntegrals$[p, q]$.

Example. The input functions are $f(x_1)$ and $g(x_2)$. Compute $I_1^G(f)I_1^G(g)$.

$In :=$ ProductTwoGaussianIntegrals$[1, 1]$

The output is a chart corresponding to

$$I_2^G (f \otimes_0 g) + I_0^G (f \otimes_1 g) =$$
$$\int_{x_1 \neq x_2} f(x_1)g(x_2)G(dx_1)G(dx_2) + \int f(x)g(x)\nu(dx).$$

A.15 Product of Two Poisson Multiple Integrals

A.15.1 ProductTwoPoissonIntegrals

Computes the product $I_p^{\widehat{N}}(f)I_q^{\widehat{N}}(g)$ of two Poisson multiple integrals, one of order $p \geq 1$ and one of order $q \geq 1$; the function f has p variables and the function g has q variables; both f and g are symmetric functions. Namely it computes

$$I_p^{\widehat{N}}(f)I_q^{\widehat{N}}(g) = \sum_{r=0}^{p \wedge q} r! \binom{p}{r} \binom{q}{r} \sum_{l=0}^{r} \binom{r}{l} I_{p+q-r-l}^{\widehat{N}}(f \star_r^l g).$$

In the displayed output of the function,

$$\text{coefficient} = r! \binom{p}{r} \binom{q}{r} \binom{r}{l}.$$

Observe that $I_{p+q-r-l}^{\widehat{N}} (f \star_r^l g))$ is equal to $I_{p+q-r-l}^{\widehat{N}} \left(\widetilde{f \star_r^l g} \right)$ where $\tilde{}$ denotes symmetrization.

▶ Please refer to Proposition 6.5.1.

Input format. ProductTwoPoissonIntegrals$[p, q]$.

Example. The input functions are $f(x_1)$ and $g(x_2)$. Compute $I_1^{\widehat{N}}(f)I_1^{\widehat{N}}(g)$.

$In :=$ ProductTwoPoissonIntegrals$[1, 1]$

The output is a chart corresponding to

$$I_2^{\widehat{N}} (f \otimes_0^0 g) + I_1^{\widehat{N}} (f \star_1^0 g) + I_0^{\widehat{N}} (f \star_1^1 g)$$
$$= \int_{x \neq y} f(x)g(y)\widehat{N}(dx)\widehat{N}(dy) + \int f(x)g(x)\widehat{N}(dx) + \int f(x)g(x)\nu(dx)$$
$$= \int_{x \neq y} f(x)g(y)\widehat{N}(dx)\widehat{N}(dy) + \int f(x)g(x)N(dx).$$

A.16 MSets and MZeroSets

These sets appear in diagram formulae involving Gaussian and Poisson multiple integrals. They have similar structure. Given a list $(n_1, \cdots, n_k)$ of positive integers, let $n = n_1 + \cdots + n_k$ and let π^* be the special partition of $[n] = \{1, ..., n\}$ defined in Section A.4.4. The MSets and MZeroSets are a collection of set partitions of $[n] = \{1, ..., n\}$.

The sets $\mathcal{M}_2([n], \pi^*)$ and $\mathcal{M}_2^0([n], \pi^*)$ appear in diagram formulae when the random measure is Gaussian (see Section 7.3). The sets $\mathcal{M}_{\geq 2}([n], \pi^*)$ and $\mathcal{M}_{\geq 2}^0([n], \pi^*)$ appear when the random measure is compensated Poisson (see Section 7.4). They are respectively subsets of $\mathcal{M}([n], \pi^*)$ (involving meet and join) and $\mathcal{M}^0([n], \pi^*)$ (involving meet only).

▶ Please refer to Section 7.2.

Input format.   FunctionName[list], where "list" is a list $(n_1, ..., n_k)$ of positive integers.

Example. Let the list be $(2, 2)$ so that $\pi^* = \{\{1, 2\}, \{3, 4\}\}$. Find the MSets and MZeroSets.

A.16.1 MSets

This function produces the same output as PiStarMeetAndJoinSolve. It provides the set partitions in $\mathcal{M}([n], \pi^*)$, namely all set partitions σ such that $\sigma \wedge \pi^* = \hat{0}$ and $\sigma \vee \pi^* = \hat{1}$.

A.16.2 MSetsEqualTwo

This function produces all solutions to PiStarMeetAndJoinSolve containing partitions only of size two. It provides the set partitions in $\mathcal{M}_2([n], \pi^*)$.

A.16.3 MSetsGreaterEqualTwo

This function produces all solutions to PiStarMeetAndJoinSolve containing no singletons. It provides the set partitions in $\mathcal{M}_{\geq 2}([n], \pi^*)$.

A.16.4 MZeroSets

This function produces the same output as PiStarMeetSolve. It provides the set partitions in $\mathcal{M}^0([n], \pi^*)$, namely all set partitions σ such that $\sigma \wedge \pi^* = \hat{0}$.

A.16.5 MZeroSetsEqualTwo

This function produces all solutions to PiStarMeetSolve containing partitions only of size two. It provides the set partitions in $\mathcal{M}_2^0([n], \pi^*)$.

A.16.6 MZeroSetsGreaterEqualTwo

This function produces all solutions to PiStarMeetSolve containing no singletons. It provides the set partitions in $\mathcal{M}_{\geq 2}^0([n], \pi^*)$.

Example. (See above.) The (full) output also indicates the number of solutions.

In :=   MSets[{2, 2}]
Out=   {{{1, 4}, {2, 3}}, {{1, 3}, {2, 4}}, {{1}, {2, 3}, {4}}, {{1}, {2, 4}, {3}}, {{1, 3}, {2}, {4}}, {{1, 4}, {2}, {3}}}

In :=   MSetsEqualTwo[{2, 2}]
Out=   {{{1, 4}, {2, 3}}, {{1, 3}, {2, 4}}}

In :=   MSetsGreaterEqualTwo[{2, 2}]
Out=   {{{1, 4}, {2, 3}}, {{1, 3}, {2, 4}}}

In :=   MZeroSets[{2, 2}]
Out=   {{{1, 4}, {2, 3}}, {{1, 3}, {2, 4}}, {{1}, {2, 3}, {4}}, {{1}, {2, 4}, {3}}, {{1, 3}, {2}, {4}}, {{1, 4}, {2}, {3}}}, {{1}, {2}, {3}, {4}}}

In :=   MZeroSetsEqualTwo[{2, 2}]
Out=   {{{1, 4}, {2, 3}},   {{1, 3}, {2, 4}}}

In :=   MZeroSetsGreaterEqualTwo[{2, 2}]
Out=   {{{1, 4}, {2, 3}},   {{1, 3}, {2, 4}}}

Example. A case where the cardinality of π^* is odd and, therefore, the output is empty when executing MZeroSetsEqualTwo.

In :=   MZeroSetsEqualTwo[{2, 2, 1}]
Out=   := {}

A.17 Hermite Polynomials

A.17.1 HermiteRho

Computes Hermite polynomials with a leading coefficient of 1 and parameter $\rho > 0$. They are orthogonal with respect to the Gaussian distribution with mean zero and variance ρ. The generating function is $e^{tx - \frac{1}{2}\rho t^2}$.

These polynomials are defined by:

$$H_n(x, \rho) = (-\rho)^n e^{\frac{x^2}{2\rho}} \frac{d^n}{dx^n} e^{-\frac{x^2}{2\rho}}, \ n \geq 0,$$

and satisfy

$$H_n(x, \rho) = \sum_{k=0}^{\lfloor n/2 \rfloor} \binom{n}{2k} (2k-1)!!(-\rho)^k x^{n-2k}, \ n \geq 0.$$

▶ Please refer to Section 8.1.

Example. Compute the Hermite-Rho polynomial of order 5 with a leading coefficient of 1.

In := HermiteRho$[5, x, \rho]$
Out= $x^5 - 10x^3\rho + 15x\rho^2$

A.17.2 HermiteRhoGrid

Display the Hermite-Rho polynomials up to order n, using a leading coefficient of 1 for each polynomial.

Example. Make a grid of the first 11 Hermite-Rho polynomials (that is, go up to order 10).

In := HermiteRhoGrid$[10]$

The output is a chart:

$$
\begin{array}{ll}
0 & 1 \\
1 & x \\
2 & -\rho + x^2 \\
3 & -3\rho x + x^3 \\
4 & 3\rho^2 - 6\rho x^2 + x^4 \\
5 & 15\rho^2 x - 10\rho x^3 + x^5 \\
6 & -15\rho^3 + 45\rho^2 x^2 - 15\rho x^4 + x^6 \\
7 & -105\rho^3 x + 105\rho^2 x^3 - 21\rho x^5 + x^7 \\
8 & 105\rho^4 - 420\rho^3 x^2 + 210\rho^2 x^4 - 28\rho x^6 + x^8 \\
9 & 945\rho^4 x - 1260\rho^3 x^3 + 378\rho^2 x^5 - 36\rho x^7 + x^9 \\
10 & -945\rho^5 + 4725\rho^4 x^2 - 3150\rho^3 x^4 + 630\rho^2 x^6 - 45\rho x^8 + x^{10}.
\end{array}
$$

A.17.3 Hermite

Computes Hermite polynomials with a leading coefficient of 1. These polynomials are obtained by setting $\rho = 1$ and are defined as

$$
H_n(x) = (-1)^n e^{\frac{x^2}{2}} \frac{d^n}{dx^n} e^{-\frac{x^2}{2}}, \quad x \in \mathbb{R}.
$$

Example. Compute the Hermite polynomial of order 5 with a leading coefficient of 1.

In :=   Hermite[5, x] // TraditionalForm
Out= $x^5 - 10x^3 + 15x$

A.17.4 HermiteGrid

Display the Hermite polynomials up to order n, using a leading coefficient of 1 for each polynomial.

Example. Make a grid of the first 11 Hermite polynomials.

In :=   HermiteGrid[10] // TraditionalForm

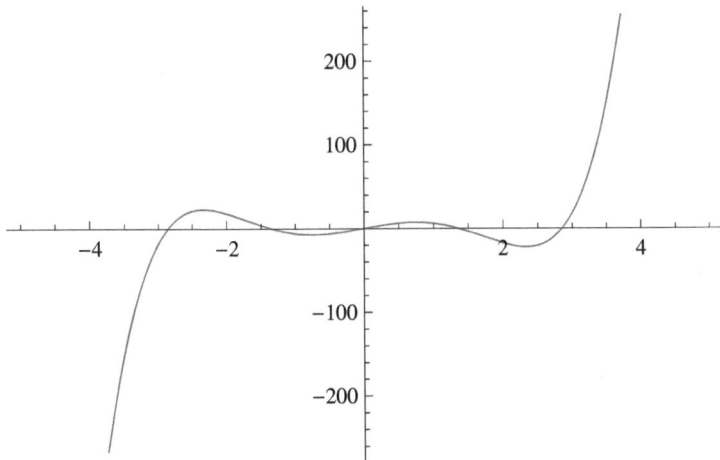

Fig. A.1. Hermite polynomial with $n = 5$

The output is a chart:

0	1
1	x
2	$x^2 - 1$
3	$x^3 - 3x$
4	$x^4 - 6x^2 + 3$
5	$x^5 - 10x^3 + 15x$
6	$x^6 - 15x^4 + 45x^2 - 15$
7	$x^7 - 21x^5 + 105x^3 - 105x$
8	$x^8 - 28x^6 + 210x^4 - 420x^2 + 105$
9	$x^9 - 36x^7 + 378x^5 - 1260x^3 + 945x$
10	$x^{10} - 45x^8 + 630x^6 - 3150x^4 + 4725x^2 - 945.$

Example. Plot $H_5(x)$ for $-5 \leq x \leq 5$. The output is Fig. A.1.

In :=   Plot[Hermite[5, x], {x, -5, 5}]

Example. Plot $H_k(x)$ for $k = 0, \cdots, 6$ and $-5 \leq x \leq 5$. The output is Fig. A.2.

In :=   Plot[Evaluate[Hermite[#, x] & /@ Range[0, 6, 1]], {x, -5, 5}, PlotStyle →
{}]

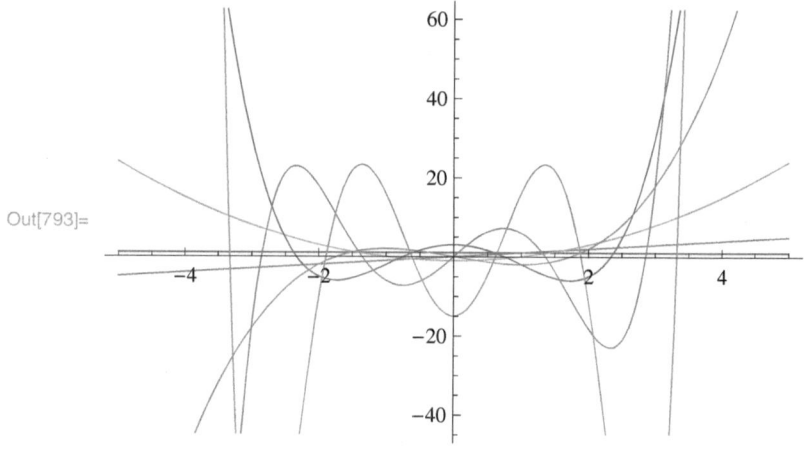

Out[793]=

Fig. A.2. Hermite polynomials with $1 \le n \le 6$

A.17.5 HermiteH

The built-in *Mathematica* function HermiteH[n,x] is defined as

$$\text{HermiteH}[n, x] = (-1)^n e^{x^2} \frac{d^n}{dx^n} e^{-x^2}.$$

One has:

$$H_n(x) = 2^{-n/2} \text{HermiteH}\left[n, \frac{x}{\sqrt{2}}\right].$$

$In :=$ HermiteH$[2, x]$
$Out=$ $-2 + 4x^2$

A.17.6 HermiteHGrid

Displays HermiteH as a chart. For example,

$In :=$ HermiteHGrid[3]

The output is a chart:

0   1
1   $2x$
2   $-2 + 4x^2$
3   $-12x + 8x^3$.

A.18 Poisson-Charlier Polynomials

A.18.1 Charlier

Computes the Poisson-Charlier Polynomials. They are orthogonal with respect to the Poisson distribution with mean a. These polynomials $c_n(x, a)$, defined for $n \geq 0$ and $a > 0$, satisfy the orthogonality relation

$$\sum_{k=0}^{\infty} c_n(k, a) c_m(k, a) e^{-a} \frac{a^k}{k!} = \frac{n!}{a^n} \delta_{nm}$$

and the recursion relation

$$c_0(x, a) = 1$$
$$c_{n+1}(x, a) = a^{-1} x c_n(k - 1, a) - c_n(x, a), \quad n \geq 1.$$

▶ Please refer to Chapter 10.

  Example. Compute the Poisson-Charlier polynomial with $n = 8$ and $a = 1$.

  In :=   Charlier[8, x, 1] // TraditionalForm
  Out= $x^8 - 36x^7 + 518x^6 - 3836x^5 + 15659x^4 - 34860x^3 + 38618x^2 - 16072x + 1$

  Example. Compute the Poisson-Charlier polynomial with $n = 2$ and $a > 0$.

  In :=   Collect[Charlier[2, x, a], x] // TraditionalForm
  Out= $\frac{x^2}{a^2} + \left(-\frac{1}{a^2} - \frac{2}{a}\right) x + 1$

  ("Collect" gathers together terms involving the same powers of x.)

A.18.2 CharlierGrid

Displays the Poisson-Charlier polynomials up to order n with Poisson mean $a > 0$.

  Example. Make a grid of the Poisson-Charlier polynomials for $0 \leq n \leq 5$.

  In :=   ChalierGrid[5, a] // TraditionalForm

Out=

0	1
1	$\frac{x}{a} - 1$
2	$\frac{x^2}{a^2} + \left(-\frac{1}{a^2} - \frac{2}{a}\right)x + 1$
3	$\frac{x^3}{a^3} + \left(-\frac{3}{a^3} - \frac{3}{a^2}\right)x^2 + \left(\frac{2}{a^3} + \frac{3}{a^2} + \frac{3}{a}\right)x - 1$
4	$\frac{x^4}{a^4} + \left(-\frac{6}{a^4} - \frac{4}{a^3}\right)x^3 + \left(\frac{11}{a^4} + \frac{12}{a^3} + \frac{6}{a^2}\right)x^2 + \left(-\frac{6}{a^4} - \frac{8}{a^3} - \frac{6}{a^2} - \frac{4}{a}\right)x + 1$
5	$\frac{x^5}{a^5} + \left(-\frac{10}{a^5} - \frac{5}{a^4}\right)x^4 + \left(\frac{35}{a^5} + \frac{30}{a^4} + \frac{10}{a^3}\right)x^3 + \left(-\frac{50}{a^5} - \frac{55}{a^4} - \frac{30}{a^3} - \frac{10}{a^2}\right)x^2 + \left(\frac{24}{a^5} + \frac{30}{a^4} + \frac{20}{a^3} + \frac{10}{a^2} + \frac{5}{a}\right)x - 1$

Example. Plot the Poisson-Charlier polynomial with $n = 5$ and $a = 1/2$ over the interval $-1 \le x \le 3$.

In :=   Plot[Charlier[5, x, 0.5], {x, -1, 3}, PlotRange → All]

Example. Plot the Poisson-Charlier polynomials with $a = 1$ and $1 \le n \le 6$ over the interval $-1 \le x \le 1$. The output is Fig. A.3.

In :=   Plot[Evaluate[Charlier[#, x, 1] & /@ Range[1, 6]], {x, -1, 1}, PlotStyle → { }]

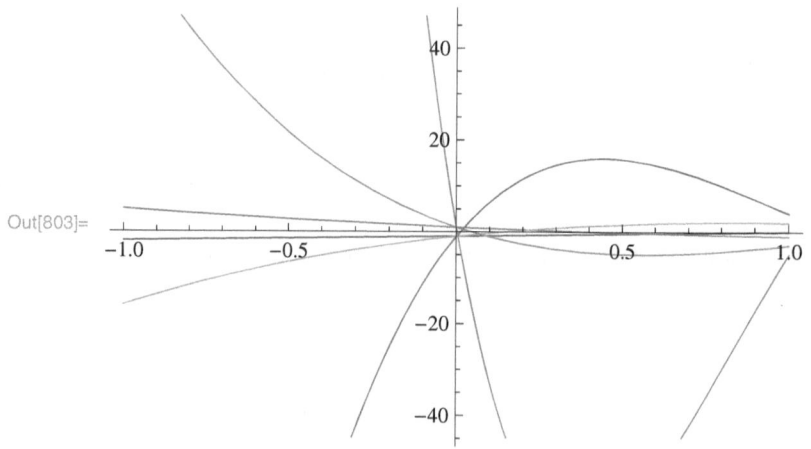

Fig. A.3. Poisson-Charlier polynomials with $1 \le n \le 6$

A.18.3 CharlierCentered

These polynomials are defined by

$$C_n(x, a) = a^n c_n(x + a, a).$$

They are orthogonal with respect to the centered Poisson distribution with parameter $a > 0$.

Example. Compute the centered Poisson-Charlier polynomial with $n = 3$.

In := Collect[CharlierCentered[3, x, a], x]
Out= $2a + (2 - 3a)x - 3x^2 + x^3$

Example. Compute the centered Poisson-Charlier polynomial with $n = 3$ and $a = 1$.

In := CharlierCentered[8, x, 1] // TraditionalForm
Out= $x^8 - 28x^7 + 294x^6 - 1428x^5 + 3059x^4 - 1428x^3 - 3326x^2 + 2904x - 7$

Example. Plot the centered Poisson-Charlier polynomials with $a = 1$ and $1 \le n \le 6$ over the interval $-1 \le x \le 1$. The output is Fig. A.4.

In := Plot[Evaluate[CharlierCentered[#, x, 1] & /@ Range[1, 6]], x, -1, 1, Plot Range → All]

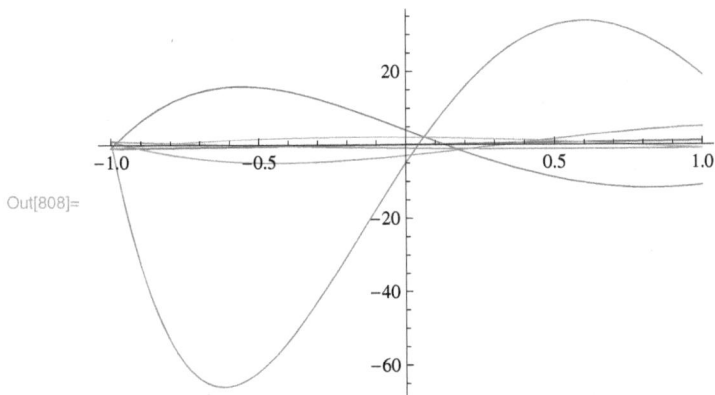

Out[808]=

Fig. A.4. Centered Charlier polynomials with $a = 1$ and $1 \le n \le 6$

A.18.4 CharlierCenteredGrid

Displays the centered Poisson-Charlier polynomials up to order n.

Example. Make a grid of the centered Poisson-Charlier polynomials for $0 \le n \le 5$.

$In :=$ ChalierCenteredGrid[5, a]
$Out=$

0 1

1 x

2 $x^2 - x - a$

3 $x^3 - 3x^2 + (2 - 3a)x + 2a$

4 $x^4 - 6x^3 + (11 - 6a)x^2 + (14a - 6)x + 3a^2 - 6a$

5 $x^5 - 10x^4 + (35 - 10a)x^3 + (50a - 50)x^2 + (15a^2 - 70a + 24)x - 20a^2 + 24a$

6 $x^6 - 15x^5 + (85 - 15a)x^4 + (130a - 225)x^3 + (45a^2 - 375a + 274)x^2$
$$+(-165a^2 + 404a - 120)x - 15a^3 + 130a^2 - 120a$$

Appendix B

Tables of moments and cumulants

In the following tables, μ_n and κ_n denote respectively the nth moment and nth cumulant of a random variable X. If X has mean zero, then $\mu_1 = \kappa_1 = 0$ and μ_n and κ_n are said to be "centered". If, in addition, X has variance one, then $\mu_2 = \kappa_2 = 1$ as well, and μ_n and κ_n are said to be "centered and scaled".

List of tables:

1. Centered and scaled Cumulant to Moments
2. Centered Cumulant to Moments
3. Cumulant to Moments
4. Centered and scaled Moment to Cumulants
5. Centered Moment to Cumulants
6. Moment to Cumulants

G. Peccati, M.S. Taqqu: Wiener Chaos: Moments, Cumulants and Diagrams –
A survey with computer implementation.
© Springer-Verlag Italia 2011

Centered and scaled Cumulant to Moments

n	$\kappa_n =$
1	0
2	1
3	μ_3
4	$-3 + \mu_4$
5	$-10\mu_3 + \mu_5$
6	$30 - 10\mu_3^2 - 15\mu_4 + \mu_6$
7	$210\mu_3 - 35\mu_3\mu_4 - 21\mu_5 + \mu_7$
8	$-630 + 560\mu_3^2 + 420\mu_4 - 35\mu_4^2 - 56\mu_3\mu_5 - 28\mu_6 + \mu_8$
9	$-7560\mu_3 + 560\mu_3^3 + 2520\mu_3\mu_4 + 756\mu_5 - 126\mu_4\mu_5 - 84\mu_3\mu_6 - 36\mu_7 + \mu_9$

$$n = 10 \quad 22680 - 37800\mu_3^2 - 18900\mu_4 + 4200\mu_3^2\mu_4 + 3150\mu_4^2 + 5040\mu_3\mu_5 - 126\mu_5^2 +$$
$$1260\mu_6 - 210\mu_4\mu_6 - 120\mu_3\mu_7 - 45\mu_8 + \mu_{10}$$

$$n = 12 \quad -1247400 + 3326400\mu_3^2 - 92400\mu_3^4 + 1247400\mu_4 - 831600\mu_3^2\mu_4 - 311850\mu_4^2 +$$
$$11550\mu_4^3 - 498960\mu_3\mu_5 + 55440\mu_3\mu_4\mu_5 + 16632\mu_5^2 - 83160\mu_6 + 18480\mu_3^2\mu_6 +$$
$$27720\mu_4\mu_6 - 462\mu_6^2 + 15840\mu_3\mu_7 - 792\mu_5\mu_7 + 2970\mu_8 - 495\mu_4\mu_8 - 220\mu_3\mu_9 -$$
$$66\mu_{10} + \mu_{12}$$

$$n = 13 \quad -32432400\mu_3 + 14414400\mu_3^3 + 21621600\mu_3\mu_4 - 1201200\mu_3^3\mu_4 - 2702700\mu_3\mu_4^2$$
$$+120120\mu_3\mu_4\mu_6 + 72072\mu_5\mu_6 - 154440\mu_7 + 34320\mu_3^2\mu_7 + 51480\mu_4\mu_7 - 1716\mu_6\mu_7$$
$$+25740\mu_3\mu_8 - 1287\mu_5\mu_8 + 4290\mu_9 - 715\mu_4\mu_9 - 286\mu_3\mu_{10} - 78\mu_{11} + \mu_{13}$$

$$n = 14 \quad 97297200 - 378378000\mu_3^2 + 33633600\mu_3^4 - 113513400\mu_4 +$$
$$151351200\mu_3^2\mu_4 + 37837800\mu_4^2 - 6306300\mu_3^2\mu_4^2 - 3153150\mu_4^3 + 60540480\mu_3\mu_5 -$$
$$3363360\mu_3^3\mu_5 - 15135120\mu_3\mu_4\mu_5 - 2270268\mu_5^2 + 252252\mu_4\mu_5^2 + 7567560\mu_6 -$$
$$5045040\mu_3^2\mu_6 - 3783780\mu_4\mu_6 + 210210\mu_4^2\mu_6 + 336336\mu_3\mu_5\mu_6 + 84084\mu_6^2 -$$
$$2162160\mu_3\mu_7 + 240240\mu_3\mu_4\mu_7 + 144144\mu_5\mu_7 - 1716\mu_7^2 - 270270\mu_8 +$$
$$60060\mu_3^2\mu_8 + 90090\mu_4\mu_8 - 3003\mu_6\mu_8 + 40040\mu_3\mu_9 - 2002\mu_5\mu_9 + 6006\mu_{10} -$$
$$1001\mu_4\mu_{10} - 364\mu_3\mu_{11} - 91\mu_{12} + \mu_{14}$$

Centered Cumulant to Moments

n	$\kappa_n =$
1	0
2	μ_2
3	μ_3
4	$-3\mu_2^2 + \mu_4$
5	$-10\mu_2\mu_3 + \mu_5$
6	$30\mu_2^3 - 10\mu_3^2 - 15\mu_2\mu_4 + \mu_6$
7	$210\mu_2^2\mu_3 - 35\mu_3\mu_4 - 21\mu_2\mu_5 + \mu_7$
8	$-630\mu_2^4 + 560\mu_2\mu_3^2 + 420\mu_2^2\mu_4 - 35\mu_4^2 - 56\mu_3\mu_5 - 28\mu_2\mu_6 + \mu_8$

9    $-7560\mu_2^3\mu_3 + 560\mu_3^3 + 2520\mu_2\mu_3\mu_4 + 756\mu_2^2\mu_5 - 126\mu_4\mu_5 - 84\mu_3\mu_6 - 36\mu_2\mu_7 + \mu_9$

10    $22680\mu_2^5 - 37800\mu_2^2\mu_3^2 - 18900\mu_2^3\mu_4 + 4200\mu_3^2\mu_4 + 3150\mu_2\mu_4^2 + 5040\mu_2\mu_3\mu_5 - 126\mu_5^2 + 1260\mu_2^2\mu_6 - 210\mu_4\mu_6 - 120\mu_3\mu_7 - 45\mu_2\mu_8 + \mu_{10}$

11    $415800\mu_2^4\mu_3 - 92400\mu_2\mu_3^3 - 207900\mu_2^2\mu_3\mu_4 + 11550\mu_3\mu_4^2 - 41580\mu_2^3\mu_5 + 9240\mu_3^2\mu_5 + 13860\mu_2\mu_4\mu_5 + 9240\mu_2\mu_3\mu_6 - 462\mu_5\mu_6 + 1980\mu_2^2\mu_7 - 330\mu_4\mu_7 - 165\mu_3\mu_8 - 55\mu_2\mu_9 + \mu_{11}$

12    $-1247400\mu_2^6 + 3326400\mu_2^3\mu_3^2 - 92400\mu_3^4 + 1247400\mu_2^4\mu_4 - 831600\mu_2\mu_3^2\mu_4 - 311850\mu_2^2\mu_4^2 + 11550\mu_4^3 - 498960\mu_2^2\mu_3\mu_5 + 55440\mu_3\mu_4\mu_5 + 16632\mu_2\mu_5^2 - 83160\mu_2^3\mu_6 + 18480\mu_3^2\mu_6 + 27720\mu_2\mu_4\mu_6 - 462\mu_6^2 + 15840\mu_2\mu_3\mu_7 - 792\mu_5\mu_7 + 2970\mu_2^2\mu_8 - 495\mu_4\mu_8 - 220\mu_3\mu_9 - 66\mu_2\mu_{10} + \mu_{12}$

13    $-32432400\mu_2^5\mu_3 + 14414400\mu_2^2\mu_3^3 + 21621600\mu_2^3\mu_3\mu_4 - 1201200\mu_3^3\mu_4 - 2702700\mu_2\mu_3\mu_4^2 + 3243240\mu_2^4\mu_5 - 2162160\mu_2\mu_3^2\mu_5 - 1621620\mu_2^2\mu_4\mu_5 + 90090\mu_4^2\mu_5 + 72072\mu_3\mu_5^2 - 1081080\mu_2^2\mu_3\mu_6 + 120120\mu_3\mu_4\mu_6 + 72072\mu_2\mu_5\mu_6 - 154440\mu_2^3\mu_7 + 34320\mu_3^2\mu_7 + 51480\mu_2\mu_4\mu_7 - 1716\mu_6\mu_7 + 25740\mu_2\mu_3\mu_8 - 1287\mu_5\mu_8 + 4290\mu_2^2\mu_9 - 715\mu_4\mu_9 - 286\mu_3\mu_{10} - 78\mu_2\mu_{11} + \mu_{13}$

Cumulant to Moments

n	$\kappa_n =$

1   μ_1

2    $-\mu_1^2 + \mu_2$

3    $2\mu_1^3 - 3\mu_1\mu_2 + \mu_3$

4    $-6\mu_1^4 + 12\mu_1^2\mu_2 - 3\mu_2^2 - 4\mu_1\mu_3 + \mu_4$

5    $24\mu_1^5 - 60\mu_1^3\mu_2 + 30\mu_1\mu_2^2 + 20\mu_1^2\mu_3 - 10\mu_2\mu_3 - 5\mu_1\mu_4 + \mu_5$

6
$$-120\mu_1^6 + 360\mu_1^4\mu_2 - 270\mu_1^2\mu_2^2 + 30\mu_2^3 - 120\mu_1^3\mu_3 + 120\mu_1\mu_2\mu_3 - 10\mu_3^2 + 30\mu_1^2\mu_4 - 15\mu_2\mu_4 - 6\mu_1\mu_5 + \mu_6$$

7
$$720\mu_1^7 - 2520\mu_1^5\mu_2 + 2520\mu_1^3\mu_2^2 - 630\mu_1\mu_2^3 + 840\mu_1^4\mu_3 - 1260\mu_1^2\mu_2\mu_3 + 210\mu_2^2\mu_3 + 140\mu_1\mu_3^2 - 210\mu_1^3\mu_4 + 210\mu_1\mu_2\mu_4 - 35\mu_3\mu_4 + 42\mu_1^2\mu_5 - 21\mu_2\mu_5 - 7\mu_1\mu_6 + \mu_7$$

8
$$-5040\mu_1^8 + 20160\mu_1^6\mu_2 - 25200\mu_1^4\mu_2^2 + 10080\mu_1^2\mu_2^3 - 630\mu_2^4 - 6720\mu_1^5\mu_3 + 13440\mu_1^3\mu_2\mu_3 - 5040\mu_1\mu_2^2\mu_3 - 1680\mu_1^2\mu_3^2 + 560\mu_2\mu_3^2 + 1680\mu_1^4\mu_4 - 2520\mu_1^2\mu_2\mu_4 + 420\mu_2^2\mu_4 + 560\mu_1\mu_3\mu_4 - 35\mu_4^2 - 336\mu_1^3\mu_5 + 336\mu_1\mu_2\mu_5 - 56\mu_3\mu_5 + 56\mu_1^2\mu_6 - 28\mu_2\mu_6 - 8\mu_1\mu_7 + \mu_8$$

9
$$40320\mu_1^9 - 181440\mu_1^7\mu_2 + 272160\mu_1^5\mu_2^2 - 151200\mu_1^3\mu_2^3 + 22680\mu_1\mu_2^4 + 60480\mu_1^6\mu_3 - 151200\mu_1^4\mu_2\mu_3 + 90720\mu_1^2\mu_2^2\mu_3 - 7560\mu_2^3\mu_3 + 20160\mu_1^3\mu_3^2 - 15120\mu_1\mu_2\mu_3^2 + 560\mu_3^3 - 15120\mu_1^5\mu_4 + 30240\mu_1^3\mu_2\mu_4 - 11340\mu_1\mu_2^2\mu_4 - 7560\mu_1^2\mu_3\mu_4 + 2520\mu_2\mu_3\mu_4 + 630\mu_1\mu_4^2 + 3024\mu_1^4\mu_5 - 4536\mu_1^2\mu_2\mu_5 + 756\mu_2^2\mu_5 + 1008\mu_1\mu_3\mu_5 - 126\mu_4\mu_5 - 504\mu_1^3\mu_6 + 504\mu_1\mu_2\mu_6 - 84\mu_3\mu_6 + 72\mu_1^2\mu_7 - 36\mu_2\mu_7 - 9\mu_1\mu_8 + \mu_9$$

10
$$-362880\mu_1^{10} + 1814400\mu_1^8\mu_2 - 3175200\mu_1^6\mu_2^2 + 2268000\mu_1^4\mu_2^3 - 567000\mu_1^2\mu_2^4 + 22680\mu_2^5 - 604800\mu_1^7\mu_3 + 1814400\mu_1^5\mu_2\mu_3 - 1512000\mu_1^3\mu_2^2\mu_3 + 302400\mu_1\mu_2^3\mu_3 - 252000\mu_1^4\mu_3^2 + 302400\mu_1^2\mu_2\mu_3^2 - 37800\mu_2^2\mu_3^2 - 16800\mu_1\mu_3^3 + 151200\mu_1^6\mu_4 - 378000\mu_1^4\mu_2\mu_4 + 226800\mu_1^2\mu_2^2\mu_4 - 18900\mu_2^3\mu_4 + 100800\mu_1^3\mu_3\mu_4 - 75600\mu_1\mu_2\mu_3\mu_4 + 4200\mu_3^2\mu_4 - 9450\mu_1^2\mu_4^2 + 3150\mu_2\mu_4^2 - 30240\mu_1^5\mu_5 + 60480\mu_1^3\mu_2\mu_5 - 22680\mu_1\mu_2^2\mu_5 - 15120\mu_1^2\mu_3\mu_5 + 5040\mu_2\mu_3\mu_5 + 2520\mu_1\mu_4\mu_5 - 126\mu_5^2 + 5040\mu_1^4\mu_6 - 7560\mu_1^2\mu_2\mu_6 + 1260\mu_2^2\mu_6 + 1680\mu_1\mu_3\mu_6 - 210\mu_4\mu_6 - 720\mu_1^3\mu_7 + 720\mu_1\mu_2\mu_7 - 120\mu_3\mu_7 + 90\mu_1^2\mu_8 - 45\mu_2\mu_8 - 10\mu_1\mu_9 + \mu_{10}$$

Centered and scaled Moment to Cumulants

n	$\mu_n =$
1	0
2	1
3	κ_3
4	$3 + \kappa_4$
5	$10\kappa_3 + \kappa_5$
6	$15 + 10\kappa_3^2 + 15\kappa_4 + \kappa_6$
7	$105\kappa_3 + 35\kappa_3\kappa_4 + 21\kappa_5 + \kappa_7$
8	$105 + 280\kappa_3^2 + 210\kappa_4 + 35\kappa_4^2 + 56\kappa_3\kappa_5 + 28\kappa_6 + \kappa_8$
9	$1260\kappa_3 + 280\kappa_3^3 + 1260\kappa_3\kappa_4 + 378\kappa_5 + 126\kappa_4\kappa_5 + 84\kappa_3\kappa_6 + 36\kappa_7 + \kappa_9$
10	$945 + 6300\kappa_3^2 + 3150\kappa_4 + 2100\kappa_3^2\kappa_4 + 1575\kappa_4^2 + 2520\kappa_3\kappa_5 + 126\kappa_5^2 + 630\kappa_6 + 210\kappa_4\kappa_6 + 120\kappa_3\kappa_7 + 45\kappa_8 + \kappa_{10}$
11	$17325\kappa_3 + 15400\kappa_3^3 + 34650\kappa_3\kappa_4 + 5775\kappa_3\kappa_4^2 + 6930\kappa_5 + 4620\kappa_3^2\kappa_5 + 6930\kappa_4\kappa_5 + 4620\kappa_3\kappa_6 + 462\kappa_5\kappa_6 + 990\kappa_7 + 330\kappa_4\kappa_7 + 165\kappa_3\kappa_8 + 55\kappa_9 + \kappa_{11}$
12	$10395 + 138600\kappa_3^2 + 15400\kappa_3^4 + 51975\kappa_4 + 138600\kappa_3^2\kappa_4 + 51975\kappa_4^2 + 5775\kappa_4^3 + 83160\kappa_3\kappa_5 + 27720\kappa_3\kappa_4\kappa_5 + 8316\kappa_5^2 + 13860\kappa_6 + 9240\kappa_3^2\kappa_6 + 13860\kappa_4\kappa_6 + 462\kappa_6^2 + 7920\kappa_3\kappa_7 + 792\kappa_5\kappa_7 + 1485\kappa_8 + 495\kappa_4\kappa_8 + 220\kappa_3\kappa_9 + 66\kappa_{10} + \kappa_{12}$
13	$270270\kappa_3 + 600600\kappa_3^3 + 900900\kappa_3\kappa_4 + 200200\kappa_3^3\kappa_4 + 450450\kappa_3\kappa_4^2 + 135135\kappa_5 + 360360\kappa_3^2\kappa_5 + 270270\kappa_4\kappa_5 + 45045\kappa_4^2\kappa_5 + 36036\kappa_3\kappa_5^2 + 180180\kappa_3\kappa_6 + 60060\kappa_3\kappa_4\kappa_6 + 36036\kappa_5\kappa_6 + 25740\kappa_7 + 17160\kappa_3^2\kappa_7 + 25740\kappa_4\kappa_7 + 1716\kappa_6\kappa_7 + 12870\kappa_3\kappa_8 + 1287\kappa_5\kappa_8 + 2145\kappa_9 + 715\kappa_4\kappa_9 + 286\kappa_3\kappa_{10} + 78\kappa_{11} + \kappa_{13}$
14	$135135 + 3153150\kappa_3^2 + 1401400\kappa_3^4 + 945945\kappa_4 + 6306300\kappa_3^2\kappa_4 + 1576575\kappa_4^2 + 1051050\kappa_3^2\kappa_4^2 + 525525\kappa_4^3 + 2522520\kappa_3\kappa_5 + 560560\kappa_3^3\kappa_5 + 2522520\kappa_3\kappa_4\kappa_5 + 378378\kappa_5^2 + 126126\kappa_4\kappa_5^2 + 315315\kappa_6 + 840840\kappa_3^2\kappa_6 + 630630\kappa_4\kappa_6 + 105105\kappa_4^2\kappa_6 + 168168\kappa_3\kappa_5\kappa_6 + 42042\kappa_6^2 + 360360\kappa_3\kappa_7 + 120120\kappa_3\kappa_4\kappa_7 + 72072\kappa_5\kappa_7 + 1716\kappa_7^2 + 45045\kappa_8 + 30030\kappa_3^2\kappa_8 + 45045\kappa_4\kappa_8 + 3003\kappa_6\kappa_8 + 20020\kappa_3\kappa_9 + 2002\kappa_5\kappa_9 + 3003\kappa_{10} + 1001\kappa_4\kappa_{10} + 364\kappa_3\kappa_{11} + 91\kappa_{12} + \kappa_{14}$

Centered Moment to Cumulants

n	$\mu_n =$
1	0
2	κ_2
3	κ_3
4	$3\kappa_2^2 + \kappa_4$
5	$10\kappa_2\kappa_3 + \kappa_5$
6	$15\kappa_2^3 + 10\kappa_3^2 + 15\kappa_2\kappa_4 + \kappa_6$
7	$105\kappa_2^2\kappa_3 + 35\kappa_3\kappa_4 + 21\kappa_2\kappa_5 + \kappa_7$
8	$105\kappa_2^4 + 280\kappa_2\kappa_3^2 + 210\kappa_2^2\kappa_4 + 35\kappa_4^2 + 56\kappa_3\kappa_5 + 28\kappa_2\kappa_6 + \kappa_8$
9	$1260\kappa_2^3\kappa_3 + 280\kappa_3^3 + 1260\kappa_2\kappa_3\kappa_4 + 378\kappa_2^2\kappa_5 + 126\kappa_4\kappa_5 + 84\kappa_3\kappa_6 + 36\kappa_2\kappa_7 + \kappa_9$

$$\mu_{10} = 945\kappa_2^5 + 6300\kappa_2^2\kappa_3^2 + 3150\kappa_2^3\kappa_4 + 2100\kappa_3^2\kappa_4 + 1575\kappa_2\kappa_4^2 + 2520\kappa_2\kappa_3\kappa_5 + 126\kappa_5^2 +$$
$$630\kappa_2^2\kappa_6 + 210\kappa_4\kappa_6 + 120\kappa_3\kappa_7 + 45\kappa_2\kappa_8 + \kappa_{10}$$

$$\mu_{11} = 17325\kappa_2^4\kappa_3 + 15400\kappa_2\kappa_3^3 + 34650\kappa_2^2\kappa_3\kappa_4 + 5775\kappa_3\kappa_4^2 + 6930\kappa_2^3\kappa_5 + 4620\kappa_3^2\kappa_5 +$$
$$6930\kappa_2\kappa_4\kappa_5 + 4620\kappa_2\kappa_3\kappa_6 + 462\kappa_5\kappa_6 + 990\kappa_2^2\kappa_7 + 330\kappa_4\kappa_7 + 165\kappa_3\kappa_8 + 55\kappa_2\kappa_9 + \kappa_{11}$$

$$\mu_{12} = 10395\kappa_2^6 + 138600\kappa_2^3\kappa_3^2 + 15400\kappa_3^4 + 51975\kappa_2^4\kappa_4 + 138600\kappa_2\kappa_3^2\kappa_4 + 51975\kappa_2^2\kappa_4^2 +$$
$$5775\kappa_4^3 + 83160\kappa_2^2\kappa_3\kappa_5 + 27720\kappa_3\kappa_4\kappa_5 + 8316\kappa_2\kappa_5^2 + 13860\kappa_2^3\kappa_6 + 9240\kappa_3^2\kappa_6 +$$
$$13860\kappa_2\kappa_4\kappa_6 + 462\kappa_6^2 + 7920\kappa_2\kappa_3\kappa_7 + 792\kappa_5\kappa_7 + 1485\kappa_2^2\kappa_8 + 495\kappa_4\kappa_8 +$$
$$220\kappa_3\kappa_9 + 66\kappa_2\kappa_{10} + \kappa_{12}$$

$$\mu_{13} = 270270\kappa_2^5\kappa_3 + 600600\kappa_2^2\kappa_3^3 + 900900\kappa_2^3\kappa_3\kappa_4 + 200200\kappa_3^3\kappa_4 + 450450\kappa_2\kappa_3\kappa_4^2 +$$
$$135135\kappa_2^4\kappa_5 + 360360\kappa_2\kappa_3^2\kappa_5 + 270270\kappa_2^2\kappa_4\kappa_5 + 45045\kappa_4^2\kappa_5 + 36036\kappa_3\kappa_5^2 +$$
$$180180\kappa_2^2\kappa_3\kappa_6 + 60060\kappa_3\kappa_4\kappa_6 + 36036\kappa_2\kappa_5\kappa_6 + 25740\kappa_2^3\kappa_7 + 17160\kappa_3^2\kappa_7 +$$
$$25740\kappa_2\kappa_4\kappa_7 + 1716\kappa_6\kappa_7 + 12870\kappa_2\kappa_3\kappa_8 + 1287\kappa_5\kappa_8 + 2145\kappa_2^2\kappa_9 + 715\kappa_4\kappa_9 +$$
$$286\kappa_3\kappa_{10} + 78\kappa_2\kappa_{11} + \kappa_{13}$$

$$\mu_{14} = 135135\kappa_2^7 + 3153150\kappa_2^4\kappa_3^2 + 1401400\kappa_2\kappa_3^4 + 945945\kappa_2^5\kappa_4 + 6306300\kappa_2^2\kappa_3^2\kappa_4 +$$
$$1576575\kappa_2^3\kappa_4^2 + 1051050\kappa_3^2\kappa_4^2 + 525525\kappa_2\kappa_4^3 + 2522520\kappa_2^3\kappa_3\kappa_5 + 560560\kappa_3^3\kappa_5 +$$
$$2522520\kappa_2\kappa_3\kappa_4\kappa_5 + 378378\kappa_2^2\kappa_5^2 + 126126\kappa_4\kappa_5^2 + 315315\kappa_2^4\kappa_6 + 840840\kappa_2\kappa_3^2\kappa_6 +$$
$$630630\kappa_2^2\kappa_4\kappa_6 + 105105\kappa_4^2\kappa_6 + 168168\kappa_3\kappa_5\kappa_6 + 42042\kappa_2\kappa_6^2 + 360360\kappa_2^2\kappa_3\kappa_7 +$$
$$120120\kappa_3\kappa_4\kappa_7 + 72072\kappa_2\kappa_5\kappa_7 + 1716\kappa_7^2 + 45045\kappa_2^3\kappa_8 + 30030\kappa_3^2\kappa_8 +$$
$$45045\kappa_2\kappa_4\kappa_8 + 3003\kappa_6\kappa_8 + 20020\kappa_2\kappa_3\kappa_9 + 2002\kappa_5\kappa_9 + 3003\kappa_2^2\kappa_{10} +$$
$$1001\kappa_4\kappa_{10} + 364\kappa_3\kappa_{11} + 91\kappa_2\kappa_{12} + \kappa_{14}$$

Moment to Cumulants

n	$\mu_n =$
1	κ_1
2	$\kappa_1^2 + \kappa_2$
3	$\kappa_1^3 + 3\kappa_1\kappa_2 + \kappa_3$
4	$\kappa_1^4 + 6\kappa_1^2\kappa_2 + 3\kappa_2^2 + 4\kappa_1\kappa_3 + \kappa_4$
5	$\kappa_1^5 + 10\kappa_1^3\kappa_2 + 15\kappa_1\kappa_2^2 + 10\kappa_1^2\kappa_3 + 10\kappa_2\kappa_3 + 5\kappa_1\kappa_4 + \kappa_5$

n = 6

$$\kappa_1^6 + 15\kappa_1^4\kappa_2 + 45\kappa_1^2\kappa_2^2 + 15\kappa_2^3 + 20\kappa_1^3\kappa_3 + 60\kappa_1\kappa_2\kappa_3 + 10\kappa_3^2 + 15\kappa_1^2\kappa_4 + 15\kappa_2\kappa_4 + 6\kappa_1\kappa_5 + \kappa_6$$

n = 7

$$\kappa_1^7 + 21\kappa_1^5\kappa_2 + 105\kappa_1^3\kappa_2^2 + 105\kappa_1\kappa_2^3 + 35\kappa_1^4\kappa_3 + 210\kappa_1^2\kappa_2\kappa_3 + 105\kappa_2^2\kappa_3 + 70\kappa_1\kappa_3^2 + 35\kappa_1^3\kappa_4 + 105\kappa_1\kappa_2\kappa_4 + 35\kappa_3\kappa_4 + 21\kappa_1^2\kappa_5 + 21\kappa_2\kappa_5 + 7\kappa_1\kappa_6 + \kappa_7$$

n = 8

$$\kappa_1^8 + 28\kappa_1^6\kappa_2 + 210\kappa_1^4\kappa_2^2 + 420\kappa_1^2\kappa_2^3 + 105\kappa_2^4 + 56\kappa_1^5\kappa_3 + 560\kappa_1^3\kappa_2\kappa_3 + 840\kappa_1\kappa_2^2\kappa_3 + 280\kappa_1^2\kappa_3^2 + 280\kappa_2\kappa_3^2 + 70\kappa_1^4\kappa_4 + 420\kappa_1^2\kappa_2\kappa_4 + 210\kappa_2^2\kappa_4 + 280\kappa_1\kappa_3\kappa_4 + 35\kappa_4^2 + 56\kappa_1^3\kappa_5 + 168\kappa_1\kappa_2\kappa_5 + 56\kappa_3\kappa_5 + 28\kappa_1^2\kappa_6 + 28\kappa_2\kappa_6 + 8\kappa_1\kappa_7 + \kappa_8$$

n = 9

$$\kappa_1^9 + 36\kappa_1^7\kappa_2 + 378\kappa_1^5\kappa_2^2 + 1260\kappa_1^3\kappa_2^3 + 945\kappa_1\kappa_2^4 + 84\kappa_1^6\kappa_3 + 1260\kappa_1^4\kappa_2\kappa_3 + 3780\kappa_1^2\kappa_2^2\kappa_3 + 1260\kappa_2^3\kappa_3 + 840\kappa_1^3\kappa_3^2 + 2520\kappa_1\kappa_2\kappa_3^2 + 280\kappa_3^3 + 126\kappa_1^5\kappa_4 + 1260\kappa_1^3\kappa_2\kappa_4 + 1890\kappa_1\kappa_2^2\kappa_4 + 1260\kappa_1^2\kappa_3\kappa_4 + 1260\kappa_2\kappa_3\kappa_4 + 315\kappa_1\kappa_4^2 + 126\kappa_1^4\kappa_5 + 756\kappa_1^2\kappa_2\kappa_5 + 378\kappa_2^2\kappa_5 + 504\kappa_1\kappa_3\kappa_5 + 126\kappa_4\kappa_5 + 84\kappa_1^3\kappa_6 + 252\kappa_1\kappa_2\kappa_6 + 84\kappa_3\kappa_6 + 36\kappa_1^2\kappa_7 + 36\kappa_2\kappa_7 + 9\kappa_1\kappa_8 + \kappa_9$$

n = 10

$$\kappa_1^{10} + 45\kappa_1^8\kappa_2 + 630\kappa_1^6\kappa_2^2 + 3150\kappa_1^4\kappa_2^3 + 4725\kappa_1^2\kappa_2^4 + 945\kappa_2^5 + 120\kappa_1^7\kappa_3 + 2520\kappa_1^5\kappa_2\kappa_3 + 12600\kappa_1^3\kappa_2^2\kappa_3 + 12600\kappa_1\kappa_2^3\kappa_3 + 2100\kappa_1^4\kappa_3^2 + 12600\kappa_1^2\kappa_2\kappa_3^2 + 6300\kappa_2^2\kappa_3^2 + 2800\kappa_1\kappa_3^3 + 210\kappa_1^6\kappa_4 + 3150\kappa_1^4\kappa_2\kappa_4 + 9450\kappa_1^2\kappa_2^2\kappa_4 + 3150\kappa_2^3\kappa_4 + 4200\kappa_1^3\kappa_3\kappa_4 + 12600\kappa_1\kappa_2\kappa_3\kappa_4 + 2100\kappa_3^2\kappa_4 + 1575\kappa_1^2\kappa_4^2 + 1575\kappa_2\kappa_4^2 + 252\kappa_1^5\kappa_5 + 2520\kappa_1^3\kappa_2\kappa_5 + 3780\kappa_1\kappa_2^2\kappa_5 + 2520\kappa_1^2\kappa_3\kappa_5 + 2520\kappa_2\kappa_3\kappa_5 + 1260\kappa_1\kappa_4\kappa_5 + 126\kappa_5^2 + 210\kappa_1^4\kappa_6 + 1260\kappa_1^2\kappa_2\kappa_6 + 630\kappa_2^2\kappa_6 + 840\kappa_1\kappa_3\kappa_6 + 210\kappa_4\kappa_6 + 120\kappa_1^3\kappa_7 + 360\kappa_1\kappa_2\kappa_7 + 120\kappa_3\kappa_7 + 45\kappa_1^2\kappa_8 + 45\kappa_2\kappa_8 + 10\kappa_1\kappa_9 + \kappa_{10}$$

n = 11

$$\kappa_1^{11} + 55\kappa_1^9\kappa_2 + 990\kappa_1^7\kappa_2^2 + 6930\kappa_1^5\kappa_2^3 + 17325\kappa_1^3\kappa_2^4 + 10395\kappa_1\kappa_2^5 + 165\kappa_1^8\kappa_3 + 4620\kappa_1^6\kappa_2\kappa_3 + 34650\kappa_1^4\kappa_2^2\kappa_3 + 69300\kappa_1^2\kappa_2^3\kappa_3 + 17325\kappa_2^4\kappa_3 + 4620\kappa_1^5\kappa_3^2 + 46200\kappa_1^3\kappa_2\kappa_3^2 + 69300\kappa_1\kappa_2^2\kappa_3^2 + 15400\kappa_1^2\kappa_3^3 + 15400\kappa_2\kappa_3^3 + 330\kappa_1^7\kappa_4 + 6930\kappa_1^5\kappa_2\kappa_4 + 34650\kappa_1^3\kappa_2^2\kappa_4 + 34650\kappa_1\kappa_2^3\kappa_4 + 11550\kappa_1^4\kappa_3\kappa_4 + 69300\kappa_1^2\kappa_2\kappa_3\kappa_4 + 34650\kappa_2^2\kappa_3\kappa_4 + 23100\kappa_1\kappa_3^2\kappa_4 + 5775\kappa_1^3\kappa_4^2 + 17325\kappa_1\kappa_2\kappa_4^2 + 5775\kappa_3\kappa_4^2 + 462\kappa_1^6\kappa_5 + 6930\kappa_1^4\kappa_2\kappa_5 + 20790\kappa_1^2\kappa_2^2\kappa_5 + 6930\kappa_2^3\kappa_5 + 9240\kappa_1^3\kappa_3\kappa_5 + 27720\kappa_1\kappa_2\kappa_3\kappa_5 + 4620\kappa_3^2\kappa_5 + 6930\kappa_1^2\kappa_4\kappa_5 + 6930\kappa_2\kappa_4\kappa_5 + 1386\kappa_1\kappa_5^2 + 462\kappa_1^5\kappa_6 + 4620\kappa_1^3\kappa_2\kappa_6 + 6930\kappa_1\kappa_2^2\kappa_6 + 4620\kappa_1^2\kappa_3\kappa_6 + 4620\kappa_2\kappa_3\kappa_6 + 2310\kappa_1\kappa_4\kappa_6 + 462\kappa_5\kappa_6 + 330\kappa_1^4\kappa_7 + 1980\kappa_1^2\kappa_2\kappa_7 + 990\kappa_2^2\kappa_7 + 1320\kappa_1\kappa_3\kappa_7 + 330\kappa_4\kappa_7 + 165\kappa_1^3\kappa_8 + 495\kappa_1\kappa_2\kappa_8 + 165\kappa_3\kappa_8 + 55\kappa_1^2\kappa_9 + 55\kappa_2\kappa_9 + 11\kappa_1\kappa_{10} + \kappa_{11}$$

References

1. Adler R.J. (1990). *An Introduction to Continuity, Extrema, and Related Topics for General Gaussian Processes*. Lecture Notes-Monograph Series **12**, Institut of Mathematical Statistics, Hayward, California.
2. Aigner M. (1979). *Combinatorial theory*. Springer-Verlag, Berlin Heidelberg New York.
3. Anshelevich M. (2001). Partition-dependent stochastic measures and q-deformed cumulants. *Documenta Mathematica* **6**, 343-384.
4. Anshelevich M. (2005). Linearization coefficients for orthogonal polynomials using stochastic processes. *The Annals of Probabability* **33**(1), 114-136.
5. Baldi P., Kerkyacharian G., Marinucci D. and Picard D. (2008). High-frequency asymptotics for wavelet-based tests for Gaussianity and isotropy on the torus. *Journal of Multivariate Analysis* **99**, 606-636.
6. Barbour A.D. (1986). Asymptotic expansions based on smooth functions in the central limit theorem. *Probab. Theory Rel. Fields* **72**(2), 289-303.
7. Barndorff-Nielsen O., Corcuera J., Podolskij M. and Woerner J. (2009). Bipower variations for Gaussian processes with stationary increments. *Journal of Applied Probability* **46**, 132-150.
8. Billingsley P. (1995). *Probability and Measure*, 3rd edition. Wiley, New York.
9. Bitcheler K. (1980). Stochastic integration and L^p theory of semimartingales. *The Annals of Probability* **9**(1), 49-89.
10. Breuer P. and Major P. (1983). Central limit theorems for non-linear functionals of Gaussian fields. *Journal of Multivariate Analysis* **13**, 425-441.
11. Brodskii M.S. (1971). *Triangular and Jordan Representations of Linear Operators*. Transl. Math. Monographs **32**, AMS, Providence.
12. Chambers D. and Slud E. (1989). Central limit theorems for nonlinear functionals of stationary Gaussian processes. *Probability Theory and Related Fields* **80**, 323-349.
13. Charalambides C.A. (2002). *Enumerative Combinatorics*. Chapman and Hall, New York.
14. Chen L.H.Y. and Shao Q.-M. (2005). Stein's method for normal approximation. In: A.D. Barbour and L.H.Y. Chen (eds), *An Introduction to Stein's Method*. Lecture Notes Series No. **4**, Institute for Mathematical Sciences, National University of Singapore, Singapore University Press and World Scientific 2005, 1-59.
15. Chen L.H.Y., Goldstein L. and Shao Q.-M. (2001). *Normal Approximation by Stein's Method*. Springer-Verlag, Berlin Heidelberg New York.
16. Cohen S. and Taqqu M.S. (2004). Small and large scale behavior of the Poissonized Telecom process. *Methodology and Computing in Applied Probability* **6**, 363-379.

17. Corcuera J.M., Nualart D. and Woerner J.H.C. (2006). Power variation of some integral long memory process. *Bernoulli* **12**(4), 713-735.

18. de Blasi P., Peccati G. and Prünster I. (2008). Asymptotics for posterior hazards. *The Annals of Statistics* **37**(4), 1906-1945.

19. Deheuvels P., Peccati G. and Yor M. (2006). On quadratic functionals of the Brownian sheet and related processes. *Stochastic Processes and their Applications* **116**, 493-538.

20. Dellacherie C., Maisonneuve B. and Meyer P.-A. (1992). *Probabilités et Potentiel (Chapitres XVII à XXIV)*. Hermann, Paris.

21. Di Nunno G., Øksendal B. and Proske F. (2009). *Malliavin calculus for Lévy processes with applications to finance*. Springer-Verlag, Berlin.

22. Dobiński G. (1877). Summirung der Reihe $\sum n^m/n!$ für $m = 1, 2, 3, 4, 5, ...$ *Grunert Archiv (Arch. für Mat. und Physik)* **61**, 333-336.

23. Dobrushin R.L. (1979). Gaussian and their subordinated self-similar random generalized fields. *The Annals of Probability* **7**, 1–28.

24. Dudley R.M. (1967). The sizes of compact subsets of Hilbert space and continuity of Gaussian processes. *Journal of Functional Analysis* **1**, 290-330.

25. Embrechts P. and Maejima M. (2003). *Selfsimilar Processes*. Princeton University Press, Princeton.

26. Engel D.D. (1982). The multiple stochastic integral. *Memoirs of the AMS* **38**, 1-82.

27. Erhardsson T. (2005). Poisson and compound Poisson approximations. In: A.D. Barbour and L.H.Y. Chen (eds.), *An Introduction to Stein's Method*. Lecture Notes Series No.4, Institute for Mathematical Sciences, National University of Singapore, Singapore University Press and World Scientific 2005, 61-115.

28. Farré M., Jolis M. and Utzet F. (2008). Multiple Stratonovich integral and Hu-Meyer formula for Lévy processes. To appear in: *The Annals of Probability*.

29. Feigin P.D. (1985). Stable convergence of semimartingales. *Stochastic Processes and their Applications* **19**, 125-134.T.S.

30. Ferguson (1973). A Bayesian analysis of some non-parametric problems. *The Annals of Statistics* **1**(2), 209-230.

31. Fox R. and Taqqu M.S. (1987). Multiple stochastic integrals with dependent integrators. *Journal of Multivariate Analysis* **21**(1), 105-127.

32. Fristedt B. and Gray L. (1997). *A Modern Approach to Probability theory*. Birkhäuser, Boston.

33. Giné E. and de la Peña V.H. (1999). *Decoupling*. Springer-Verlag, Berlin Heidelberg New York.

34. Giraitis L. and Surgailis D. (1985). CLT and other limit theorems for functionals of Gaussian processes. *Zeitschrift für Wahrsch. verw. Gebiete* **70**, 191-212.

35. Goldberg J. and Taqqu M.S. (1982). Regular multigraphs and their applications to the Monte Carlo evaluation of moments of non-linear functions of Gaussian processes. *Stochastic Processes and their Applications* **13**, 121-138.

36. Gripenberg G. and Norros I. (1996). On the prediction of fractional Brownian motion. *Journal of Applied Probability* **33**, 400-410.

37. Handa K. (2005). Sampling formulae for symmetric selection. *Electronic Communications in Probabability* **10**, 223-234 (Electronic).

38. Handa K. (2009). The two-parameter Poisson-Dirichlet point process. *Bernoulli* **15**(4), 1082-1116.

39. Hu Y. and Nualart D. (2005). Renormalized self-intersection local time for fractional Brownian motion. *The Annals of Probabability* **33**(3), 948-983.

40. Itô K. (1951). Multiple Wiener integral. *J. Math. Soc. Japan* **3**, 157–169.

41. Jacod J., Klopotowski A. and Mémin J. (1982). Théorème de la limite centrale et convergence fonctionnelle vers un processus à accroissements indépendants: la méthode des martingales. *Annales de l'Institut H. Poincaré (PR)* **1**, 1-45.

42. Jacod J. and Shiryaev A.N. (1987). *Limit Theorems for Stochastic Processes*. Springer-Verlag, Berlin Heidelberg New York.

43. Jakubowski A. (1986). Principle of conditioning in limit theorems for sums of random variables. *The Annals of Probability* **11**(3), 902-915.

44. James L.F., Lijoi A. and Prünster I. (2005). Conjugacy as a distinctive feature of the Dirichlet process. *Scandinavian Journal of Statistics* **33**, 105-120.

45. James L., Roynette B. and Yor M. (2008). Generalized Gamma Convolutions, Dirichlet means, Thorin measures, with explicit examples. *Probability Surveys* **5**, 346-415.

46. Janson S. (1997). *Gaussian Hilbert Spaces*. Cambridge University Press, Cambridge.

47. Jeulin T. (1980). *Semimatingales et grossissement d'une filtration*. Lecture Notes in Mathematics 833, Springer-Verlag, Berlin Heidelberg New York.

48. Julià O. and Nualart D. (1988). The distribution of a double stochastic integral with respect to two independent Brownian Sheets. *Stochastics* **25**, 171-182.

49. Kabanov Y. (1975). On extended stochastic integrals. *Theory of Probability and its applications* **20**, 710-722.

50. Kailath T. and Segall A. (1976). Orthogonal functionals of independent increment processes. *IEEE Trans. Inform. Theory* IT-22, 287-298.

51. Kallenberg O. and Szulga J. (1991). Multiple integration with respect to Poisson and Lévy processes. *Probability Theory and Related Fields* **83**, 101-134.

52. Karatzas I. and Shreve S.E. (1991). *Brownian Motion and Stochastic Calculus*, 2nd edition. Graduate Texts in Mathematics, Volume 113, Springer-Verlag, Berlin Heidelberg New-York.

53. Kemp T., Nourdin I., Peccati G. and Speicher R. (2010). Wigner chaos and the fourth moment. Preprint.

54. Kendall W.S. (2005). Stochastic integrals and their expectations. *The Mathematica Journal.* **9**(4), 757-767.

55. Kingman J.F.C. (1967). Completely random measures. *Pacific Journal of Mathematics* **21**, 59-78.

56. Kung J.P.S., Rota G.-C. and Yan C.H. (2009). *Combinatorics. The Rota Way*. Cambridge University Press, Cambridge.

57. Kuo H.-H. (1975). *Gaussian measures in Banach spaces*. LNM **463**. Springer-Verlag, Berlin Heidelberg New-York.

58. Kuo H.-H. (1986). *Introduction to Stochastic Integration*. Springer-Verlag, Berlin Heidelberg New York.

59. Kussmaul A.U. (1977). *Stochastic integration and generalized martingales*. Pitman research notes in mathematic, **11**. London.

60. Kwapień S. and Woyczyński W.A. (1991). Semimartingale integrals via decoupling inequalities and tangent processes. *Probability and Mathematical Statisitics* **12**(2), 165-200.

61. Kwapień S. and Woyczyński W.A. (1992). *Random Series and Stochastic Integrals: Single and Multiple*. Birkhäuser, Basel.

62. Leonov V.P. and Shiryayev A.N. (1959). On a method of calculations of semi-invariants. *Theory of Probability and its Applications* **4**, 319-329.

63. Lifshits M.A. (1995). *Gaussian Random Functions*. Kluwer, Dordrecht.

64. Linde W. (1986). *Probability in Banach spaces: stable and infinitely divisible ditributions*. Wiley, New York.

65. Ma J., Protter Ph. and San Martin J. (1998). Anticipating integrals for a class of martingales. *Bernoulli*, **4**, 81-114.

66. Major P. (1981). *Multiple Wiener-Itô Integrals*. LNM **849**. Springer-Verlag, Berlin Heidelberg New York.
67. Malyshev V.A. (1980). Cluster expansion in lattice models of statistical physics and quantum fields theory. *Uspehi Mat. Nauk* **35**, 3-53.
68. Marinucci D. (2006). High resolution asymptotics for the angular bispectrum of spherical random fields. *The Annals of Statistics* **34**, 1-41.
69. Marinucci D. and Peccati G. (2007). High-frequency asymptotics for subordinated stationary fields on an Abelian compact group. *Stochastic Processes and their Applications* **118**(4), 585-613.
70. Marinucci D. and Peccati G. (2007). Group representations and high-frequency central limit theorems for subordinated random fields on a sphere. To appear in: *Bernoulli*.
71. Maruyama G. (1982). Applications of the multiplication of the Itô-Wiener expansions to limit theorems. *Proc. Japan Acad.* **58**, 388-390.
72. Maruyama G. (1985). Wiener functionals and probability limit theorems, I: the central limit theorem. *Osaka Journal of Mathematics* **22**, 697-732.
73. Masani P.R. (1995). The homogeneous chaos from the standpoint of vector measures, *Phil. Trans. R. Soc. Lond, A* **355**, 1099-1258.
74. McCullagh P. (1987). *Tensor Methods in Statistics*. Chapman and Hall, London.
75. Mauldin R.D., Sudderth W.D. and Williams S.C. (1992). Pólya trees and random distributions. *The Annals of Statistics* **20**(3), 1203-1221.
76. Meyer P.-A. (1976). Un cours sur les intégrales stochastiques. *Séminaire de Probabilités* *X*, LNM **511**. Springer-Verlag, Berlin Heidelberg New York, pp. 245-400.
77. Meyer P.-A. (1992). *Quantum Probability for Probabilists*. LNM **1538**. Springer-Verlag, Berlin Heidelberg New York.
78. Neuenkirch A. and Nourdin I. (2007). Exact rate of convergence of some approximation schemes associated to SDEs driven by a fractional Brownian motion. *Journal of Theoretical Probability* **20**(4), 871-899.
79. Neveu J. (1968). *Processus Aléatoires Gaussiens*. Presses de l'Université de Montréal, Montréal.
80. Nica A. and Speciher R. (2006). *Lectures on the combinatorics of free probability*. London Mathematical Society Lecture Notes Series **335**. Cambridge University Press, Cambridge.
81. Nourdin I. (2005). Schémas d'approximation associés à une équation différentielle dirigée par une fonction höldérienne; cas du mouvement Brownien fractionnaire. *C.R.A.S.* **340**(8), 611-614.
82. Nourdin I. and Nualart D. (2010). Central limit theorems for multiple Skorohod integrals. *Journal of Theoretical Probability* **23**(1), 39-64.
83. Nourdin I., Nualart D. and Tudor C.A. (2010). Central and non-central limit theorems for weighted power variations of fractional Brownian motion. To appear in: *Annales de l'Institut H. Poincarté (PR)*.
84. Nourdin I. and Peccati G. (2007). Non-central convergence of multiple integrals. *The Annals of Probability* **37**(4), 1412–1426.
85. Nourdin I. and Peccati G. (2008). Weighted power variations of iterated Brownian motion. *The Electronic Journal of Probability* **13**, n. 43, 1229-1256 (Electronic).
86. Nourdin I. and Peccati G. (2008). Stein's method on Wiener chaos. *Probability Theory and Related Fields* **145**(1), 75-118.
87. Nourdin I. and Peccati G. (2008). Stein's method and exact Berry-Esséen asymptotics for functionals of Gaussian fields. *The Annals of Probabnility* **37**(6), 2231-2261.
88. Nourdin I. and Peccati G. (2010). Cumulants on the Wiener space. *The Journal of Functional Analysis* **258**, 3775-3791.

89. Nourdin I., Peccati G. and Reinert G. (2010). Invariance principles for homogeneous sums: universality of the Gaussian Wiener chaos. *The Annals of Probability* **38**(5), 1947-1985.

90. Nourdin I., Peccati G. and Réveillac A. (2008). Multivariate normal approximation using Stein's method and Malliavin calculus. *Annales de l'Institut H. Poincaré (PR)* **46**(1), 45-58.

91. Nourdin I. and Réveillac A. (2008). Asymptotic behavior of weighted quadratic variations of fractional Brownian motion: the critical case $H = 1/4$. *The Annals of Probability* **37**(6), 2200-2230.

92. Nualart D. (1983). On the distribution of a double stochastic integral. *Z. Wahrscheinlichkeit verw. Gebiete* **65**, 49-60.

93. Nualart D. (1998). Analysis on Wiener space and anticipating stochastic calculus. *Lectures on Probability Theory and Statistics. École de probabilités de St. Flour XXV (1995)*, LNM **1690**. Springer-Verlag, Berlin Heidelberg New York, pp. 123-227.

94. Nualart D. (2006). *The Malliavin Calculus and Related Topics*, 2nd edition. Springer-Verlag, Berlin Heidelberg New York.

95. Nualart D. and Ortiz-Latorre S. (2007). Intersection local times for two independent fractional Brownian motions. *Journal of Theoretical Probability* **20**(4), 759-767.

96. Nualart D. and Ortiz-Latorre S. (2008). Central limit theorems for multiple stochastic integrals and Malliavin calculus. *Stochastic Processes and their Applications* **118**(4), 614-628.

97. Nualart D. and Pardoux E. (1988). Stochastic calculus with anticipating integrands. *Probability Theory Related Fields*, **78**, 535-581.

98. Nualart D. and Peccati G. (2005). Central limit theorems for sequences of multiple stochastic integrals. *The Annals of Probability* **33**(1), 177-193.

99. Nualart D. and Schoutens W. (2000). Chaotic and predictable representation for Lévy processes. *Stochastic Processes and their Applications* **90**, 109-122.

100. Nualart D. and Vives J. (1990). Anticipative calculus for the Poisson space based on the Fock space. *Séminaire de Probabilités XXIV*, LNM **1426**. Springer-Verlag, Berlin Heidelberg New York, pp. 154-165.

101. Ogura H. (1972). Orthogonal functionals of the Poisson process. *IEEE Transactions on Information Theory* **18**(4), 473-481.

102. Peccati G. (2001). On the convergence of multiple random integrals. *Studia Sc. Mat. Hungarica* **37**, 429-470.

103. Peccati G. (2007). Gaussian approximations of multiple integrals. *Electronic Communications in Probability* **12**, 350-364 (Electronic).

104. Peccati G. (2008). Multiple integral representation for functionals of Dirichlet processes. *Bernoulli* **14**(1), 91-124.

105. Peccati G. and Prünster I. (2008). Linear and quadratic functionals of random hazard rates: an asymptotic analysis. *The Annals of Applied Probability* **18**(5), 1910-1943.

106. Peccati G., Solé J.-L., Utzet F. and Taqqu M.S. (2008). Stein's method and Gaussian approximation of Poisson functionals. *The Annals of Probability* **38**(2), 443-478.

107. Peccati G. and Taqqu M.S. (2007). Stable convergence of generalized L^2 stochastic integrals and the principle of conditioning. *The Electronic Journal of Probability* **12**, 447-480, n. 15 (Electronic).

108. Peccati G. and Taqqu M.S. (2008). Limit theorems for multiple stochastic integrals. *ALEA* **4**, 393-413.

109. Peccati G. and Taqqu M.S. (2008). Central limit theorems for double Poisson integrals. *Bernoulli* **14**(3), 791-821.

110. Peccati G. and Taqqu M.S. (2007). Stable convergence of multiple Wiener-Itô integrals. *The Journal of Theoretical Probability* **21**(3), 527-570.

111. Peccati G. and Tudor C.A. (2005). Gaussian limits for vector-valued multiple stochastic integrals. *Séminaire de Probabilités XXXVIII*, LNM **1857**. Springer-Verlag, Berlin Heidelberg New York, pp. 247-262.

112. Peccati G. and Yor M. (2004). Hardy's inequality in L^2 ([0, 1]) and principal values of Brownian local times. *Asymptotic Methods in Stochastics*, AMS, Fields Institute Communications Series, 49-74.

113. Peccati G. and Yor M. (2004). Four limit theorems for quadratic functionals of Brownian motion and Brownian bridge. *Asymptotic Methods in Stochastics*, AMS, Fields Institute Communication Series, 75-87.

114. Peccati G. and Zheng C. (2010). Multi-dimensional Gaussian fluctuations on the Poisson space. *The Electronic Journal of Probability*, to appear.

115. Pipiras V. and Taqqu M.S. (2000). Integration questions related to fractional Brownian motion. *Probability Theory and Related Fields* **118**(2), 251-291.

116. Pipiras V. and Taqqu M.S. (2001). Are classes of deterministic integrands for fractional Brownian motion complete? *Bernoulli* **7**(6), 873-897.

117. Pipiras V. and Taqqu M.S. (2010). Regularization and integral representations of Hermite processes. To appear in *Statistics and Probability Letters*.

118. Pipiras V. and Taqqu M.S. (2003). Fractional calculus and its connection to fractional Brownian motion. In: *Long Range Dependence*, 166-201, Birkhäuser, Basel.

119. Pitman J. (2006). *Combinatorial Stochastic Processes*. LNM **1875**. Springer-Verlag, Berlin Heidelberg New York.

120. Privault N. (1994). Chaotic and variational calculus in discrete and continuous time for the Poisson process. *Stochastics and Stochastics Reports* **51**, 83-109.

121. Privault N. (1994). Inégalités de Meyer sur l'espace de Poisson. *C.R.A.S.* **318**, 559-562.

122. Privault N. (2009). Stochastic Analysis in Discrete and Continuous Settings: With Normal Martingales. Springer-Verlag, Berlin Heidelberg New York.

123. Privault N. (2010). Invariance of Poisson measures under random transformations. Preprint.

124. Privault N., Solé J.L. and Vives J. (2001). Chaotic Kabanov formula for the Azéma martingales. *Bernoulli* **6**(4), 633-651.

125. Privault N. and Wu J.-L. (1998). Poisson stochastic integration in Hilbert spaces. *Ann. Math. Blaise Pascal* **6**(2), 41-61.

126. Protter P. (2005). *Stochastic Integration and Differential Equations*, 2nd edition. Springer-Verlag, Berlin Heidelberg New York.

127. Rajput B.S. and Rosinski J. (1989). Spectral representation of infinitely divisible processes. *Probability Theory and Related Fields* **82**, 451-487.

128. Revuz D. and Yor M. (1999). *Continuous martingales and Brownian motion*. Springer-Verlag, Berlin Heidelberg New York.

129. Rider B. and Virág B. (2007). The noise in the circular law and the Gaussian free field. *Int. Math. Res. Not.* 2, Art. ID rnm006.

130. Roman S. (1984). *The Umbral Calculus*. Academic press, New York.

131. Rota G.-C. and Shen J. (2000). On the combinatorics of cumulants. *Journal of Combinatorial Theory Series A* **91**, 283-304.

132. Rota G.-C. and Wallstrom C. (1997). Stochastic integrals: a combinatorial approach. *The Annals of Probability* **25**(3), 1257-1283.

133. Rosińsky J. and Woyczyński W.A. (1984). Products of random measures, multilinear random forms and multiple stochastic integrals. *Proc. Conference of Measure Theory*, Oberwolfach 1983, LNM 1089. Springer-Verlag, Berlin Heidelberg New York, pp. 294-315.

134. Russo F. and Vallois P. (1998). Product of two multiple stochastic integrals with respect to a normal martingale. *Stochastic Processes and their Applications* **73**(1), 47-68.

135. Samorodnitsky G. and Taqqu M.S. (1994). *Stable Non-Gaussian Random Processes.* Chapman and Hall, New York.

136. Sato K.-I. (1999). *Lévy Processes and Infinitely Divisible Distributions.* Cambridge Studies in Advanced Mathematics **68**. Cambridge University Press, Cambridge.

137. Schoutens W. (2000). Stochastic processes and orthogonal polynomials. Lecture Notes in Staistics 146. Springer-Verlag, Berlin Heidelberg New York.

138. Schreiber M. (1969). Fermeture en probabilité de certains sous-espaces d'un espace L^2. *Zeitschrift Warsch. verw. Gebiete* **14**, 36-48.

139. Sheffield S. (1997). Gaussian free field for mathematicians. *Probab. Theory Related Fields* **139**(3-4), 521-541.

140. Shigekawa I. (1980). Derivatives of Wiener functionals and absolute continuity of induced measures. *J. Math. Kyoto Univ.* **20**(2), 263-289.

141. Shiryaev A.N. (1984). *Probability.* Springer-Verlag, Berlin Heidelberg New York.

142. Slud E.V. (1993). The moment problem for polynomial forms in normal random variables. *The Annals of Probability* **21**(4), 2200-2214.

143. Solé J.-L. and Utzet F. (2008). Time-space harmonic polynomials associated with a Lévy process. *Bernoulli* **14**(1), 1-13.

144. Solé J.-L. and Utzet F. (2008). On the orthogonal polynomials associated with a Lévy process. *The Annals of Probability* **36**(2), 765-795.

145. Speed T. (1983). Cumulants and partitions lattices. *Australian Journal of Statistics* **25**(2), 378-388.

146. Stanley R. (1997). *Enumerative combinatorics, Vol. 1.* Cambridge University Press, Cambridge.

147. Stein Ch. (1986). *Approximate computation of expectations.* Institute of Mathematical Statistics Lecture Notes - Monograph Series **7**. Institute of Mathematical Statistics, Hayward, CA.

148. Steutel F.W. and van Ham K. (2004). *Infinite divisibility of probability distributions on the real line.* Marcel Dekker, New York.

149. Stroock D.W. (1987). Homogeneous Chaos Revisited. *Séminaire de Probabilités XXI*, LNM **1247**, Springer-Verlag, Berlin Heidelberg New York, pp. 1-8.

150. Surgailis D. (1984). On multiple Poisson stochastic integrals and associated Markov semigroups. *Probab. Math. Statist.* **3**(2), 217-239.

151. Surgailis D. (2000). CLTs for Polynomials of Linear Sequences: Diagram Formulae with Applications. In: *Long Range Dependence.* Birkhäuser, Basel, pp. 111-128.

152. Surgailis D. (2000). Non-CLT's: U-Statistics, Multinomial Formula and Approximations of Multiple Wiener-Itô integrals. In: *Long Range Dependence.* Birkhäuser, Basel, pp. 129-142.

153. Taqqu M.S. (1979). Convergence of integrated processes of arbitrary Hermite rank. *Zeitschrift für Wahrscheinlichkeitstheorie und verwandte Gebiete* **50**, 53-83.

154. Taqqu M.S. (2003). Fractional Brownian motion and long-range dependence. In: P. Doukhan, G. Oppenheim and M.S. Taqqu (eds.), *Theory and Applications of Long-range Dependence.* Birkhäuser, Boston, pp. 5-38.

155. Tocino A. (2009). Multiple stochastic integrals with Mathematica. *Mathematics and Computers in Simulations.* **79**, 1658-1667.

156. Tsilevich N.V. and Vershik A.M. (2003). Fock factorizations and decompositions of the L^2 spaces over general Lévy processes. *Russian Math. Surveys* **58**(3), 427-472.

157. Tsilevich N., Vershik A.M. and Yor M. (2001). An infinite-dimensional analogue of the Lebesgue measure and distinguished properties of the gamma process. *J. Funct. Anal.* **185**(1), 274-296.

158. Tudor C. (1997). Product formula for multiple Poisson-Itô integrals. *Revue Roumaine de Math. Pures et Appliquées* **42**(3-4), 339-345.
159. Tudor C.A. and Vives J. (2002). The indefinite Skorohod integral as integrator on the Poisson space. *Random Operators and Stochastic Equations* **10**, 29-46.
160. Üstünel A.S. and Zakai M. (1997). The construction of filtrations on Abstract Wiener Space. *Journal of Functional Analysis* **143**, 10-32.
161. Vitale R.A. (1990). Covariances of symmetric statistics. *Journal of Multivariate Analysis* **41**, 14-26.
162. Wiener N. (1938). The homogeneous chaos. *Amer. J. Math.* **60**, 879-936.
163. Wolfe S.J. (1971). On Moments of Infinitely Divisible Distribution Functions. *Ann. Math. Statist.* **42**(6), 2036-2043.
164. Wolpert R.L. and Taqqu M.S. (2005). Fractional Ornstein-Uhlenbeck Lévy Processes and the Telecom Process: Upstairs and Downstairs. *Signal Processing* **85**(8), 1523-1545.
165. Wu L.M. (1990). Un traitement unifié de la représentation des fonctionnelles de Wiener. *Séminaire de Probabilités XXIV*, LNM **1426**, Springer-Verlag, Berlin Heidelberg New York, pp. 166-187.
166. Xue X.-H. (1991). On the principle of conditioning and convergence to mixtures of distributions for sums of dependent random variables. *Stochastic Processes and their Applications* **37**(2), 175-186.
167. Yosida K. (1980). *Functional analysis*. Springer-Verlag, Berlin Heidelberg New York.

Index

$\Delta_n^\varphi(C)$, 85
B_π, 57
B_n, 18
$B_n(x_1, ..., x_n)$, 17
$B_{n,k}(x_1, ..., x_{n-k+1})$, 17
$Bf(x)$, 171
$C^{\otimes j}$, 87
$C_n(x, a)$, 172
C_n^φ, 83
$I_n^\varphi(f)$, 82
$I_1^\varphi(h)$, 64
$L_s^2(\nu^n)$, 81
$S(n, k)$, 15
$T_n(x)$, 17
Z_π^n, 57
$[\sigma, \pi]$, 12
$[n]$, 13
$[x, y]$, 24
$\begin{bmatrix} n \\ \lambda \end{bmatrix}$, 13
$\Delta_n^G(A)$, 93
$\Delta_n^{\widehat{N}}(A)$, 98
$\Gamma(\pi, \sigma)$, 45
$\Lambda(n)$, 7
α_B, 67
$\chi(\mathbf{X}_b)$, 31
$\chi_n(X)$, 32
$\delta(x, y)$, 21
$\hat{0}$, 12
– minimal element, 12
$\hat{1}$, 12
– maximal element, 12
$\hat{\Gamma}(\pi, \sigma)$, 54

$\int''$, 163
$\int'$, 107
$\int_Z h(z)\, \varphi(dz)$, 64
$\int_Z h d\varphi$, 64
$\lambda \vdash n$, 7
$\lambda(\pi, \sigma)$, 12
$\langle \cdot \rangle$ (as expectation), 128
$(1^{r_1} 2^{r_2} \cdots n^{r_n})$, 7
$|\pi|$, 9
$\hat{0}$, 12
$\mathbf{B}_1(\sigma)$, 97
$\mathbf{B}_2(\sigma)$, 97
$\mathbf{PB}_2(\sigma)$, 97
$\mathbf{X}^b$, 31
$\mathbf{X}_{[n]}$, 31
$\mathbf{X}_b$, 31
$\mathbf{e}(\alpha)$, 154
$\mathcal{A}_\infty$, 152
$\mathcal{A}_{\infty, q}$, 153
$\mathcal{E}(\nu^n)$, 77
$\mathcal{E}_0(\nu^n)$, 81
$\mathcal{E}_{s,0}(\nu^n)$, 81
$\mathcal{I}(P)$, 21
$\mathcal{M}([n], \pi^*)$, 132
$\mathcal{M}^0([n], \pi^*)$, 132
$\mathcal{M}_{\geq 2}^0([n], \pi^*)$, 133
$\mathcal{M}_2([n], \pi^*)$, 133
$\mathcal{M}_2^0([n], \pi^*)$, 133
$\mathcal{M}_{\geq 2}([n], \pi^*)$, 133
$\mathcal{P}(b)$, 9
$\mathcal{Z}_\nu$, 59
$\mathcal{Z}_\nu^n$, 61
$\mathcal{Z}_{s,\nu}^n = \mathcal{Z}_s^n$, 79

$\mathfrak{H}^{\odot q}$, 153
$\mathfrak{H}^{\otimes q}$, 153
$\mathfrak{S}_n$, 79
$\mu(\sigma, \pi)$, 19
$\mu(x, y)$, 21
$\otimes_r (= \star_r^r)$, 116
ρ_ν, 68
$\sigma < \pi$, 22
$\sigma \geq \pi$, 76
$\sigma \leq \pi$, 9
$\sigma \nleq \pi$, 19
$\sigma \vee \pi$, 10
$\sigma \wedge \pi$, 10
$\mathrm{St}_\pi^{\varphi,[n]}(f)$, 78
$\tilde{T}_n(x)$, 43
$\varphi(h)$, 64
$\underline{\varphi}^{[n]}$, 61
$\hat{f}$, 79
$\zeta(x, y)$, 21
$c_n(x, a)$, 171
$f * g$, 21
$f \star_r^l g$, 115
$f_1 \otimes_0 f_2 \otimes_0 \cdots \otimes_0 f_k$, 110
$f_{\sigma,k}$, 133
$g\mathbf{x}_b$, 31
$i \sim_\pi j$, 9
$m_n(\nu)$, 39
$\mathrm{St}_{\geq \pi}^{\varphi,[n]}(C)$, 77
$\mathrm{St}_\pi^{\varphi,[n]}(C)$, 76

Backward shift operator, 171
Bell numbers, 18
– generating function, 41
Binomial type, 19
Block, 9
Borel isomorphism, 60
Borel space, 57
Brownian motion, 88

Canonical isomorphism, 155, 157
Charlier polynomials, 171, 172, 175
– as multiple Poisson integrals, 173, 174
– centered, 172
Chen-Stein Lemma, 42
Class of a segment, 12
CLT
– on the Gaussian Wiener chaos, 182, 200
– on the second Poisson chaos, 204
Coarse, 9

Compound Poisson process, 75, 76
Computational rules for Gaussian and
    Poisson measures, 92
Contraction (of two functions), 115
– symmetrization, 117
Contraction (on a Hilbert space), 156
Control measure, 59
– symmetric, 151
Convolution on the incidence algebra, 21
Convolutional inverse, 21
Cumulants
– homogeneity of, 32
– joint, 31
– of a gamma random variable, 144
– of a Poisson random variable, 33
– of a single random variable, 32
– of an element of the second Wiener chaos,
    140
– order of, 32

Delta function, 21
Diagonal measure, 85
Diagram
– circular, 49, 56
    – and contractions, 198
– connected, 47, 56
– definition, 45
– disconnected, 47
– flat, 48
– formulae, 3
    – for Gaussian measures, 133
    – for Poisson measures, 140
    – for stochastic measures, 129, 132
    – Malyshev, 50
– Gaussian, 48, 56
– non-flat, 48, 56
Dirichlet process, 75
Divergence operator, 145
Dobiński formula, 41
Double factorial, 38
Double Wiener-Itô integrals as series of
    chi-square, 178
Dudley, R.M., 149

Edge, 46
Elementary functions, 77
Engel, D.D., 1, 4, 62
Even function, 151

Fourth cumulant condition, 182
Fourth cumulant of a multiple integral
  (Gaussian case), 180
Fractional Brownian motion, 151, 166, 190
Free probability, 5, 51
Fubini, 106

Gaussian free field, 152
Gaussian is determined by its moments, 39

Hermite polynomials, 145, 146, 154, 163,
  172, 175
Hermite processes, 165
Hu-Meyer formulae, 5, 89
Hypercontractivity, 92
– for an isonormal Gaussian process, 155
Hyperdiagonal, 107, 162
Hypergraph, 46

Incidence algebra, 21
Infinitely divisible, 65
Integral notation for stochastic measures, 84
Isomorphism of partially ordered sets, 25
Isonormal Gaussian process, 3, 149, 187
– built from a covariance, 150
– spectral representation, 152, 157, 159

Join, 10

Kailath-Segall formula, 86
– on the real line, 90
Khintchine's theorem, 66

Lévy measure, 67
Lévy-Khintchine
– characteristics, 67
– exponent, 67
  – of a Gaussian single integral, 71
  – of a Poisson single integral, 72
– representation, 67
Lévy process, 73, 88
Lattice, 10, 27
Length of a partition of an integer, 7
Length of a segment, 24
Leonov and Shiryaev formulae, 34
Long memory, 148
Long-range dependence, 148

Möbius function
– of $\mathcal{P}(b)$, 19

– of a partially ordered set, 22
– recursive computation, 25
Möbius inversion formula
– for random measures, 78
– on $\mathcal{P}(b)$, 20
– on a partially ordered set, 23
Malliavin calculus, 5
Malyshev formula, 34
Mathematica, 3
Meet, 10
Method of (moments and) cumulants, 181
Moments of a Gaussian random variable, 38
Multigraph, 53
Multiindex, 153
Multiplication formulae
– for isonormal Gaussian processes, 157, 187
– Gaussian case, 118
– general, 109
  – with diagrams, 112
  – with multigraphs, 112
– Poisson case, 122

Non-crossing partition, 51
Normal martingales, 92

Partial ordering, 9
Partition of a set, 9
Partition of an integer, 7
Partition segment (interval), 12
Perfect matching, 49
Polish space, 57
Polynomials
– centered Touchard, 44
– Charlier, 171, 172, 175
  – as multiple Poisson integrals, 173, 174
  – centered, 172
– complete Bell, 17
– Hermite, 145, 154, 163, 172, 175
  – as multiple integrals, 147
  – generalized, 146
– partial Bell, 17
– Touchard (exponential), 17
Prime, 107, 163
Product of partially ordered sets, 25
Purely diagonal set, 58

Random measure
– additivity of, 60
– compensated Poisson, 63
  – diagonals, 98

– completely random (or independently scattered), 1, 59
– even, 159
– gamma, 74
– Gaussian, 63
  – diagonals, 93
– good, 61
– Hermitian, 159
– homogeneous, 73
– in $L^2(\mathbb{P})$, 60
– independently scattered (or completely random), 1
– infinite divisibility of, 65
– monotonicity of, 62
– multiplicative, 128
Rank, 197
Rota, G.-C., 1, 4, 62, 109

Segment (interval) of a partially ordered set, 24
Self-similar processes
– Fractional Brownian motion, 151
– Hermite processes, 165
Size of a partition of a set, 9
Spectral domain, 164–166
Spectral representation, 159, 163
Stein's bound, 186
Stein's Lemma, 38
Stein's method, 38, 146, 186
– and Malliavin calculus, 186
Stirling numbers of the second kind, 15
Stochastic Fubini, 106, 187
Stochastic measure (of order n), 83
Stochastic processes, 164
Symmetric σ-field, 79

Symmetrization of a function, 79

Tensor product, 153
– and symmetric functions, 153
– symmetric, 153
Tensor product of functions, 110
Teugels martingales, 89
Time domain, 164–166
Total variation distance, 185
Touchard polynomials as Poisson moments, 40

Variation process, 89

Wallstrom, T.J., 1, 4, 62, 109
Weak composition, 8
Wick formula, 38
Wick products, 5
Wiener chaos, 2
– definition, 83
– isonormal Gaussian process, 154
Wiener-Itô chaotic decomposition, 91
– Poisson case, 174
– as a Hilbert space decomposition, 92
– Gaussian case, 148
– unicity, 91
Wiener-Itô integral
– absolute continuity of the law, 178
– isometry, 64, 83
– multiple, 82, 83
– non-Gaussianity, 178
– single, 64
  – infinite divisibility of, 67

Zeta function, 21

B&SS – Bocconi & Springer Series

THE ONLINE VERSION OF THE BOOKS PUBLISHED IN THE SERIES IS AVAILABLE ON SpringerLink

1. G. Peccati, M.S. Taqqu
    Wiener Chaos: Moments, Cumulants and Diagrams – A survey with computer implementation
    2011, XIV+274 pp, ISBN 978-88-470-1678-1

2. A. Pascucci
    PDE and Martingale Methods in Option Pricing
    2011, XVIII+720 pp, ISBN 978-88-470-1780-1

3. P. Hirsch, C. Profeta, B. Roynette, M. Yor
    Peacocks and Associated Martingales, with Explicit Constructions
    TO BE PUBLISHED, ISBN 978-88-470-1907-2

For further information, please visit the following link:
http://www.springer.com/series/8762